BIOCHEMICAL NEUROLOGY

BIOCHEMICAL NEUROLOGY

M. J. Eadie and J. H. Tyrer
Department of Medicine
University of Queensland
Brisbane, Australia

Springer-Science+Business Media, B.V.

British Library Cataloguing in Publication Data

Eadie, M. J.
    Biochemical neurology.
    1. Neurochemistry
    I. Title   II. Tyrer, J. H.
    612'.8042   QP356.3

ISBN 978-94-017-0898-2   ISBN 978-94-017-0896-8 (eBook)
DOI 10.1007/978-94-017-0896-8

Filmset by Northumberland Press Ltd, Gateshead
Printed by McCorquodale (Scotland) Ltd.

Contents

Preface

'... it is probable that by the aid of chemistry many derangements of the brain and mind, which are at present obscure, will become accurately definable.'

J. L. W. Thudichum (1884) *Chemical Constitution of the Brain.*

Dysfunction of the human nervous system may be interpreted at different levels of the system's own organizational complexity. For centuries neurological disorders had to be assessed mainly by means of the unaided human senses. In the present century, light and electron microscopy have opened the way to an understanding of neurological disorder at a cellular and subcellular level, while electronic amplification devices have permitted the study of nervous system dysfunction at an electrophysiological level. In recent years there has been increasing investigation of both normal and abnormal neural function at its ultimate level, the molecular or chemical one. In the last analysis, all dysfunction of the nervous system must somewhere involve alterations in its component molecules. At present it is not possible to provide a complete account of all molecular changes which must occur in the various neurological diseases. However, biochemical knowledge of neurological disorders has grown rapidly and sufficient material is now available for a moderately coherent biochemical interpretation of many neurological diseases. In the past, when neurological disorder could be explained at a molecular level, the material has often been presented as if the main aim was to illuminate normal neural function. In the present work, the subject of biochemical neurology is presented with the main aim of interpreting in molecular terms, as far as seems possible, the disorders which comprise the content of clinical neurological practice. This approach may lead to a result which may appear unbalanced to the scientific neurochemist, but it may prove more relevant and useful to the clinical neurologist. The latter, looking at familiar clinical facts from a relatively unfamiliar biochemical viewpoint, may gain new insights and become aware of gaps in neurochemical knowledge. Thus research may be stimulated. A chemically-based approach to clinical neurology should include the study of neuromuscular as well as purely neural diseases. Muscle disease is customarily regarded as part of the field of clinical neurology, though there are some major chemical differences between neural and muscular tissue.

To what extent can the content of contemporary clinical neurology be regarded as biochemical? All neural and neuromuscular disorders ultimately involve molecular changes, and all genetic disorders are determined by abnormalities of desoxyribosenucleic acids. Therefore it might be argued that the scope of biochemical neurology includes all clinical neurology. However, any attempt at present to deal with all neurological disorders at a biochemical level would show substantial areas for which little biochemical information is available. This book therefore deals only with those neurological conditions where contemporary biochemical knowledge contributes significantly to their understanding. Where the relation between a disorder and a described biochemical abnormality appears too tenuous, e.g. the question of altered central cholinergic neurotransmission in Alzheimer's disease, we have generally felt it inappropriate to discuss the matter, bearing in mind the level at which the book is written. We have also omitted from consideration disorders in which described biochemical abnormalities appear to be purely consequences of a disease process which itself has no known biochemical basis or mechanism. Because of this policy we do not discuss a disorder such as subarachnoid haemorrhage, even though this may lead to clinical biochemical abnormalities, e.g. glycosuria.

Thus we have attempted to write a book interpreting dysfunction of neural and neuromuscular tissues at the molecular level, directed towards the interests of those who investigate and treat such disorders clinically, be they neurologists, physicians, paediatricians, psychiatrists or pathologists. Such an approach may also be of value to laboratory-based neurochemists and biochemists, despite its obvious shortcomings from their point of view.

Throughout the book references have not been supplied to material widely available in standard texts on biochemistry and clinical neurology. Except for recently published work unlikely to have yet been included in the review-type literature, we have preferred to cite reviews of subjects or papers with extensive bibliographies, rather than original research contributions. Thus it is hoped that the reader will be provided with an entry to the relevant literature, without the text becoming unduly cluttered with references.

We are indebted to many persons for help in preparing this book and in particular would wish to express our gratitude to Mrs Janet Wickham for patiently transforming so many pages of largely indecipherable handwriting into a highly professional typescript, to Mr D. Sheehy for drawing the numerous chemical formulae and other illustrations, and to Mr G. Jurott for photographing them. We would also wish to indicate our thanks to Dr J. Marks, of Girton College, Cambridge, for instigating this project, and to Mr D. Bloomer of MTP Press, for his encouragement and help throughout its course.

M. J. Eadie
J. H. Tyrer
Brisbane
May, 1982

1
The arrangement of the material

In writing for clinicians, it is difficult to know how much biochemical detail to present, and what basic knowledge of chemistry to assume. Generally in this book we have tried to give only enough chemistry to show the reader the structures of the various molecules considered and to follow the different chemical reaction sequences discussed. Reaction mechanisms as such have not been considered. Illustration devices such as shading have been used to allow the reader's eye to follow the fates of relevant portions of molecules more easily, and to emphasize aspects of molecular structure which are pertinent in particular contexts. We have assumed relatively little background chemical knowledge on the part of the reader, though we realize this policy may irritate those chemically more sophisticated. For simplicity we have often not followed all the ramifications of a particular chemical reaction at the one time, but have concentrated on those aspects germane to the disturbance currently under discussion. However, we have attempted by cross-referencing to bring out the interrelations between the various reactions of the one molecule which can yield different chemical products and be responsible for different diseases. Thus it is hoped that material presented as individual chemical facets can later be synthesized by the reader into a biochemical overview.

The material has been arranged around a concept of biochemical functioning of nervous tissue in a way which it is hoped may be relevant to the clinician, but which may appear unconventional and perhaps inappropriate to the biochemist. We have tried to fit the individual biochemical neurological disorders into this arrangement, even when the fit has sometimes not been a particularly comfortable one, because we wished to present the individual disturbances in relation to an over-riding concept of neurochemical function, rather than describing the disorders as a series of self-contained and sometimes apparently unrelated chemical entities.

1.1 THE APPROACH TO THE SUBJECT

It seemed possible to deal with biochemical neurology in two main ways. First, we could follow the conventional biochemist's method of considering in turn the various classes of molecule found in neural tissue and muscle, and then discuss the neurological disorders associated with disturbed metabolism of each of these classes of molecule. Secondly, we could take the various diseases in the order in which they are dealt with in most textbooks of clinical neurology, and describe the abnormal chemistry of each condition in turn. The latter approach had the advantage that the book's intended reader, the clinician, would move from the more familiar to the less familiar. He would see the chemistry in relation to clinical data that he already knew. Unfortunately, however, it would mean that chemically unrelated conditions might be dealt with side by side, and chemically related conditions might be divorced. It would therefore be more difficult to develop any comprehensive view of normal and abnormal neurochemistry, unless what might appear to the clinician as an indigestible and doubtfully relevant body of normal neurochemistry were presented first, and the diseases were then considered seriatim as in textbooks of clinical neurology. To avoid this undesirable arrangement we have dealt with neurological disease in an order that has been determined chiefly by chemical considerations. We believe this approach allows the greater opportunity for the clinician to develop chemical insights. However, we have not simply taken chemical class as the primary criterion for determining the order in which neurological diseases are considered. Rather, we have seen in the complex patterns of chemical activities that occur in cells, groupings of reactions that serve particular functional purposes that appear especially germane to the interests of clinical neurology. We have built the book around an analysis of disorders of these functional biochemical reaction sequences (which, to a considerable extent, do correlate with chemical class).

1.2 THE FUNCTIONAL SIGNIFICANCE OF CERTAIN NEUROCHEMICAL REACTION SEQUENCES

The nervous system is chiefly concerned with the rapid transmission of signals from point to point within the body. This transmission involves electrical conduction along preformed anatomical pathways which are kept in a state of excitability, and also chemical transmission across synaptic clefts between nerve cells and, at the periphery, between nerve cells and effector cells. One can conceive the main biochemical functions of neural tissue as involving the production of energy, required to drive chemical reactions serving various purposes, including:

(1) The synthesis and degradation of large molecules, required to maintain the structural integrity on which the function of neural tissue depends,

(2) The production of ionic concentration gradients, required to sustain nerve cell excitability, and

(3) The maintenance of the synaptic transmission process.

2

In addition, there are reactions of small metabolic intermediate molecules involved in various synthetic activities. The energy yielding reaction sequence and the reaction of small metabolic intermediate molecules are so intimately linked that they are conveniently considered together. Thus the functional concept

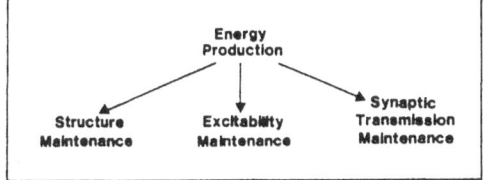

may be correlated with a concept of neurochemical processes.

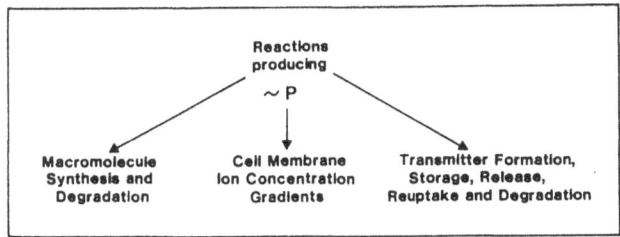

Such a schematic overview embraces many important aspects of neural activity, though it does not bring out the mutual interdependence of the various processes. When one considers skeletal muscle function, certain different macromolecules become relevant, the whole matter of synaptic transmission is largely inapplicable, and high energy phosphate ($\sim$ P) use is heavily directed towards muscle contraction rather than towards impulse conduction.

In this book, the disorders to be considered have been grouped according to which of the major neurochemical functions is primarily disturbed in each disorder. A disorder may, of course, produce consequent disturbance of biochemical mechanisms in a different functional category or categories, and the primary disturbance of molecular mechanisms (usually an enzyme defect) may not always be known with certainty. Consequently the classification of disorders here adopted may not prove valid in all instances, as knowledge grows.

It may now be useful to amplify to a limited extent the more important chemical reaction sequences involved in the major groups of neurological functions mentioned in Figure 1.1. These reaction sequences determine the subdivision of the contents of the various chapters of this book. In general, for each subdivision (i.e. a major group of related reactions) we have attempted to provide an initial account of the chemistry in sufficient detail for the clinician's purposes. We have then related known neurological disorders to this chemistry, attempting to emphasize principles and features common to a group of disorders, where possible. Then individual clinical disorders of each reaction sequence are dealt with briefly, but in a systematic fashion, under the

3

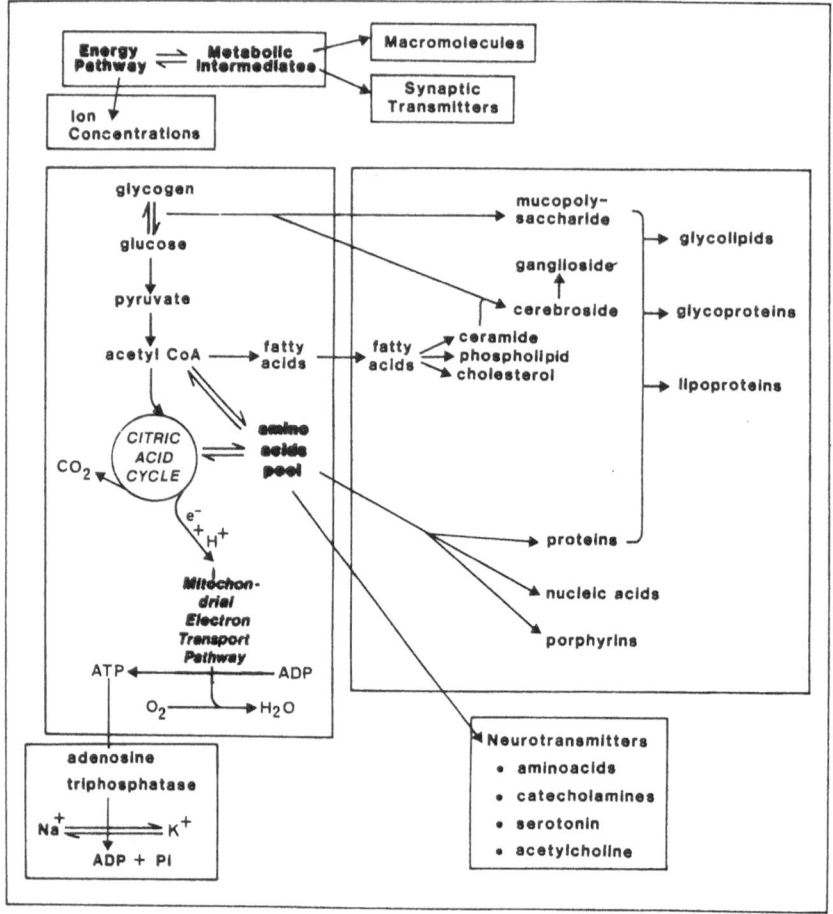

Figure 1.1 Major aspects of neurochemical activity, with the relevant chemical pathways (in outline)

following subheadings, unless the available material is too slight to warrant such subdivision:

(1) Introductory comment,
(2) Biochemical abnormality,
(3) Aetiology,
(4) Structural pathology,
(5) Clinical features,
(6) Diagnosis
   (a) clinical
   (b) laboratory,
(7) Treatment.

4

Such a structured approach may highlight gaps in knowledge which might be obscured by a less methodical presentation. Because of the interests of the intended major readership and the level at which the material is presented, details of biochemical analytical methods are not provided. However, the types of analytical method used in making biochemical diagnoses are indicated. By use of this information, and the references, those interested should be able to gain access to details of the relevant laboratory techniques. Many of the disorders considered here are rare, or comparatively rare. Therefore at times both clinicians and laboratory workers may find it preferable to use the references to make contact with those who have the appropriate chemical assays already functioning, rather than attempt to set up these assays for themselves to study one or two patients.

2

Disorders of the energy pathway and the metabolic intermediate pool

Part A: The energy pathway

Neural tissue obtains almost all its energy from the oxidative metabolism of glucose. The whole human brain consumes about 50 ml of oxygen per minute, some 20% of the body's total oxygen requirement under basal conditions. This oxygen is required to provide energy by oxidatively metabolizing the 72 mg per minute of glucose that the brain takes up over the same time interval[1]. In certain circumstances acetyl groups ($CH_3.CO—$) and keto acids ($R'—CO—R''—COOH$), derived respectively from fatty acids by β-oxidation and from amino acids by oxidative deamination, can be used to provide energy. Energy is stored in the form of high energy phosphate (adenosine triphosphate, i.e. ATP). Skeletal muscle can make more extensive use of fatty acids as an energy source than can neural tissue.

The chemical events between the entry of glucose into cells and the production of ATP can be divided into three stages:

(1) Glucose is anaerobically metabolized to form the 3 carbon derivative pyruvate ($CH_3.CO.COO^-$). This stage (Embden–Meyerhof glycolysis) occurs in the cytosol of neural tissue and muscle. In the presence of relative anoxia the pyruvate may be reduced to lactate. Ordinarily pyruvate enters mitochondria, for the subsequent stages of its metabolism.

(2) In mitochondria pyruvate is then converted to activated acetate (acetyl-coenzyme A; acetyl-CoA; $CH_3.CO.S.CoA$). This subsequently enters the Krebs tricarboxylic acid (citric acid) cycle by combining with a molecule of oxaloacetate to form citrate. During one turn of the Krebs cycle the

7

acetate moiety is completely consumed, yielding two molecules of carbon dioxide, and four hydrogen atoms. These hydrogen atoms enter the final stage of the energy metabolism pathway and reduce nicotinamide adenine dinucleotide (NAD^+) to form NADH or reduce flavine adenine dinucleotide (FAD^+) to form $FADH_2$.

(3) The four electron pairs produced from metabolism of each acetate moiety during a single turn of the Krebs cycle are carried as NADH or $FADH_2$ into the mitochondrial electron transport system. This is an assembly of enzymes that is located in the inner membranes of mitochondria. During electron movement along this electron transport pathway NADH and $FADH_2$ are oxidized back to NAD^+ and FAD^+ respectively, while at the end of the pathway oxygen is reduced to water. At stages of electron transport along the pathway, in the process of oxidative phosphorylation, adenosine triphosphate (ATP) is formed from adenosine diphosphate (ADP).

Within cells, glucose not required for the immediate production of energy is stored in a polymer form (glycogen). Glycogen is the major energy reserve for skeletal muscle, though it is less important as an energy reserve for neural tissue. Glycogen is formed from, and broken down again to glucose-6-phosphate, through different anabolic and catabolic pathways.

Thus overall energy metabolism can be regarded as falling into four stages, as shown in Figure 2.1.

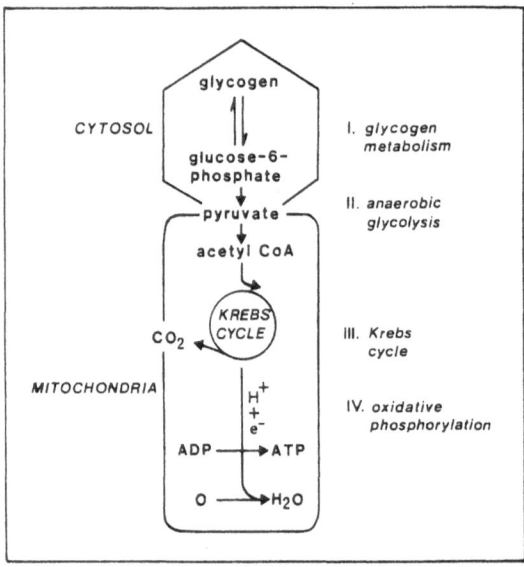

Figure 2.1 The four stages of the energy pathway

Neurological disorders involving the energy production pathway will be discusssed in relation to these four stages.

2.1 DISORDERS OF GLYCOGEN METABOLISM

Glycogen is a high molecular weight branched chain polymer comprising multiple glucose units in α-1,4-glycosidic links (straight chains), with α-1,6-glycosidic links at the points of branching. Thus it is a polyglucosan (Figure 2.2).

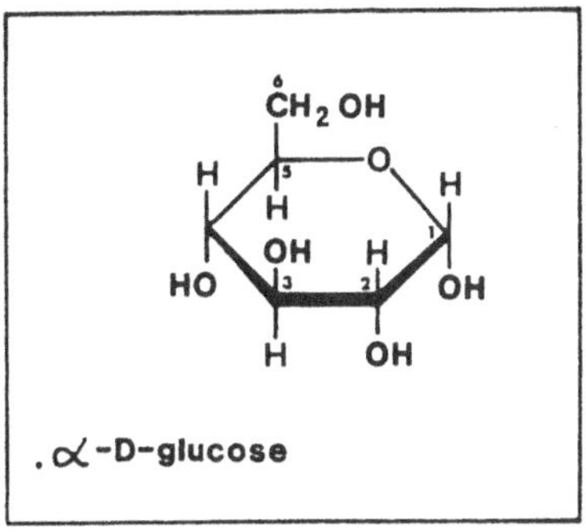

$.\alpha$-D-glucose

The synthetic path from glucose to glycogen is set out in Figure 2.3(a) (indicating how a molecule of glucose comes to be attached to the glycogen skeleton). The degradation pathway for a molecule of glucose detached from the polymer is shown in Figure 2.3(b).

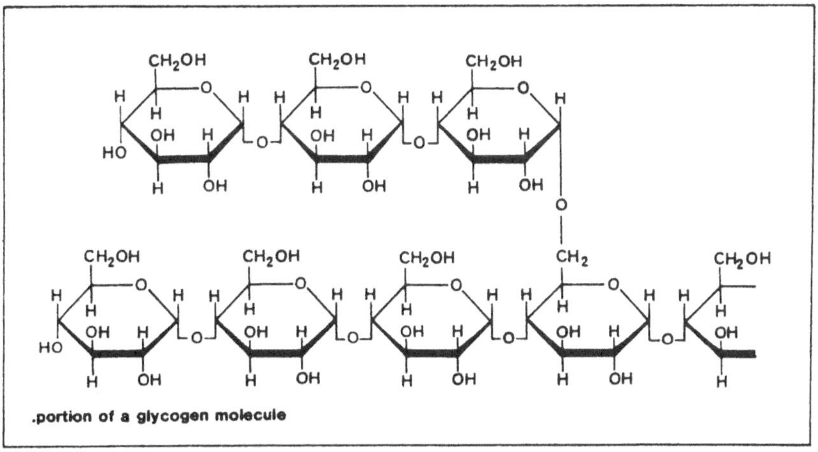

.portion of a glycogen molecule

Figure 2.2 Portion of a glycogen molecule showing two straight chains with α-1,4-glucosidic bonds, with a branching via an α-1,6-bond linking two chains. The glucose units are shown in Howarth projection formulae. The —OH group on the glucose residues is in the α-configuration

9

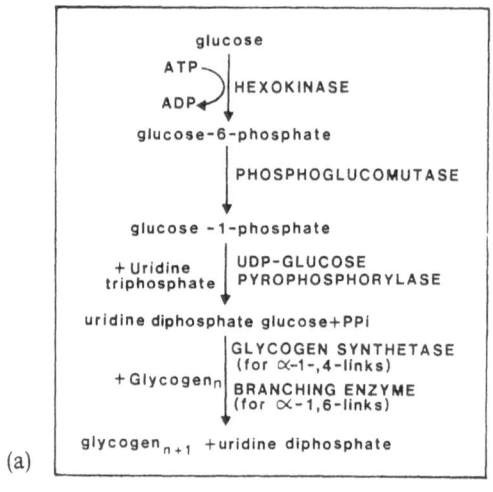

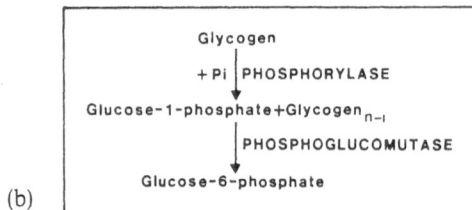

Figure 2.3 (a) The biosynthetic pathway from glucose to glycogen (b) the degradative pathway

The enzyme glucose-6-phosphatase catalyses the conversion of glucose-6-phosphate to glucose in liver and certain other tissues. The enzyme phosphorylase will not cleave the α-1,6-glucosidic bonds in glycogen, or the α-1,4-bonds in the 3 glucose residues nearest to an α-1,6-glycosidic linkage. During glycogen catabolism these three α-1,4 linked glucose residues must be removed to another polyglucosan chain before glycogen can break down further. A transferase is required to effect this intramolecular translocation, after which α-1,6-glucosidase (debranching enzyme) can catalyse the splitting of the exposed α-1,6-linkage (Figure 2.4) allowing phosphorylase to act again. Glycogen may also be broken down, at a slower rate, by the hydrolytic enzymes α-amylase and acid maltase (α-1,4-glucosidase). The latter is a lysosomal rather than a cytosol enzyme, unlike the other enzymes of glycogen catabolism.

Entry of glucose into cells is accelerated by insulin. Therefore it could be regarded as logical to consider the effects of excess or deficient insulin in relation to glycogen synthesis. However, it is equally logical, and more convenient, to consider insulin effects in relation to anaerobic glycolysis, the set of metabolic reactions to be considered after glycogen synthesis and catabolism are dealt with. With the exception of glycogen synthetase deficiency

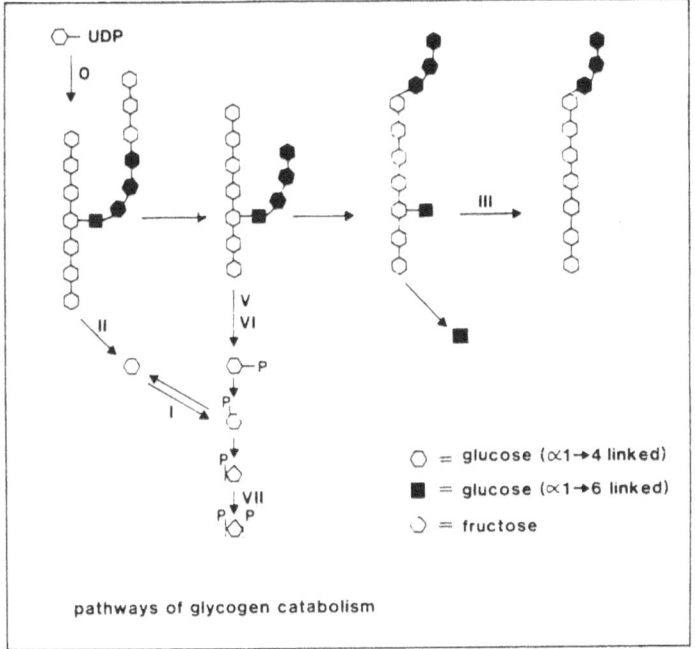

Figure 2.4 Schematic representation of routes of glycogen catabolism. The Roman numerals correspond with the types of glycogen storage disease listed in Table 2.1, and indicate sites where failure of metabolism causes glycogen accumulation in tissues

(Section 2.1.2.3) there are no known neural or muscular diseases proven due to abnormalities of glycogen synthesis. Corpora amylacea and Lafora bodies, as occur in normal ageing and in a peculiar brain degeneration respectively, are both polyglucosan collections reasonably similar in molecular structure to glycogen. The chemical factors responsible for their occurrences are unknown. A number of disorders of glycogen breakdown exist (the glycogenoses). These disorders result in glycogen accumulation in various tissues. Several of these glycogen accumulation disorders, in particular deficiency of the lysosomal enzyme acid maltase (α-1,4-glucosidase), affect the nervous system clinically. Others, e.g. McArdle's syndrome (myophosphorylase deficiency), affect skeletal muscle function by limiting the energy available for sustained or repeated contraction.

2.1.1 Disorders involving accumulation of abnormal polyglycosans

2.1.1.1 Corpora amylacea formation

Corpora amylacea occur in fibrous astrocytes and neurons with increasing frequency as the nervous system ages. Their chemical composition is described by Austin[2] and Austin and Sakai[3]. These bodies consist chiefly of polyglucosans (glucose polymers containing both α-1,4- and α-1,6-glycosidic links), though some acid phosphate groups are present. Corpora amylacea do not contain amyloid, which is an abnormal protein. Ultrastructurally, and histochemically, corpora amylacea are relatively similar to but not identical with Lafora bodies (see below). The biochemical factors responsible for the formation of corpora amylacea are unknown. The presence of corpora amylacea does not itself appear to cause neural dysfunction.

2.1.1.2 Lafora body disease

Lafora body disease is an uncommon inherited disorder of later childhood and adolescence. It causes progressive dementia and myoclonic epileptic seizures[3]. Lafora bodies accumulate in cells in neural and several extraneural tissues. The biochemical pathogenesis of Lafora body disease is not understood.

Biochemical abnormality – Lafora bodies occur in the perikarya of neurons, in the liver and the myocardium. In composition these bodies are almost pure glucose polymer and contain both α-1,4- and α-1,6-glucosidic links. There is a tendency for α-1,6- branching points to occur near the surface of the polyglucosan macromolecules of Lafora bodies. This polyglucosan differs in some ways from that of normal glycogen. Lafora bodies contain some acidic groups (probably phosphate) and also sulphur molecules and a little protein. It is not known whether the biochemical fault responsible for Lafora body formation lies in the defective synthesis of glucose polymer or in its defective degradation[4]. How the presence of Lafora bodies disturbs tissue function is unclear.

Aetiology – The disorder is inherited as an autosomal recessive trait. The detailed biochemical genetics are unknown.

Structural pathology – The neuropathological findings in Lafora body disease are described in detail by Seitelberger[5]. The brain in Lafora body disease shows some neuronal loss, with gliosis. The substantia nigra may be depigmented and there is neuronal fall-out in the inferior olives, pontine tegmentum, dentate nuclei and thalami. Lafora bodies occur in these structures, and in the neocortex, pallidum and retina. Lafora bodies may be as large as 30 μm in diameter. They have a pale periphery which stains with Best's carmine (for glycogen) and a basophil core, which stains with the PAS reagent. Lafora bodies and multiple small PAS positive granules occur in liver cells, and basophil deposits with an acidophil periphery may occur in myocardial fibres.

Clinical features – Myoclonus epilepsy (Lafora body disease: Unverricht–Lundborg disease) usually begins between the ages of 6 to 20 years, and runs an average course of 3 to 10 years. Very rare adult onset cases are known[6]. Initially there are personality alterations and brief bilateral myoclonic seizures.

The e.e.g. often shows the bilateral high amplitude discharges of an acquired type of generalized epilepsy. Progressive dementia with increasing myoclonic and convulsive seizures develops. In the later stages features of parkinsonism and cerebellar deficiency may be superimposed on the dementia and epilepsy.

Robitale et al.[4] have described four cases of a somewhat different syndrome associated with the presence of enormous numbers of Lafora bodies in neuronal processes and astrocytes. There was progressive upper and lower motor neuron involvement, with marked sensory loss in the lower limbs, neurogenic bladder dysfunction and, in two cases, dementia.

Diagnosis
(1) Clinical: The disorder should be differentiated from other causes of progressive dementia with bilateral myoclonic seizures which begin in later childhood or early adult life. Such disorders include (a) the ceroid lipofuscinoses, (b) the cherry-red macula myoclonus syndrome (sialidosis), (c) sequelae of brain hypoxia, and (d) subacute sclerosing panencephalitis.

(2) Laboratory: Definitive diagnosis depends on the identification of Lafora bodies in brain biopsy specimens or characteristic inclusions in liver biopsies[7] using light and electron microscopy, and histochemical techniques.

Therapy – The epilepsy may be treated symptomatically, with anticonvulsants. No curative therapy is available. The disorder pursues a fatal course despite all known treatments.

Table 2.1   Varieties of glycogen storage disease (after Hug[9])

Glycogenosis type	Deficient enzyme
0	uridine diphosphate glucose-glycogen transferase (glycogen synthetase)
I	glucose-6-phosphatase (hepatic)
II	α-1,4-glucosidase (lysosomal), i.e. acid maltase
III	amylo-1,6-glucosidase
IV	amylo-1,4→1,6-transglucosidase (branching enzyme)
V	phosphorylase (muscle)
VI	phosphorylase (liver)
VII	phosphofructokinase (muscle)
VIII	phosphorylase (inactive, not deficient in liver)
IX	phosphorylase kinase (liver)
X	cyclic 3′5′-AMP dependent kinase
XI	?

13

2.1.2   Disorders involving impaired catabolism of glycogen

Impaired glycogen catabolism leads to glycogen accumulation in tissues. Eleven types of glycogen storage disease (Figure 2.4, Table 2.1) are known[8,9]. Only six appear to involve the nervous system and/or skeletal muscle. Of these six disorders, one (phosphofructokinase deficiency) is due to a defect in anaerobic glycolysis and is considered later (Section 2.2.1.1), in relation to that stage of glucose metabolism.

2.1.2.1   Acid α-1,4-glucosidase (acid maltase) deficiency: glycogenosis Type II (Pompe's disease)

Pompe's disease is an uncommon inherited disorder in which glycogen accumulates in the brain and viscera of infants, older children or adults. The disorder is due to deficiency, or defective function, of the lysosomal enzyme acid maltase (acid α-1,4-glucosidase). The disorder is sometimes classified among the lysosomal storage disorders.

Biochemical abnormality – In Pompe's disease, lysosomal glycogen (taken up by autophagy) accumulates due to catalytic inactivity (glycogenosis Type IIa) or deficiency (glycogenosis Type IIb) of acid maltase. Extralysosomal glycogen breakdown via debranching enzyme and phosphorylase is not affected. Acid maltase catalyses the cleavage of both α-1,4- and α-1,6-glucosidic links[10].

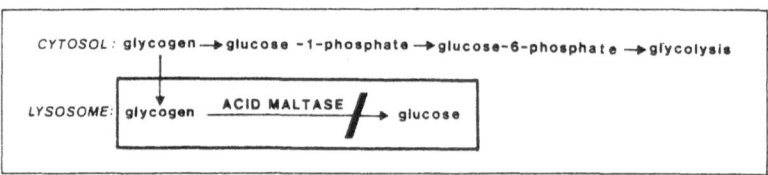

Apart from its mechanical effects, which might disturb other lysosomal functions, it is difficult to see how intralysosomal glycogen accumulation could disturb function of nervous tissue and skeletal muscle when extralysosomal glycogen degradation is normal. Conceivably, reduced cleavage of some polysaccharide or oligosaccharide other than glycogen is the cause of the altered tissue function in Pompe's disease.

Two instances have been reported of a clinically and ultrastructurally similar disorder with intralysosomal glycogen accumulation in muscle, but with no acid maltase deficiency[11].

Aetiology – The disorder is inherited as an autosomal recessive trait. In Type IIa glycogenosis, acid maltase is catalytically inactive, suggesting the synthesis of abnormal enzyme in this variant. In Type IIb glycogenosis there appears to be deficient synthesis of normal enzyme.

Structural pathology – There may be enlargement of the heart, skeletal muscle (including the tongue and diaphragm) and the liver, but not of the brain, in Pompe's disease. There is excess normally-structured glycogen in nearly all tissues (demonstrated by PAS or Best's carmine staining). There are ballooned neurons in the brain stem, deep cerebellar nuclei and spinal cord (anterior horn cells), with glycogen accumulation in the glia. The cerebral and cerebellar cortices usually appear intact, though there may be gliosis of the white matter. Metachromatic material (possibly acid mucopolysaccharide) occurs in skeletal muscle, and sometimes in the brain and heart. At electron microscopy, cells from various tissues show lysosomal storage vacuoles packed with glycogen granules, though there may also be extravacuolar glycogen accumulation in cardiac and skeletal muscle.

Clinical features – In the infantile form of glycogenosis Type II, symptoms appear in the first few months of life. There is failure to thrive, profound hypotonia, reduced motor activity, dyspnoea, cyanosis and sometimes febrile periods. The heart is enlarged and the e.c.g. may show giant QRS complexes in all leads, with a short PR interval. Despite the weakness there is no muscle wasting. The e.m.g. may show activity resembling myotonia. Intelligence tends to be normal. Death from cardiac failure often occurs in early life, but if the heart is not enlarged longer survival is possible. Juvenile and adult cases do occur. In the latter, the presentation is typically that of a chronic myopathy, first affecting the lower limbs, with e.m.g. changes suggesting the presence of myotonia, though there is no clinical myotonia[12]. Some adult cases may present with diaphragmatic paralysis and dyspnoea[13].

Diagnosis
(1) Clinical: The clinical features of the fully developed infantile form of Pompe's disease, with an enlarged heart and tongue, are highly suggestive of the diagnosis. A myopathic distribution of weakness in the late onset form, without much muscle wasting, might raise the possibility of the diagnosis. The question of Duchenne or other type of muscle dystrophy might arise.

(2) Laboratory: In the infantile form of the disorder the e.c.g. may be characteristic. Glycogen accumulation in muscle or liver biopsies can be detected histochemically, biochemically, or by electron microscopy. The ultrastructural appearance of the intravacuolar (lysosomal) glycogen deposits helps differentiate Pompe's disease from other forms of glycogenosis. The diagnosis can also be made by electron microscopy of skin biopsies[14]. Definitive diagnosis depends on demonstration of reduced, or absent, acid α-1,4-glucosidase activity in liver or muscle biopsies, in peripheral lymphocytes[15] or in cultured skin fibroblasts or in amniotic fluid cells. The enzyme is also deficient in the placenta and umbilical cord of affected infants. Heterozygous carriers can be detected by finding reduced enzyme activity in cultured skin fibroblasts. Although acid α-1,4-glucosidase activity is

15

present in normal urine, this enzyme activity is not deficient in acid maltase deficiency.

Therapy – No useful therapy for Pompe's disease is known. Purified acid maltase administration has not helped sufferers.

2.1.2.2 Myophosphorylase deficiency (glycogenosis Type V – McArdle's syndrome)

This uncommon disorder of childhood and adult life causes severe muscle pain, cramp and swelling on exercise. It is due to a deficiency of muscle phosphorylase. Brain function is unaltered. Howell[8] provides a detailed description of the disorder.

Biochemical abnormality – Glycogen synthesis proceeds normally in McArdle's syndrome, but the polyglucosan that is formed cannot be degraded to glucose-1-phosphate in the cytosol of skeletal muscle, because of deficiency of the muscle enzyme phosphorylase. Liver phosphorylase is a different protein which is not deficient in the disorder, so that liver glycogen can be degraded.

The inability of skeletal muscle to utilize its energy reserves stored as glycogen causes an inadequate production of high energy phosphate (adenosine triphosphate and creatine phosphate) when muscle contracts repeatedly. The contractile mechanism then becomes starved of ATP, and fails. However, for less vigorous exercise, muscle may synthesize sufficient ATP from the aerobic metabolism of glucose and fatty acids to continue satisfactory functioning.

Aetiology – The disorder is usually inherited as an autosomal recessive trait, though rarely as an autosomal dominant.

Structural pathology – Subsarcolemmal and intermyofibrillar glycogen vacuoles occur in skeletal muscle fibres[16].

Clinical features – From early adult life, affected patients experience muscle pain, cramp, swelling, fatigue and weakness on attempting repeated or sustained muscle contraction. There is no muscle wasting or reflex abnormality. After exercise, the swollen, apparently contracted, muscle is electrically silent to concentric needle electromyography. Repeated attempts at exercise may cause muscle injury leading to myoglobinaemia and myoglobinuria. Ultimately permanent weakness of proximal muscles may occur. Rarely, the disorder may cause generalized weakness in infancy, with early death[17]. After exercise which produces painful cramps, plasma levels of lactate dehydrogenase, aldolase and creatine kinase may rise considerably.

Diagnosis
(1) Clinical: The occurrence of electrically silent contracture at electromyography after ischaemic exercise in a patient with a complaint of muscle pain

and stiffness on exertion is very suggestive of the diagnosis. However, defects in other enzymes involved in glycogenolysis and the earlier stages of anaerobic glycolysis may produce a similar clinical picture.

(2) Laboratory: Ischaemic exercise fails to produce a rise in blood lactate level in McArdle's syndrome[18]. A similar phenomenon occurs in related disorders in which there is an insufficient availability of glucose metabolites in skeletal muscle to meet energy requirements. The deficiency in phosphorylase activity may be demonstrated histochemically or biochemically in muscle biopsy specimens.

Therapy – No effective therapy is known. Severe exercise should be avoided. In theory a high fructose diet might increase the availability of energy sources to skeletal muscle (see Figure 2.5). In practice, this proves of little value, and may simply promote weight gain.

2.1.2.3 *Other glycogenoses*

The known types of glycogenosis are summarized in Table 2.1 and Figure 2.4.

Occasional patients with Type III glycogenosis, i.e. amylo-1,6-glucosidase deficiency (debranching enzyme deficiency) may develop muscle weakness[19,20]. In a series of five such cases in adults[21] there was slowly progressive weakness beginning in adult life. In two of these subjects there was distal wasting, in two there was heart failure, and in three hepatomegaly. The e.m.g. showed fibrillation, pseudomyotonic discharges and infrequent polyphasic units of prolonged duration. The muscle glycogen content was increased and the muscle contained vacuoles at microscopy. At least one patient with Type IV glycogenosis (amylo-1,4→1,6-transglucosidase deficiency, brancher enzyme deficiency) is known to have had hypotonia and muscle atrophy[22]. In this condition glycogen spheroids are found in the brain stem and spinal cord.

In Type VIII glycogenosis (due to inactive liver phosphorylase) giant glycogen particles occur in the cerebral cortex, and there is progressive neurological deterioration leading to decerebration and death[9].

Epileptic seizures have been reported in relation to glycogenosis Type I (glucose-6-phosphatase deficiency), glycogenosis Type III, and glycogenosis Type 0 (glycogen synthetase, i.e. uridine diphosphate glucose glycogen glucosyl transferase deficiency), according to Menkes[23]. In both Type 0 and Type I glycogenosis, hypoglycaemia may occur due to impaired hepatic formation of glucose from glycogen. Possibly this hypoglycaemia causes, or contributes to, the epilepsy that has been reported in these particular disorders.

Type VII glycogenosis (muscle phosphofructokinase deficiency) is a disorder of Embden–Meyerhof glycolysis, as was mentioned earlier, and is therefore described in the following section.

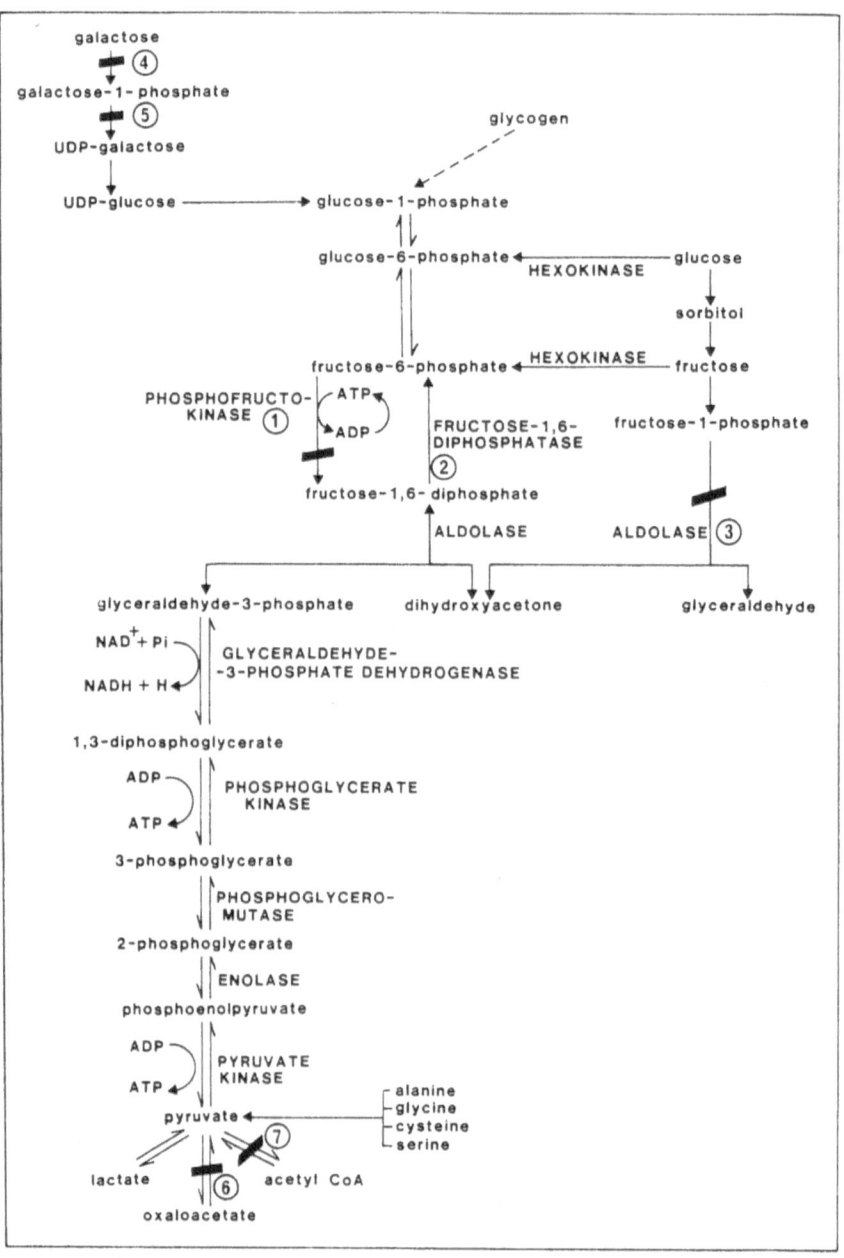

Figure 2.5 The Embden–Meyerhof glycolytic pathway, and relevant associated reactions. Sites of enzyme defects which cause known human disease are indicated. (1) phosphofructokinase, (2) fructose-1,6-diphosphatase, (3) aldolase, (4) galactose kinase, (5) galactose-1-phosphate uridyl transferase, (6) pyruvate carboxylase, (7) pyruvate dehydrogenase complex

2.2 DISORDERS OF EMBDEN–MEYERHOF GLYCOLYSIS

The series of reactions that comprises the glycolytic pathway permits the conversion of one molecule of glucose to two molecules of pyruvate, with the net production of two molecules of adenosine triphosphate and the reduction of one molecule of NAD^+. There are certain reactions associated with this pathway which are also relevant to the concerns of the clinical neurologist. This anaerobic glycolysis pathway, with the relevant associated reactions, is shown in Figure 2.5.

Simply to bring it into proximity to the other glycogenoses, phosphofructokinase deficiency is the first disorder dealt with in this section. Then disorders involving reactions, or reactants, feeding into the pathway are considered, followed by disorders of constituents of the more distal glycolytic pathway. In practice this last category comprises disorders of pyruvate metabolism.

2.2.1 Disorders of the proximal direct glycolytic pathway

2.2.1.1 Phosphofructokinase deficiency: glycogenosis Type VII

This very rare inherited disorder, causing exercise-related weakness and pain in skeletal muscle, is due to deficiency of the enzyme phosphofructokinase.

Biochemical abnormality – Phosphofructokinase catalyses the forward reaction phosphorylating fructose-6-phosphate to form fructose-1,6-diphosphate. The enzyme deficiency leads to reduced flux of glucose derivatives through the glycolytic pathway. Glycogen accumulates and there is a shortage of the metabolites required for high energy phosphate production. As in McArdle's syndrome, the clinical features of the disorder seem related to insufficient energy being available to skeletal muscle to permit repeated or sustained muscle contraction.

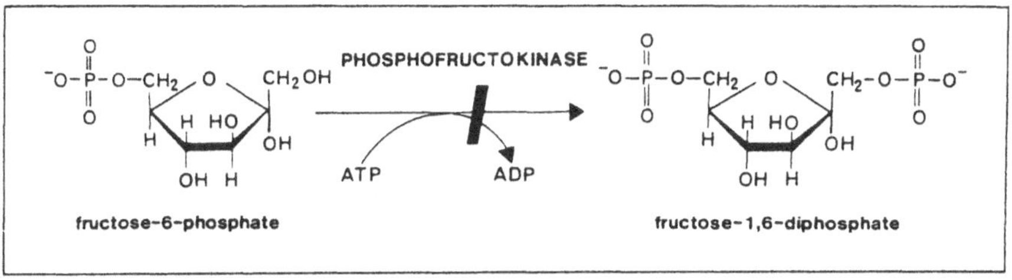

Aetiology – Phosphofructokinase deficiency is inherited as an autosomal recessive trait. The molecule of phosphofructokinase is a tetramer. The muscle enzyme comprises four identical subunits, all of which are absent in glycogenosis Type VII. Levels of the erythrocyte enzyme, which contains two different subunits, are reduced by 50% in the disorder.

19

Structural pathology – Skeletal muscle in the disorder contains subsarcolemmal glycogen blebs.

Clinical features – The clinical presentation of the disorder[24] is very similar to that of myophosphorylase activity, namely fatigue and aching pain in skeletal muscle with more severe pain, contracture and sometimes myoglobinuria on strenuous muscular exercise.

Diagnosis
(1) Clinical: The clinical features suggest the consequences of a deficient energy supply to skeletal muscle.

(2) Laboratory: The failure of ischaemic exercise to produce a rise in venous lactate concentration supports the diagnostic category, but exact diagnosis requires demonstration of deficient phosphofructokinase activity in a muscle biopsy specimen.

Therapy – No effective therapy is available.

2.2.1.2 *Hyperglycaemia*

Hyperglycaemia, nearly always due to diabetes mellitus, is a relatively common metabolic disturbance which may alter the functions of both central and peripheral nervous systems. The biochemical mechanisms involved are not fully understood though a certain amount of information is available[25]. Diabetes mellitus has other chemical effects on the nervous system (e.g. those due to ketoacidosis and dehydration with hyperosmolality) apart from the consequences of pure hyperglycaemia. Only the latter are considered here. Certain other consequences of diabetes mellitus are considered in Chapter 4. The effects of hyperglycaemia have been reviewed by Prockop[25].

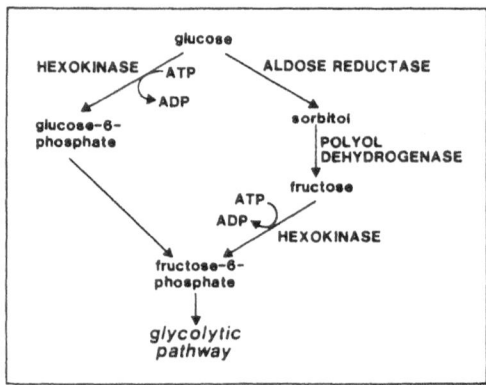

Figure 2.6 Alternative metabolic pathways for glucose

Biochemical abnormality – After entry into nervous tissue from the bloodstream, following active transport through the blood–brain barrier[26], glucose may be metabolized along one of two metabolic pathways, as set out in Figure 2.6.

The enzyme hexokinase has a low K_m for glucose, whereas aldose reductase has a high K_m for this hexose. Consequently glucose at physiological concentrations is preferentially phosphorylated to its -6-phosphate, and little sorbitol (D-glucitol) forms. However, if glucose concentrations rise, there is a relative diversion of glucose towards sorbitol formation, along the glucuronate (or polyol, i.e. polyhydric alcohol) pathway. If this diversion occurs, sorbitol and fructose tend to accumulate in neural tissue because hexokinase has a much higher K_m towards fructose than towards glucose. There is therefore rather less flux of glucose metabolite through the glycolytic pathway than might have been expected purely from the degree of hyperglycaemia. Fructose levels in CSF rise in parallel with the blood glucose concentration.

Details of the polyol pathway (with the hexoses shown in open chain structure rather than in the cyclic form in which they have been represented earlier in this chapter) are shown in Figure 2.7. In solution the hexoses coexist in

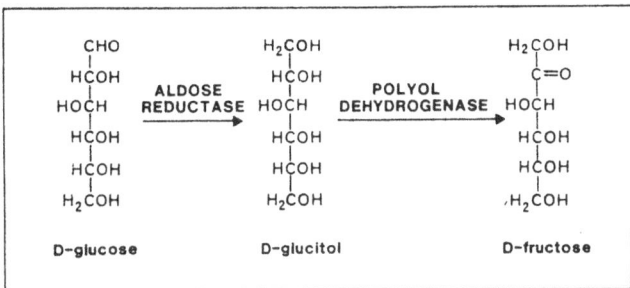

Figure 2.7 Structural details of the polyol pathway

open chain and in cyclic form. Polyols do not cross cell membranes readily. Hence sorbitol, formed intracellularly in increased amounts in the presence of hyperglycaemia, and, with its metabolite fructose, only slowly metabolized (by hexokinase), tends to accumulate *in situ*. By its osmotic effect, sorbitol increases intracellular water content and cell volume. (A similar set of events may occur in the lens of the eye and cause cataract.) If a dehydrated diabetic patient is treated with intravenous fluids, as plasma osmolality is restored toward normal it is possible that sufficient water dislocation may occur into the intracellular space with its raised polyol concentration for cerebral oedema and raised intracranial pressure to develop. In peripheral nerve it is possible that the fluid shifts may separate the myelin lamellae[25]. However, in a study of sural nerve biopsies from diabetic patients Dyck *et al.*[27] found that sorbitol and fructose content correlated better with severity of hyperglycaemia than with severity of neuropathy.

In diabetes mellitus, reduced tissue concentrations of myoinositol (i.e. inositol, another polyol) may occur. Myoinositol levels fall as sorbitol levels

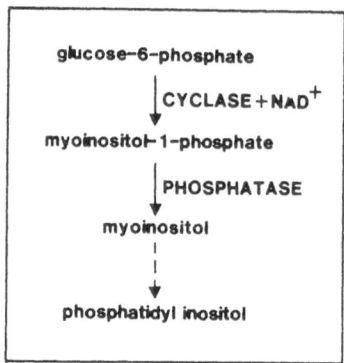

rise. The biochemical mechanisms involved in the changed inositol levels are unclear. The consequence of reduced inositol levels may be decreased formation of phosphatidyl inositol, a myelin constituent. In chronic experimental diabetes mellitus in rats, sciatic nerve phosphatidylinositol concentration was reduced[28]. This alteration in inositol derivatives may contribute to the peripheral neuropathy that occurs in chronic hyperglycaemia. However, myoinositol levels were not consistently reduced in human diabetic sural nerve[27], and were more often elevated in diabetic than in control nerves.

.inositol (cyclohexanehexanol)

Another factor which may conceivably contribute to nervous system dysfunction in hyperglycaemia is the increased non-enzymatic glucosylation of free amino groups in proteins (mainly lysine and arginine residues) which may occur because of the high concentration of glucose that is present. Such N-glucosidation of haemoglobin in diabetes mellitus is well known, and it seems likely that similar reactions may occur on other proteins (Figure 2.8), though the question has not been extensively explored.

Aetiology – Temporary hyperglycaemia may be due to glucose infusion or high glucose intake, but in the longer term hyperglycaemia almost always occurs as part of the metabolic disturbances of diabetes mellitus.

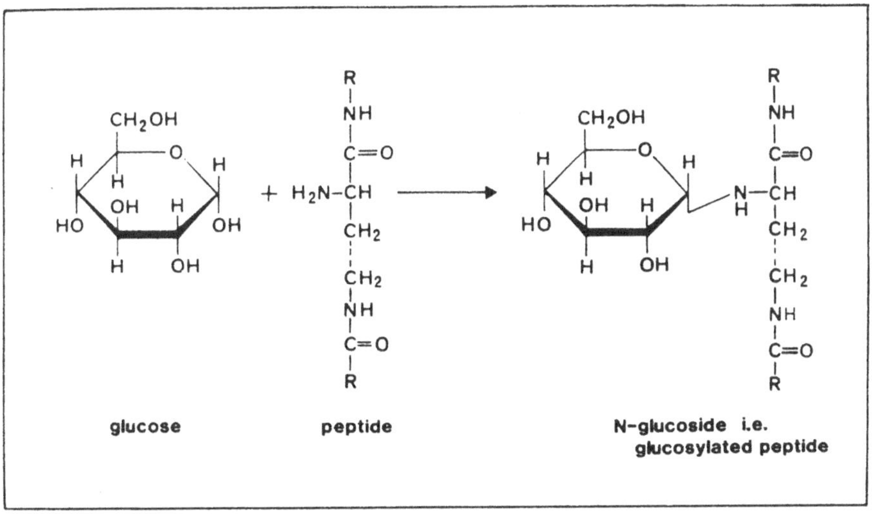

Figure 2.8 Formation of a glucosylated peptide

Structural pathology – Cerebral oedema may develop in hyperglycaemia. Peripheral nerves in diabetes mellitus show segmental demyelination with axon degeneration. Autonomic nerves may also be affected. The extent to which these latter changes depend on hyperglycaemia itself is uncertain. Associated arterial disease and acidosis may make some contribution to the peripheral neuropathy of diabetes mellitus, and to other patterns of nervous system damage in the disorder.

Clinical features – The effects of diabetes mellitus on brain function, namely precoma and coma, are due to the combined consequences of dehydration, acidosis and the biochemical effects of hyperglycaemia itself (see above). Occasional cases of diabetic coma occur whose clinical state at first improves when insulin and fluids are given as treatment. Subsequently the level of consciousness in these patients deteriorates as they develop irreversible cerebral oedema through the biochemical mechanisms mentioned above (namely raised intracellular osmolality with lowered extracellular osmolality.) Severe hyperglycaemia, without ketoacidosis, may occur in diabetes and may cause sufficient dehydration, particularly in the elderly, to produce hyperosmolar, hyperglycaemic, non-ketotic diabetic coma, usually in the presence of plasma osmolalities[29] exceeding 350 mOsm/l.

Peripheral nerve involvement in diabetes mellitus may take the form of (1) a peripheral polyneuropathy, (2) a mononeuropathy (often femoral or cranial), and (3) an autonomic neuropathy, often causing hypotension, diarrhoea and impaired sphincter control, but sometimes affecting the innervation of the pupils.

Diagnosis
(1)   Clinical: The possibility of hyperglycaemia is suggested by the occurrence of polyuria and polydypsia, particularly in the presence of weight loss, evidence of dehydration and ketoacidosis, and a diminished level of consciousness.

(2)   Laboratory: The diagnosis is established by blood glucose measurement. Glycosuria and ketonuria are suggestive.

Treatment – Diabetes mellitus is treated by insulin replacement therapy. In rehydrating patients in diabetic coma or precoma the dangers of too rapid a restoration of plasma osmolality should be remembered. The peripheral neuropathic manifestations may not be completely reversed by insulin therapy.

2.2.1.3 Hypoglycaemia

A blood glucose level below 40 mg/100 ml, i.e. 2.5 μmol/l (30 mg/100 ml in full term infants, 20 mg/100 ml in premature babies), however caused, is likely to be associated with disturbed neurological function. Such an event is not particularly common in clinical neurological practice, but its recognition is most important. The effects of hypoglycaemia on the nervous system are reviewed by Wilkinson and Prockop[30].

Biochemical abnormality – Glucose is the brain's chief source of energy, though the brain can obtain some energy by oxidizing ketone bodies (acetoacetate and D-3-hydroxybutyrate) which are formed mainly in the liver and are carried to the brain in the blood. Unlike muscle, brain cannot obtain significant amounts of energy from fatty acid oxidation. Hypoglycaemia leads to decreased metabolic flux through the glycolytic pathway and Krebs cycle. Hence hypoglycaemia is likely to cause decreased energy availability within cells. Neural tissue contains very little stored glycogen which can be used to yield energy when there is a shortage of glucose. In rats if blood glucose levels fall below 3 μmol/l there is no measurable intracellular glucose[31]. In hypoglycaemia amino acids may be deaminated to keto acids which enter the Krebs cycle to help it continue turning. This mechanism is of limited compensatory capacity[32] and distorts amino acid metabolism. In the neonatal dog's brain there may be increased lactate utilization, a mechanism which may contribute to the neonatal brain's better tolerance of hypoglycaemia[33]. The potential compensatory mechanisms are summarized in Figure 2.9. During hypoglycaemia in humans cerebral blood flow does not increase[34].

As hypoglycaemia develops, brain function becomes disturbed earlier than the function of any other organ. In the presence of hypoglycaemia, skeletal muscle can continue functioning considerably longer because of its capacity to obtain energy from fatty acid oxidation and from its glycogen reserve.

Experimental work in animals suggests that brain function in hypoglycaemia may become disturbed before there is any fall in high energy phosphate

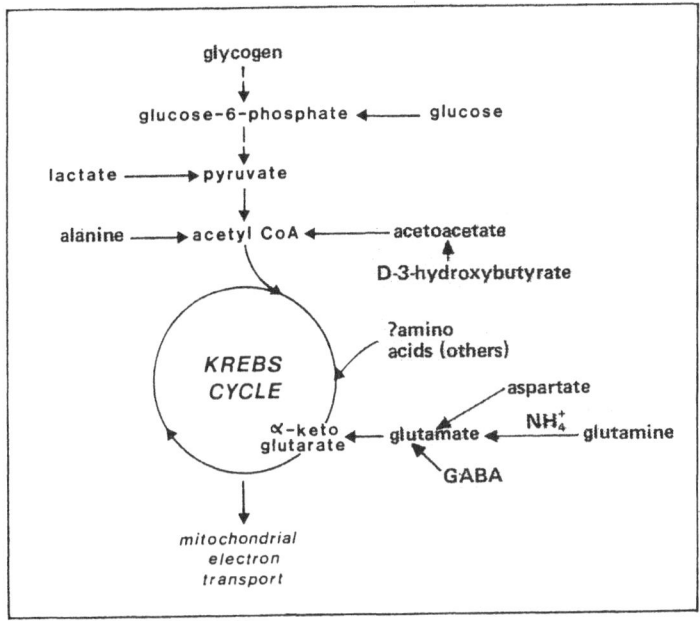

Figure 2.9 Alternative sources of energy (shaded over), and relevant metabolic pathways (in outline), in hypoglycaemia

stores[31,35-37]. However, by the time the animal's e.e.g. becomes isoelectric there is profound energy failure[31]. Before there is any shortage of ATP, brain levels of acetylcholine[38], glutamate, glutamine, γ-aminobutyrate and alanine all fall, and aspartate levels rise, while NH_4^+ accumulates[34,35,39]. However, Dirks *et al.*[37] found that γ-aminobutyrate levels were preserved in these circumstances. These various changes probably reflect effects of diversion of amino acids and acetyl derivatives into the Krebs cycle (Figure 2.9). There are consequent altered concentrations of acetylcholine and of amino-acid neurotransmitters. These neurotransmitter changes may be the initial cause of altered brain function in hypoglycaemia. Actually, in hypoglycaemia the brain makes decreased use of its available glucose, due to inhibition which develops at the phosphofructokinase stage of glycolysis. The fact that brain function in hypoglycaemic animals may return to normal within 45 s of intravenous glucose administration, by which time only brain levels of glucose and aspartate have recovered, also suggests that the earlier stages of the neurological disturbance in hypoglycaemia may be due to altered neurotransmission[35]. Further, in animals there is evidence that reduced brain glucose level may in some way rapidly switch off use of ATP[36], thereby conserving energy. Thus there is now some basis for calling into question the simplistic view that all manifestations of hypoglycaemia are due purely to progressive limitation of available energy.

25

However the later stage manifestations of hypoglycaemia probably are due to insufficient energy availability.

Pre-existing clinical or subclinical brain abnormality from other causes may be present in patients who subsequently develop hypoglycaemia. Therefore the clinical expression of hypoglycaemia may vary somewhat from person to person, since hypoglycaemia may bring abnormal brain areas to their threshold of clinical dysfunction before function of normal brain becomes disturbed.

Aetiology – The causes of hypoglycaemia include the following:

(1) The presence of excess circulating insulin, which may be
    (a) Iatrogenic, as in the treatment of diabetes mellitus,
    (b) Due to insulinoma or to pancreatic islet cell hyperplasia,
    (c) Reactive, as in
       – leucine sensitivity
       – alimentary hypoglycaemia
       – mild, maturity onset, diabetes mellitus.

(2) Glucose production failure, which may occur in
    (a) Severe liver disease,
    (b) Various glycogenoses (failure to break down preformed glycogen),
    (c) Metabolite inhibition of glycogen breakdown
       – fructose-1-phosphate aldolase ⎫
       – galactose-1-phosphate uridyl transferase ⎬ deficiencies
       – familial fructose or galactose intolerance, ⎭
    (d) Failure to form glucose-6-phosphate, required for gluconeogenesis, due to
       – pyruvate carboxylase ⎫
       – fructose-1,6-diphosphatase ⎬ deficiencies.
       – phosphoenolpyruvate carboxykinase ⎭

(3) Deficiency of hormones which normally oppose the effects of insulin
       – panhypopituitarism
       – adrenal cortical insufficiency
       – hypothyroidism
       – isolated growth hormone deficiency
       – isolated ACTH deficiency.

(4) Decreased supply of gluconeogenic material
       – ketotic hypoglycaemia in childhood
       – fasting hypoglycaemia of late pregnancy
       – starvation.

(5) Renal glycosuria.

(6) Retroperitoneal tumours.

(7) Overdosage with oral hypoglycaemic agents.

(8) Salicylate overdose.

(9)   Idiopathic hypoglycaemia.

Of this considerable range of causes, the most important in ordinary clinical neurological practice is pancreatic insulinoma.

Structural pathology – Probably milder degrees of hypoglycaemia produce no structural change in the nervous system. In more severe and prolonged hypoglycaemia which terminates fatally[40], there may be focal or laminar necrosis of the cerebral cortex with relative preservation of the visual cortex. Ammon's horn is involved and there may be striatal necrosis, particularly affecting small neurons, though the globus pallidus usually remains intact. The lateral nucleus of the thalamus is commonly damaged and there may be a relatively selective loss of cerebellar Purkinje cells. Ischaemic necrotic type changes have been reported in anterior horn cells. If the patient survives long enough, gliosis occurs at sites of neuronal damage in hypoglycaemia.

Clinical features – The manifestations of hypoglycaemia are protean[41]. They are often episodic, and tend to occur some hours after the most recent intake of food. Symptoms, if not present for too long, are rapidly relieved by the intake of food or glucose. Hypoglycaemia may cause epileptic seizures or episodes of confusion, stupor or altered behaviour which may raise the possibility of psychiatric disorder. There may be manifestations of peripheral adrenergic overactivity, e.g. sweating, pallor. Focal motor or cerebellar disturbances may occur, and in more severe hypoglycaemic episodes there may be coma. The coma may be fatal, if severe enough hypoglycaemia persists for long enough. It is said that chronic hypoglycaemia may cause a clinical picture resembling pure progressive muscular atrophy, presumably resulting from anterior horn cell damage.

Diagnosis
(1)   Clinical: Hypoglycaemia enters the differential diagnosis of many episodic disturbances of nervous system function, and is particularly likely when the manifestations tend to occur some hours after the last intake of food. Rapid reversal of the disturbance after glucose intake is highly suggestive diagnostically.

(2)   Laboratory: The diagnosis of hypoglycaemia is made by blood glucose measurement. Sometimes it may be necessary to carry out a prolonged (e.g. three day) fast, with serial blood glucose estimations, to diagnose the development of significant episodic hypoglycaemia. Simultaneous plasma insulin assays may permit the strong suspicion that an insulinoma is present. Appropriate tests should otherwise be done to seek out the aetiology of the hypoglycaemia once its presence is demonstrated (see the list of causes, above).

Therapy – Treatment of hypoglycaemia comprises the oral or intravenous administration of sufficient glucose to correct clinical symptoms, with appro-

27

priate treatment for the causative disorder whenever possible, e.g. surgical removal of a pancreatic insulinoma.

2.2.2 Disorders of fructose metabolism

There are two known disturbances of fructose metabolism[42]. Both typically produce symptoms by causing hypoglycaemia.

2.2.2.1 Fructose-1,6-diphosphatase deficiency

This very rare inherited disorder leads to defective gluconeogenesis, with consequent hypoglycaemia. The disorder results from fructose-1,6-diphosphatase deficiency.

Biochemical abnormality – Fructose-1,6-diphosphate catalyses the hydrolysis of fructose-1,6-diphosphate to fructose-6-phosphate (Figure 2.5), the reverse

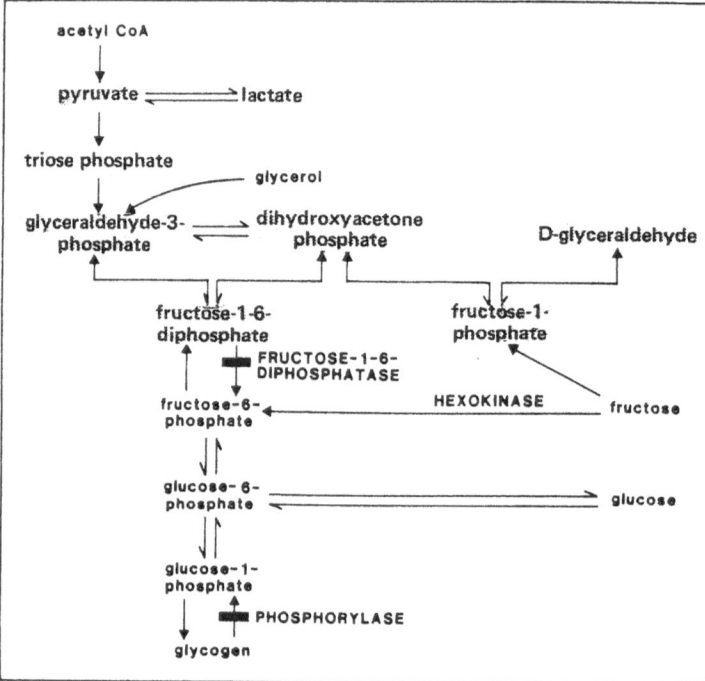

Figure 2.10 Metabolic (gluconeogenic) pathways relevant to fructose-1,6-diphosphatase deficiency. Reactants prior to the relevant enzyme in the gluconeogenetic pathway accumulate, and the triose phosphate esters inhibit phosphorylase. The latter inhibition results in hypoglycaemia. Substances which accumulate in the disorder are shaded

28

reaction to that catalysed by phosphofructokinase. Deficiency of the latter enzyme (Section 2.2.1.1) produces a quite different clinical and biochemical syndrome.

A metabolic block at the fructose-1,6-diphosphatase-catalysed reaction stage of glycolysis produces a deficiency of inorganic phosphate and an accumulation of various glycolytic diphosphate esters (Figure 2.10). Several of these esters appear to inhibit the enzyme phosphorylase a. Inhibition of phosphorylase a reduces the breakdown of glycogen to glucose, causing hypoglycaemia and metabolic (lactic) acidosis. (The acidosis probably occurs because (1) keto-acid formation increases to provide an alternative source of 'fuel' for the Krebs cycle when there is a shortage of glucose metabolites, and (2) more pyruvate converts to lactate when there is decreased Krebs cycle activity and decreased gluconeogenesis.) In fructose-1,6-diphosphatase deficiency, blood glucose levels can be maintained at normal values only by a continuing dietary intake of glucose or galactose. Dietary fructose or glycerol intake fails to maintain blood glucose levels in these circumstances.

Aetiology – The disorder is inherited as an autosomal recessive trait.

Structural pathology – No details are available.

Clinical features – Symptoms usually appear in the first week of life in the affected neonate. Hypoglycaemia and metabolic acidosis lead to trembling, hypotonia, convulsions and coma. These symptoms occur in episodes which appear when food is not taken sufficiently often. Hepatomegaly often occurs.

A single instance of central core myopathy associated with a selective deficiency of muscle fructose-1,6-diphosphatase has been recorded[43].

Diagnosis
(1) Clinical: The presence of appropriate clinical features in an infant may lead to suspicion of hypoglycaemia.

(2) Laboratory: In the disorder hypoglycaemia, together with lactate accumulation in the blood, develop on prolonged fasting. The hypoglycaemia is refactory to glucagon, which normally raises blood glucose levels by increasing hepatic gluconeogenesis. The enzyme defect can be measured in liver biopsy specimens.

Therapy – Hypoglycaemic episodes may be treated with glucose, and any acidosis should be corrected. Frequent feeding with glucose or lactose helps prevent episodes. Fructose and sorbitol intake in the diet should be avoided.

2.2.2.2 Hereditary fructose intolerance

This is an uncommon inherited disorder of early life which results in hypoglycaemia and its consequences[42,44]. Fructose intolerance is due to a deficiency of the enzyme aldolase (the liver variety of the enzyme only).

Biochemical abnormality – Aldolase catalyses the reactions in which (1) fructose-1,6-diphosphate forms dihydroxyacetone phosphate and glyceraldehyde-3-phosphate, and (2) fructose-1-phosphate forms glyceraldehyde and dihydroxyacetone phosphate (Figures 2.5, 2.11). Hepatic aldolase has a K_m

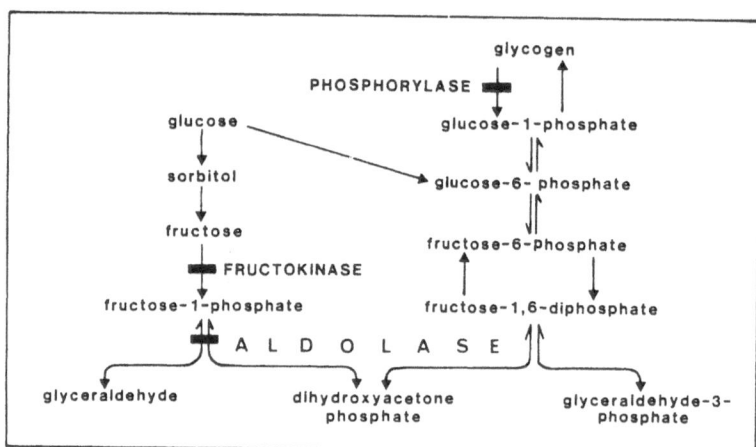

Figure 2.11 Metabolic pathways relevant to aldolase deficiency. The accumulating fructose-1-phosphate inhibits phosphorylase

towards fructose-1-phosphate about 100 times greater than its K_m towards fructose-1,6-diphosphate. Consequently, when aldolase is deficient, after a fructose or sorbitol load fructose-1-phosphate accumulates very much more than does fructose-1,6-diphosphate. Effectively, the defect becomes almost entirely one of fructose-1-phosphate aldolase activity. The accumulating fructose-1-phosphate inhibits phosphorylase, as mentioned earlier. As a result glycogen breakdown is impaired. Hypoglycaemia occurs after fructose or sorbitol intake, which results in increased fructose-1-phosphate formation. The accumulating fructose-1-phosphate also inhibits fructokinase, thus causing fructose accumulation.

Aetiology – The disorder is inherited as an autosomal recessive trait.

Structural pathology – Data are scanty. The liver may show early cirrhotic changes.

Clinical features – Symptoms appear when the infant is first exposed to dietary fructose intake (commonly taken in the form of the disaccharide sucrose). The intake of sweetened foods is followed by manifestations of hypoglycaemia, e.g. nausea, vomiting, sweating, impaired consciousness and seizures. Small infants may develop hepatomegaly, jaundice, ascites, hyperammonaemia, lactic acidosis, hypoproteinaemia, and renal tubular glycosuria and aminoaciduria.

The serum phosphate level is reduced during attacks. Affected persons develop a revulsion for sweetened foods, and remain healthy if they avoid such food.

Diagnosis
(1) Clinical: Episodes in infants consistent with hypoglycaemia, beginning when sucrose is introduced into the diet and following each intake of sweetened food, may suggest the diagnosis.

(2) Laboratory: Blood fructose levels may be raised after intake of this sugar, and fructosuria may occur. Glucagon-resistant hypoglycaemia occurs. The enzyme deficiency can be demonstrated in liver biopsy specimens.

Treatment – Avoidance of sucrose, fructose and sorbitol intakes will prevent symptoms.

2.2.3 Disorders of galactose metabolism

Galactose, an isomer of glucose, enters the glycolytic pathway through the series of reactions set out in Figure 2.12. The main source of galactose is the disaccharide lactose, present in milk. Lactose is hydrolysed in the intestine to form glucose and galactose. There are two variants of galactosaemia, depending

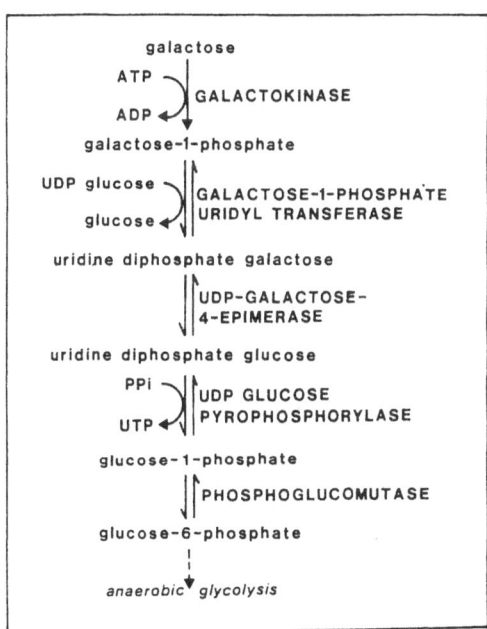

Figure 2.12 Metabolic pathways of galactose

on whether galactokinase or galactose-1-phosphate uridyl transferase is deficient[45]. Cataract occurs in both types and is probably due to polyol accumulation. However, in the transferase deficiency there are additional manifestations, probably due to accumulation of a toxic metabolite which has not yet been identified.

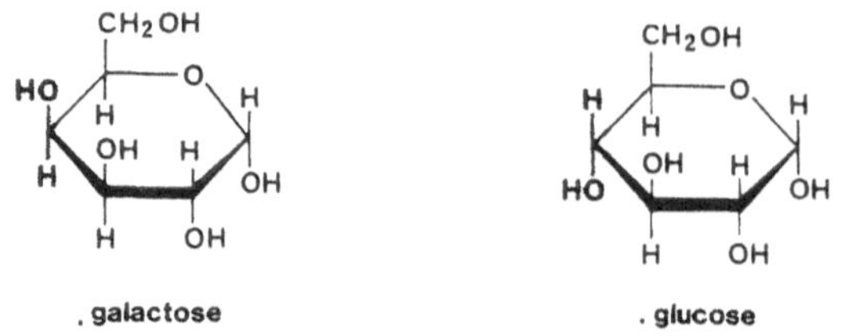

. galactose · glucose

2.2.3.1 Galactokinase deficiency

This is a rare inherited disorder causing galactosaemia and cataracts. It is due to a deficiency of the enzyme galactokinase[44]. The nervous system does not itself appear to be affected in the disorder. It is worth discussing this condition largely to set the other variant of galactosaemia, with its neurological involvement, in perspective.

Biochemical abnormality – Galactokinase deficiency causes galactose, derived from milk in the infant's diet, to accumulate in blood and other tissues. Galactose metabolism is diverted towards the polyol galacticol (dulcitol) which tends to accumulate within cells at its sites of formation (as does sorbitol, in hyperglycaemia. See Section 2.2.1.2). This galacticol accumulation raises the local tissue osmotic pressure, leading to imbibition of extracellular water and cellular swelling. These events, occurring in the lens of the eye, appear related to the development of cataract in galactosaemia. Galactose itself appears to be non-toxic.

Aetiology – The disorder is inherited as an autosomal recessive trait.

Structural pathology – Cataract is the only pathological manifestation.

Clinical features – The only clinical manifestations are galactosuria and cataract.

Diagnosis
(1) Clinical: The possibility of galactosaemia should be suspected in all instances of early-life cataract.

(2) Laboratory: The increased blood galactose content can be measured, and

32

the enzyme deficiency itself demonstrated in red blood cells. Galactosuria can also be detected. Its presence would suggest the possibility of the diagnosis.

Therapy – A diet low in galactose (and lactose), which usually involves the intake of a milk substitute, will prevent progression of the disorder. Genetic counselling is important for families in which the disorder has occurred. Heterozygotes have half normal levels of galactokinase in their peripheral blood cells.

2.2.3.2 Galactose-1-phosphate uridyl transferase deficiency: classical galactosaemia

This is a less uncommon hereditary disorder of the neonatal period than galactokinase deficiency. Galactosaemia is due to a deficiency of the enzyme galactose-1-phosphate uridyl transferase. The nervous system, and other organs, are affected[44].

Biochemical abnormality – Because of inadequate transferase activity, galactose derived from milk in the diet cannot be metabolized beyond galactose-1-phosphate. This substance, and galactose itself, accumulate in blood and tissues, and galactose metabolism is diverted towards galacticol (dulcitol) formation. Galacticol accumulates in tissues, and appears responsible for the cataracts which occur in both varieties of galactosaemia. It appears that galactose-1-phosphate itself, in excessive quantity, or some as yet unidentified metabolite derived from it, is toxic to the nervous system and other tissues, including the liver and kidneys. The manifestations of classical galactosaemia do not appear due to shortage of galactose metabolites subsequent to the deficient enzyme (Figure 2.12) since these metabolites can still be synthesized in the reverse direction from glucose. There is no deficiency of galactose for cerebroside or other galactolipid formation[46].

Aetiology – The disorder is inherited as an autosomal recessive trait.

Structural pathology – The disorder causes cataract, diffuse fatty infiltration of the liver proceeding to cirrhosis and, in the brain, slight microcephaly, loss of cerebral cortical neurons and cerebellar Purkinje cells, and gliosis of the cerebral white matter[47,48].

Clinical features – Symptoms usually appear in the first few days of postnatal life, commencing once milk is introduced into the diet. Affected infants feed poorly, and tend to have diarrhoea and vomiting. Jaundice develops, with increasing hepatomegaly and ascites. Proteinuria and generalized amino-aciduria occur, and cataracts develop (though at first they may be seen only with the slit lamp). Infants who survive the neonatal period, perhaps because they have been recognized as milk intolerant without a formal diagnosis of galactosaemia being made, are mentally retarded.

Diagnosis

(1) Clinical: The clinical picture described above, occurring in the appropriate age group, is highly suggestive of the diagnosis, particularly if the presence of cataract is noted.

(2) Laboratory: Galactosuria (the occurrence of a reducing substance in urine which does not react to a glucose oxidase test) is suggestive. A raised blood galactose level confirms the diagnosis. Enzyme activity measurement and raised galactose-1-phosphate levels in red blood cells will indicate the presence of the transferase deficiency. Enzyme levels are also reduced in leukocytes, skin fibroblasts and liver biopsy specimens. Galactose loading tests in such patients may be dangerous because they may cause severe hypoglycaemia. Prenatal diagnosis of the disorder is possible, by measuring the defective enzyme in cultured amniotic fluid cells.

Therapy – A diet with a very low galactose content, with the use of a milk substitute, is effective if commenced early enough. Such a diet may also be necessary throughout pregnancy in women who have already borne galactosaemic children. In later childhood and adult life, the patient may not need to avoid galactose so carefully. Heterozygous carriers within a family may be detected by their having a 50% reduction in red cell galactose-1-phosphate uridyl transferase activity, and genetic counselling may then be given. However, not all persons with reduced red cell enzyme levels prove to be carriers.

2.2.4 Glycerol kinase deficiency

Two families have been reported in which there has been glycerol kinase deficiency with raised blood glycerol levels and glyceroluria. The phenotypic expression has been so variable that it is impossible to know if the enzyme defect is more than fortuitous[49]. However, in one family mental retardation, spasticity and myopathy were present.

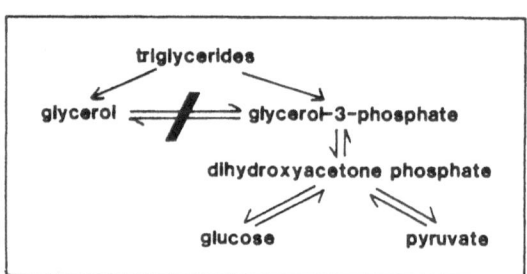

2.2.5 Disorders of pyruvate metabolism

Pyruvate is a key metabolic intermediate. It and acetyl coenzyme A constitute the major links between anaerobic glycolysis and the Krebs citric acid cycle.

$$CH_3.CO.COOH$$
.pyruvic acid

The major pathways of pyruvate metabolism are as set out in Figure 2.13.

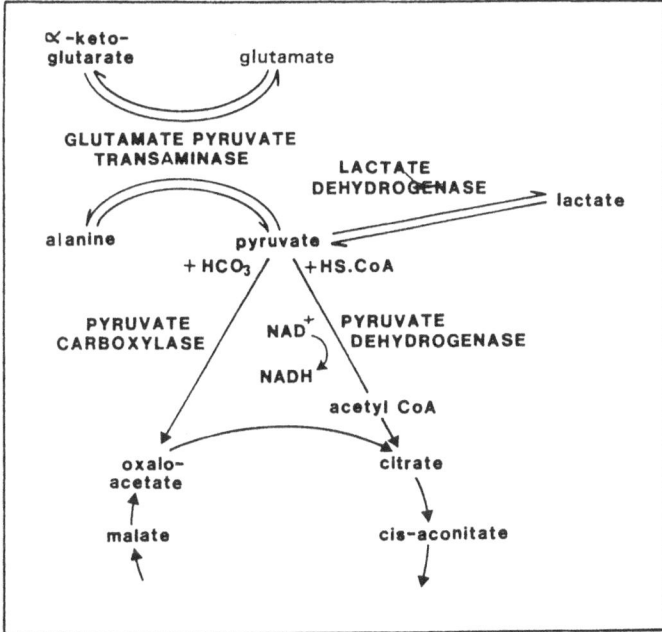

Figure 2.13 Major pathways of pyruvate metabolism

The pyruvate dehydrogenase system, catalysing the conversion of pyruvate to acetyl CoA, is a complex of three enzymes (1) pyruvate dehydrogenase (pyruvate decarboxylase) to which thiamine pyrophosphate is a cofactor, (2) dihydrolipoyl transacetylase (lipoate acetyltransferase), and (3) dihydrolipoyl dehydrogenase (lipoamide dehydrogenase). Details of the chemical mechanisms involved in the activity of this enzyme complex are available in textbooks of biochemistry. The net effect of the reactions catalysed by the complex is:

$$CH_3.CO.COOH + HSCoA + NAD^+ \longrightarrow CH_3.CO.SCoA + NADH + CO_2$$
acetic acid   acetyl CoA

Defects of several of the enzymes and cofactors involved in pyruvate metabolism, when the metabolic disturbance is severe, may lead to neurological disorders with features similar to those of Wernicke's encephalopathy and Leigh's subacute necrotizing encephalopathy. In milder degrees of metabolic defect the clinical picture may resemble those of some of the spinocerebellar degenerations[50,51].

2.2.5.1 Pyruvate carboxylase deficiency

This is a very rare disorder which in some, but not all, cases have been associated with a picture resembling that of Leigh's subacute necrotizing encephalopathy (Section 2.2.5.5).

Biochemical abnormality – Deficiency of pyruvate carboxylase causes decreased production of oxaloacetate, needed to 'prime' the Krebs cycle. Thus the carboxylase deficiency would lead to decreased Krebs cycle flux, ultimately producing decreased electron and H^+ entry into the mitochondrial electron transport pathway, and therefore reduced ATP production. Defects in both pyruvate decarboxylase and pyruvate dehydrogenase activity may lead to similar clinical and pathological pictures, and both enzyme defects also reduce Krebs cycle flux. Therefore the production of symptoms in these conditions may relate to biochemical events some distance beyond the primary enzymatic defect.

Pyruvate carboxylase utilizes biotin as a cofactor, and there are reports of pyruvate carboxylase deficiency occurring as part of a more generalized biotin-dependent carboxylase deficiency[52,53] (see Section 2.6.3.5).

Aetiology – The enzyme deficiency appears to be inherited.

Structural pathology – No information is available, except in one case who showed the haemorrhages around the third and fourth ventricles characteristic of Leigh's disease[51].

Clinical features – In three of five cases the clinical picture resembled that of Leigh's disease (Section 2.2.5.5). Another had convulsions, hypoglycaemia and metabolic acidosis on the first day of postnatal life. The final case was mentally retarded. Possibly the clinical picture of pyruvate decarboxylase deficiency varies with the severity of the enzyme deficiency, as seems to be the case for pyruvate dehydrogenase deficits.

Diagnosis
(1) Clinical: A clinical presentation similar to that of Leigh's disease would raise the possibility of pyruvate carboxylase deficiency.

(2) Laboratory: The enzyme defect can be demonstrated in liver biopsies and in cultured skin fibroblasts.

36

Therapy – A low carbohydrate diet, plus additional thiamine, appears helpful. Why thiamine is of benefit is obscure, since biotin, rather than thiamine, is the cofactor for pyruvate carboxylase. Pyruvate carboxylase deficiency is also said to respond to aspartate and pyridoxine, a combination which might be expected to increase oxaloacetate formation, thus correcting one consequence of pyruvate carboxylase deficiency.

2.2.5.2   *Pyruvate dehydrogenase deficiency*

Both inherited and acquired disorders of the pyruvate dehydrogenase complex are known. Pyruvate dehydrogenase deficiency may occur as a primary inherited enzyme defect of different degrees of severity, or it may be due to cofactor (thiamine pyrophosphate) deficiency. The latter deficiency may be of dietary origin, or due to the presence of an inhibitor of thiamine phosphorylation. Pyruvate dehydrogenase deficiency due to cofactor defects is dealt with in subsequent sections (2.2.5.4 and 2.2.5.5). Only the primary enzymatic defect is considered here. The clinical features appear to depend on the severity of the enzyme deficiency.

The recent assertion that about 40% of cases of spinocerebellar degeneration have defective function of the pyruvate dehydrogenase complex[54] has aroused interest in this area of neural metabolism.

Biochemical abnormality – Pyruvate dehydrogenase deficiency leads to decreased acetyl CoA formation from pyruvate. Acetyl CoA deficiency causes (1) decreased metabolic flux through the Krebs cycle and therefore decreased ATP production, (2) decreased acetylcholine formation, and (3) decreased biosynthetic activity, e.g. decreased fatty acid and protein production. The pyruvate which cannot be metabolized accumulates in blood and tissues. Some is reduced to lactic acid. It should be noted that when deficient pyruvate dehydrogenase activity is a consequence of thiamine pyrophosphate deficiency rather than of primary enzyme deficiency, the effects of defective function of other thiamine-dependent enzymes (transketolase, α-ketoglutarate dehydrogenase) are added to those of the pyruvate dehydrogenase deficiency itself.

The exact biochemical mechanisms involved in the pathological and clinical consequences of pyruvate dehydrogenase deficiency are incompletely understood. Possibly both insufficient energy availability (perhaps responsible for the Leigh-like picture which may occur when enzyme activity is greatly reduced, and which also occurs in pyruvate carboxylase deficiency), and chronically reduced acetylcholine production (which may correlate with partial enzyme deficiency and a clinical picture like that of Friedreich's ataxia), may be involved.

Aetiology – Pyruvate dehydrogenase complex deficiency is usually inherited as an autosomal dominant trait in the spinocerebellar degenerations (mainly Friedreich's ataxia). However, in a broader sense pyruvate dehydrogenase complex deficiency probably arises from a heterogeneous group of genetic defects. In some instances it is probably pyruvate decarboxylase which is defective, in others dihydrolipoyl transacetylase, while in Friedreich's ataxia-like syndromes the deficiency appears to be in lipoamide dehydrogenase[55]. This enzyme, at least in platelets, may have an abnormal conformation in the disorder[56]. The latter defect is not present in all cases[57] and some workers[58-60] have failed to find it in any of their cases of Friedreich's ataxia. Classification of the various clinical disorders in this area may be revised as time passes, and the relevant genetic defects may then become better defined. It may be worth pointing out in passing that Pliatakis *et al.*[61] have found evidence of glutamate dehydrogenase deficiency in cases of spinocerebellar degeneration.

Structural pathology – Some cases of pyruvate dehydrogenase deficiency which present in early life have shown widespread severe loss of neurons and myelin, and others the structural changes of Leigh's disease. Autopsy data do not yet appear available for those cases of spinocerebellar degeneration which have shown defective pyruvate dehydrogenase activity, though muscle biopsies have shown so-called 'ragged red' fibres in some of these cases. One cannot yet say whether the typical neuropathological changes of Friedreich's ataxia will prove to be present in such cases.

Clinical features – The manifestations of pyruvate dehydrogenase deficiency appear to depend on the severity of the defect in enzyme activity. When activity is reduced by 85%, or more, there is a syndrome of intermittent cerebellar ataxia, psychomotor retardation, choreoathetosis, optic atrophy and microcephaly which begins in the first year of life (compare Leigh's disease, Section 2.2.5.5). Seizures and lactic acidosis may occur. With a 65–80% decrease in enzyme activity, a similar but milder syndrome occurs, though cerebellar disturbance is more prominent. With a 50–65% loss of enzyme activity, a clinical picture resembling the various spinocerebellar degenerations, particularly Friedreich's ataxia[62], occurs.

Williams[63] found deficient pyruvate dehydrogenase activity in skin fibroblasts from cases of Charcot–Marie–Tooth disease (a chronic hereditary motor peripheral polyneuropathy often considered related to Friedreich's ataxia), but raised the possibility that the enzyme defect was secondary.

Diagnosis
(1) Clinical: The possibility of pyruvate dehydrogenase deficiency should be considered in all cases of diffuse cerebellar disturbance developing in infancy and childhood, and in all cases that would be diagnosed clinically as spinocerebellar degeneration.

(2) Laboratory: Blood pyruvate and lactate levels are raised in the more severe cases, particularly after a glucose load[57]. However, these findings are not diagnostic. Defective enzyme activity can be demonstrated in platelets[54] or in liver biopsy specimens or cultured fibroblasts. However, there appear to be technical difficulties in these assays so that replicated measurements may be necessary to obtain reliable data.

Therapy – A ketogenic diet may be helpful, possibly by providing an alternative energy source to pyruvate. Dietary thiamine supplements in high dose have helped occasional patients, and corticosteroids appear to have been beneficial in some cases[64]. In the heredo-ataxias an attempt has been made to correct possible acetylcholine deficiency by giving centrally-acting cholinergic agents, e.g. physostigmine. This is said to have been of some benefit[65]. Providing choline, the rate limiting factor in acetylcholine formation, appears of little benefit[66,67], as does providing lecithin, i.e. phosphatidyl choline[68].

2.2.5.3 *Carnitine acetyltransferase deficiency*

Di Donato *et al.*[69] described a child with a fatal illness comprising an intermittent ataxic syndrome with oculomotor paresis, hypotonia, confusion and disturbed consciousness, a clinical picture akin to that associated with a moderately severe reduction in pyruvate oxidation capacity. There was reduced carnitine acetyltransferase activity in the liver, brain, kidney and in cultured skin fibroblasts. A moderate increase in blood pyruvate concentrations occurred during exacerbations of symptoms. Acetylcarnitine can serve as an intramitochondrial store for acetate groups, freeing coenzyme A for other metabolic purposes, and can also serve as a means of transferring acetyl groups from within the mitochondria to the cytosol, where the acetyl groups may be involved in acetylcholine synthesis. Thus carnitine acetyltransferase deficiency might produce biochemical (and clinical) effects like those of decreased pyruvate oxidation by impeding the further metabolism of acetate, and by thus diminishing acetylcholine formation. It should be emphasized that in the case reported by Di Donato *et al.*[69] the defective enzyme was carnitine acetyltransferase: there was no decrease in carnitine acyltransferase activity. This latter enzyme is required for transfer of medium and longer chain fatty acids into mitochondria (see Section 2.7.2).

2.2.5.4 *Thiamine deficiency*

Thiamine deficiency, occurring mostly in chronic alcoholics, affects the nervous

39

system to produce three main patterns of disturbance, namely peripheral poly-neuropathy, Wernicke's encephalopathy and Korsakoff's psychosis. Such events are comparatively common in certain social sections of the community.

Biochemical abnormality – Thiamine, as its pyrophosphate, is a cofactor for at least four enzymes: (1) pyruvate decarboxylase, part of the pyruvate dehydro-genase complex, (2) α-ketoglutarate dehydrogenase, one of the enzymes of the Krebs cycle, (3) the branched chain keto-acid decarboxylase system (see branched chain ketoaciduria, Section 2.5.2.1), and (4) transketolase, one of the enzymes of the biosynthetic hexose monophosphate shunt (pentose phosphate pathway). Thiamine pyrophosphate deficiency reduces the conversion of

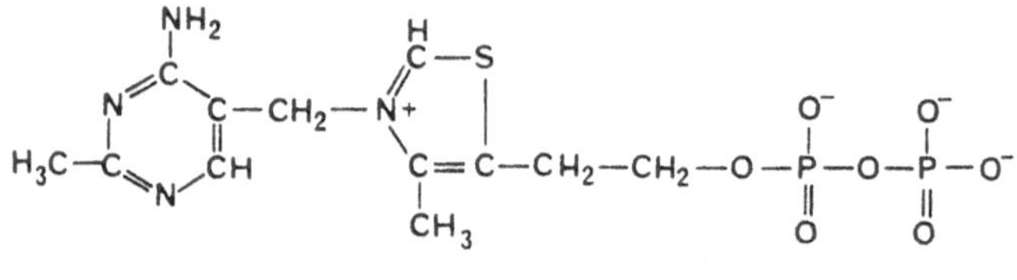

.thiamine pyrophosphate

pyruvate to acetyl CoA and causes accumulation of pyruvate in blood and tissues. This effect, and the interference with the conversion of α-ketoglutarate to succinyl CoA, both decrease molecular flux through the Krebs cycle and hence limit ATP production. These effects will also lead to a shortage of acetyl groups for various biosynthetic activities, including acetylcholine formation. Impaired transketolase activity will interfere with the following reactions:

$$\text{xylulose-5-phosphate} + \text{erthrose-4-phosphate} \rightleftharpoons$$
$$\text{glyceraldehyde-3-phosphate} + \text{fructose-6-phosphate}$$

$$\text{xylulose-5-phosphate} + \text{ribose-5-phosphate} \rightleftharpoons$$
$$\text{glyceraldehyde-3-phosphate} + \text{sedoheptulose-7-phosphate}$$

It is not clear if these impaired transketolations lead to altered neurological function. However, decreased ATP availability, and acetylcholine depletion, may be expected to interfere with nervous system activity. The brunt of the

40

damage might be expected to fall most heavily on parts of the nervous system which utilize acetylcholine as a neurotransmitter and/or have the highest energy requirement or the most precarious energy supply. The brain has little reserve capacity for pyruvate decarboxylation and the decarboxylase is unevenly distributed through the brain[70]. Hence one might anticipate a selective regional distribution of tissue injury in thiamine deficiency.

In experimental animals, thiamine deprivation causes raised brain α-keto-glutarate levels, with reduced glutamate and γ-aminobutyrate levels[71]. The mechanism involved in the latter change is not clear.

There is also some evidence of altered central noradrenergic function in cases of Korsakoff's psychosis[72]. Cerebrospinal fluid levels of the noradrenaline metabolite methoxyhydroxyphenylglycol (see page 229) are reduced, and administration of the α-noradrenergic receptor agonist clonidine seems to improve the memory defect in the disorder. In animals with induced thiamine deficiency there is reduced serotonin uptake into synaptosomes obtained from the cerebellum, brain stem and thalamus[73].

Aetiology – Thiamine deficiency is usually due to reduced thiamine intake resulting from a poor diet and/or chronic alcoholism. Rarely thiamine deficiency may be due to frequent vomiting, however caused.

Structural pathology – The neuropathological changes are described by Loken[74] and Smith[75]. In the brain there is capillary proliferation with small haemorrhages around the third ventricle, in the hypothalamus (classically in the mamillary bodies), the periaqueductal grey matter and in the floor of the fourth ventricle, particularly in the region of the vestibular and dorsal vagal nuclei. Neurons in these regions show some degenerative changes and there may be local gliosis. Peripheral nerves also show degenerative changes, particularly affecting large diameter fibres, and sometimes there is associated anterior horn cell chromatolysis.

Clinical features – There are three main components to the neurological defect caused by thiamine deficiency: (1) sensori-motor peripheral polyneuropathy, (2) Wernicke's encephalopathy, with ataxia of gait, nystagmus and paresis of external eye muscles causing diplopia, and (3) Korsakoff's psychosis, with its characteristic confabulatory dysmnesia. Elements of these three patterns of neurological defect often coexist in the thiamine deficient individual; the hyperkinetic circulatory state of beriberi heart disease may also be present.

Diagnosis
(1) Clinical: The combination of nystagmus or eye muscle paresis, peripheral polyneuropathy and Korsakoff's type psychosis in a chronic alcoholic is virtually diagnostic of thiamine deficiency.

(2) Laboratory: Blood pyruvate levels are raised, but the diagnosis is based on the finding of reduced transketolase activity.

41

Therapy – High dosage thiamine administration will halt deterioration in the disorder and may lead to at least some improvement as reversible components of the neurological damage recover. Clonidine therapy may be worth a trial in patients with Korsakoff's psychosis[72]. If patients with Wernicke–Korsakoff syndrome are given parenteral glucose without thiamine, they may become comatose, and die. The biochemical mechanisms involved in this effect are uncertain. Possibly the glucose load cannot be handled metabolically, and hyperosmolality may develop and add to the neurological damage (see Section 2.2.1.2).

2.2.5.5 Leigh's subacute necrotizing encephalopathy

This is a rare inherited disorder of early life in which neurological manifestations of thiamine deficiency occur despite adequate dietary thiamine intake[76].

Biochemical abnormality – Some cases of the disorder appear to involve decreased activity of both pyruvate carboxylase and pyruvate decarboxylase. *In vivo* defective activation of pyruvate dehydrogenase is another (rare) cause of the disorder[77]. These enzyme deficiencies cause raised plasma levels of pyruvate, lactate and alanine. In other cases of Leigh's disease there appears to be present a circulating inhibitor of the enzyme thiamine pyrophosphate-ATP phosphoryl transferase. This enzyme catalyses the phosphorylation of thiamine to thiamine pyrophosphate which is the biologically active cofactor for pyruvate decarboxylase. Thiamine triphosphate levels are reduced in histologically abnormal brain regions in Leigh's disease[78]. The enzyme inhibitor also occurs in urine[79], and appears to be a protein with a molecular weight of around 37 000 daltons[80].

The biochemical consequences of these enzyme abnormalities are as discussed in previous sections. De Vivo *et al.*[77] have suggested that in pyruvate decarboxylase deficiency, there may be partially reversed flux through the Krebs cycle (Figure 2.14).

Aetiology – Infantile Leigh's disease is inherited as an autosomal recessive trait. Sporadic cases may also occur in childhood and adult life[81].

Structural pathology – The lesions of Leigh's syndrome[75,82] are very similar to those of Wernicke's encephalopathy, though they are not identical. The mamillary bodies are less likely to be involved[81].

Clinical features – The disorder typically begins in the first one or two years of life. It presents with failure to thrive, hypotonia, peripheral neuropathy, ataxia, dysphagia, optic atrophy, nystagmus, epileptic seizures and mental retardation. Sufferers often die within one or two years of the onset of symptoms. Occasional chronic cases occur[83]. Rarely, the disorder begins in adult life[81,84].

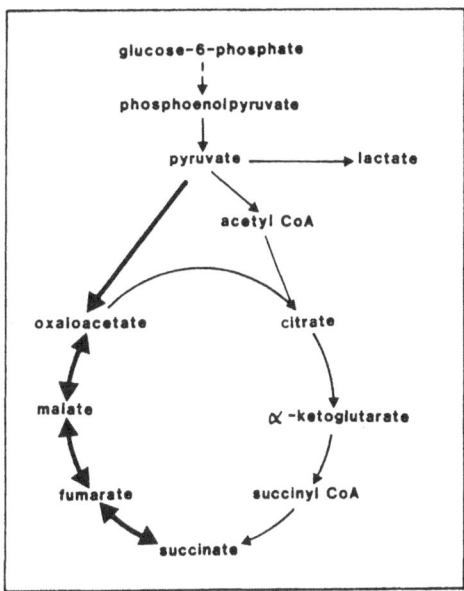

Figure 2.14 The suggested partial reversal of flux through the Krebs cycle that may occur when pyruvate decarboxylation is impaired in Leigh's disease[77]

Diagnosis
(1) Clinical: The development of a worsening ataxic syndrome in early life may raise suspicion of a disorder of the pyruvate dehydrogenase complex, though progressive structural brain disease is also possible. The diagnosis of Leigh's disease becomes more probable if there have been similarly affected siblings.

(2) Laboratory: Structural posterior fossa lesions should be excluded by appropriate investigations. Raised blood pyruvate, lactate and alanine levels may suggest the biochemical category of the disorder, but exact diagnosis requires enzyme activity measurements in a tissue specimen. The enzyme activities should also be measured in the presence and absence of the patient's plasma or urine[79,83] to ascertain if a circulating inhibitor of the enzyme is present. It should be noted that the inhibitor could not be found in the amniotic fluid in one pregnancy which resulted in the birth of an affected child[79].

Therapy – No curative treatment of the disorder is known. One has seen a patient derive useful temporary benefit from very high dosage thiamine intake, and there have been occasional similar reports in the literature.

2.3 DISORDERS OF THE KREBS CYCLE

Details of the Krebs cycle are shown in Figure 2.15. So far, primary disorders of the Krebs cycle itself do not seem to have been described in man. The Krebs cycle occupies a central role in cellular metabolism. Various amino acids, after deamination, can enter the cycle at different points (Figure 2.17). Therefore it is possible that metabolic blocks which may perhaps exist at certain

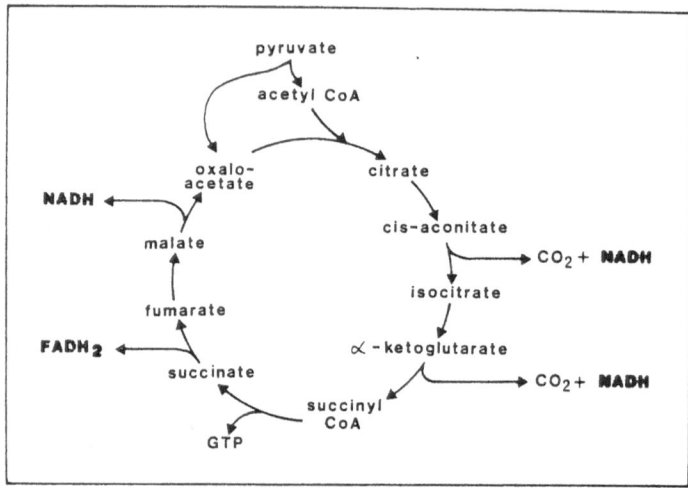

Figure 2.15   The Krebs tricarboxylic acid cycle

stages in the cycle may be bypassed by diversion of amino acids, while enzyme defects causing severe impairment of flux through the cycle at other points may be incompatible with life.

2.4 DISORDERS OF MITOCHONDRIAL ELECTRON TRANSPORT AND OXIDATIVE PHOSPHORYLATION

Electron transport and oxidative phosphorylation are thought to occur in respiratory assemblies located in the inner membranes of mitochondria, adjacent to sites of Krebs cycle reactions. In the respiratory assemblies, electrons are transferred from NADH or $FADH_2$ to oxygen via a number of intermediate stages (Figure 2.16). Stages at which ATP formation occurs are indicated in Figure 2.16. ADP concentration determines rate of electron flow along the pathway, and hence controls ATP concentration. Most of the NADH and the $FADH_2$ entering the mitochondrial electron transport chain arises from the Krebs cycle (see Figure 2.15, in which sites of NADH and $FADH_2$ production are also indicated). NADH formed in the cytosol (e.g. from the oxidation of glyceraldehyde-3-phosphate to 1,3-diphosphoglycerate during anaerobic glycolysis), can transfer its electrons via shuttle reactions across the mitochondrial membrane. Consequently these electrons can also subsequently pass down the mitochondrial electron transport pathway.

There do not appear to be well recognized human diseases due to specific

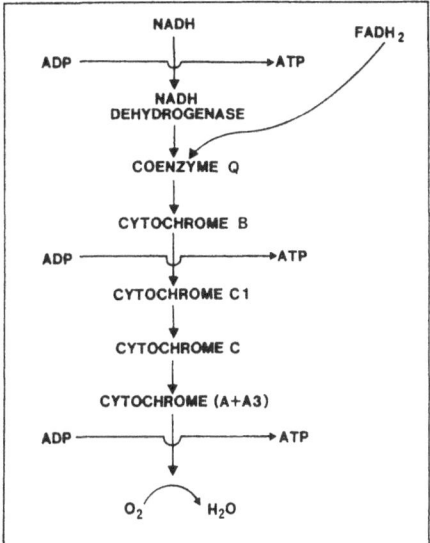

Figure 2.16 Mitochondrial electron transport pathway

primary disorders of the mitochondrial electron transport pathway itself, though stages of the pathway may be affected as consequences of several disorders of other chemical mechanisms. Sometimes coupling of oxidative phosphorylation to the pathway may be rendered defective, as in the malignant hyperthermia syndrome, thyrotoxicosis and porphyria (see also Section 2.6.2). Oxygen lack may impair electron flow down the pathway, since there is then a deficiency of the final electron acceptor of the pathway. Among the rarely reported seemingly primary disorders of function of the mitochondrial electron transport pathway, Di Mauro et al.[85] have described cytochrome-c oxidase deficiency. This enzyme deficiency was associated with the presence of aggregates of abnormal mitochondria in skeletal muscle and occurred in an infant with seizures, failure to thrive, hypotonia, lactic acidosis, proteinuria, glycosuria and generalized aminoaciduria. Di Mauro et al.[85] provided references to a few recorded instances of deficiencies of other cytochromes. Land et al.[86] have described instances of myopathy associated with exercise-induced lactic acidaemia which were probably due to deficient NADH-cytochrome b reductase activity.

2.4.1 Hypoxia

Hypoxia, variously caused, is a comparatively common occurrence in clinical medicine. It may severely disturb brain function. Often tissue hypoxia occurs in association with ischaemia, in which case impaired delivery of glucose to neural tissue, and incomplete removal of acid metabolites, are likely to coexist with the effects of pure hypoxia. However, only hypoxia itself will be discussed at this stage.

Biochemical abnormality – Tissue oxygen depletion impairs electron flux down the mitochondrial electron transport pathway. NADH levels rise relative to those of NAD^+. Intracellular H^+ concentration also rises[87]. Raised NADH levels inhibit the decarboxylation of pyruvate to acetyl CoA and the oxidations of α-ketoglutarate to succinyl CoA, and of isocitrate to α-ketoglutarate. Consequently pyruvate accumulates, and is reduced to lactate, with parallel re-oxidation of NADH to NAD^+.

Decreased acetyl CoA formation leads to reduced acetylcholine synthesis[38,88]. Reduced acetylcholine levels occur before there is any fall in ATP levels or in ATP:ADP ratios and prior to any lactate accumulation. Hypoxia activates pyruvate kinase and hexokinase, but does not activate the other rate-limiting enzyme in glycolysis, phosphofructokinase[87]. With severe enough hypoxia, high energy phosphate stores become depleted, and lactate and NH_4^+ accumulate. Hypoxia also impairs mitochondrial capacity[88] to concentrate Ca^{2+}.

Muscle, with its capacity to obtain some energy from fatty acids, is likely to tolerate hypoxia better than neural tissue.

Aetiology – Pure hypoxia of neural tissue is most likely to occur in conditions when there is (1) a decreased partial pressure of oxygen in inspired air, (2) impaired oxygen diffusing capacity across the pulmonary membranes, or (3) when there is decreased oxygen carrying capacity of the blood, due to anaemia or the presence of abnormal haemoglobins, e.g. methaemoglobin, carboxy-haemoglobin.

Structural pathology – Pathological changes [40,89,90] in fatal cases of hypoxia occur mainly in the cerebral and cerebellar cortices, in the Sommer sector of Ammon's horn, and in the striatum. In these regions neuronal degeneration occurs, followed by gliosis. The white matter may also be involved.

Clinical features – Brain hypoxia may cause confusion, stupor, irritability, disturbed behaviour, convulsions, coma and finally death. Consciousness is usually lost once the arterial oxygen tension falls below 30 mmHg. If the hypoxia is of short duration full recovery is possible. If hypoxia is prolonged there may be permanent neurological defects. To some extent the pattern of the neurological disorder which develops during and after hypoxia is determined by pre-existing local factors, e.g. earlier neuronal injury, local alteration in blood flow due to atheroma.

Diagnosis
(1) Clinical: The presence of cerebral hypoxia may be suspected when there is altered neural function (especially diffusely altered function) in circumstances when decreased oxygen availability to the circulation, or decreased oxygen carriage by the circulation, appears likely. Cyanosis may be present.

(2) Laboratory: Hypoxia is demonstrated by measuring the arterial pO_2 (normally 75–100 mmHg) and the haemoglobin content of the circulating blood.

Therapy – Treatment of hypoxia is based on correction of the cause of the hypoxia.

2.4.2 Ischaemia

It is now convenient to discuss ischaemia since two of its main biochemical consequences, oxygen and glucose lack, have already been dealt with individually. Ischaemia, due to decreased arterial blood flow, may affect brain and muscle diffusely, or focally. Consequences of ischaemia are very common in clinically neurological practice.

Biochemical abnormality – The consequences of ischaemia result from the combined effects of hypoxia, glucose deficiency and accumulation of metabolites, in particular lactic acid. It seems likely that failure of acetylcholine production occurs soon after ischaemia develops[91]. Through mechanisms already discussed, namely decreased formation of acetate from pyruvate, both hypoxia and hypoglycaemia may affect acetylcholine levels rapidly. Experimental work in animals suggests that brain ATP production seems to fail suddenly at a critical degree of ischaemia[92]. ATP levels can fall by 75% in

47

15–60 seconds, and lactate levels quadruple over this period[93]. There may be temporary hypermetabolism after ischaemia passes off[94], if irreversible metabolic and structural consequences have not occurred. After several minutes of experimental circulatory arrest, cerebral blood flow may not be restored when circulation resumes. Factors suggested as responsible for this 'no reflow' phenomenon include capillary endothelial or perivascular glial swelling, intravascular sludging and altered blood viscosity[29].

Aetiology – Generalized brain ischaemia is usually due to circulatory failure, though it also occurs when intracranial pressure becomes high enough to impair arterial blood flow. Local ischaemia is usually a consequence of local narrowing or occlusion of cerebral arteries.

Structural pathology – Generalized ischaemia, in the absence of any further regional reduction in brain perfusion, has a predilection to damage the cerebral cortex, striatum, globus pallidus, thalamus, Ammon's horn and Purkinje cell layer of the cerebellar cortex[40,90]. However, in more severe ischaemia the brain is damaged more widely. The spinal cord and medulla are more resistant to ischaemia than is cerebral tissue.

In local ischaemia, changes may range in severity from minor alterations in neuronal cytology to massive local necrosis with subsequent gliosis and cavitation.

Clinical features – Diffuse brain ischaemia causes a generalized disturbance in cerebral function ranging in severity from mild behavioural alterations and blunting of alertness to profound coma, possibly with convulsing. Local ischaemia produces local disturbances of brain or spinal cord function. Local muscle ischaemia usually produces cramping pain on use of the affected muscle or muscles (e.g. claudication).

Diagnosis
(1) Clinical: Diffuse disturbance of brain function, in the presence of systemic arterial hypotension, suggests the presence of diffuse cerebral ischaemia. The presence of local ischaemia produces the familiar, but varied, clinical patterns of neurological deficit of the various stroke syndromes.

(2) Laboratory: Cerebral blood flow studies and angiography can demonstrate decreased brain perfusion.

Therapy – In the acute phase, treatment is directed to restoring the arterial circulation, if possible by correcting the primary cause of the ischaemia. Thus far, attempts to correct the consequent biochemical disturbances themselves do not seem to have produced much clinical benefit.

48

Part B: The metabolic intermediate pool

2.5 THE AMINO-ACID POOL AND RELATED SMALL MOLECULES

Amino acids of dietary and endogenous metabolic origin enter neural tissue from the circulation by means of active uptake processes. Within neural tissues these amino acids may undergo two main types of metabolic fate:

(1) Amino acids may be deaminated, and the keto-acid residues then enter the glycolytic pathway at the levels of pyruvate or acetyl CoA, or else enter the Krebs cycle. The urea cycle, which takes up the NH_4^+ produced from amino-acid deamination in other tissues, is not fully developed in the brain. In neural tissue, NH_4^+ is taken up by α-ketoglutarate or glutamate, or else is transported in the circulation for entry to the urea cycle in the liver.

(2) Amino acids may be involved in various biosyntheses, forming:
   (a) Proteins,
   (b) Nucleic acids,
   (c) Porphyrin derivatives,
   (d) Various metabolic cofactors, and
   (e) Various neurotransmitters.

Essential amino acids cannot by synthesized within the body and must be ingested in the diet. Non-essential amino acids can be synthesized in man. The majority of abnormalities of metabolism within the amino-acid pool which disturb neurological function in man appear to involve reactions leading to the ultimate entry of amino-acid derivatives into the Krebs cycle. Mostly, the essential amino acids are involved in these disorders. Metabolic disorders usually involve a block at a single step in a metabolic sequence. The inter-conversions that can occur between the non-essential amino acids and other molecules may allow such a block to be bypassed if it occurs in their metabolic pathways whereas a block in essential amino-acid metabolism may not be bypassed easily.

The disorders of amino-acid metabolism will be here considered in relation to groups of structurally related amino acids, which often enter the energy pathway at a common point. The points of entry of various amino acids into the energy pathway are shown in Figure 2.17.

The essential amino acids are:

arginine	methionine
histidine	phenylalanine
isoleucine	threonine
leucine	tryptophan
lysine	valine

Only disorders of amino-acid metabolism which disturb neurological

49

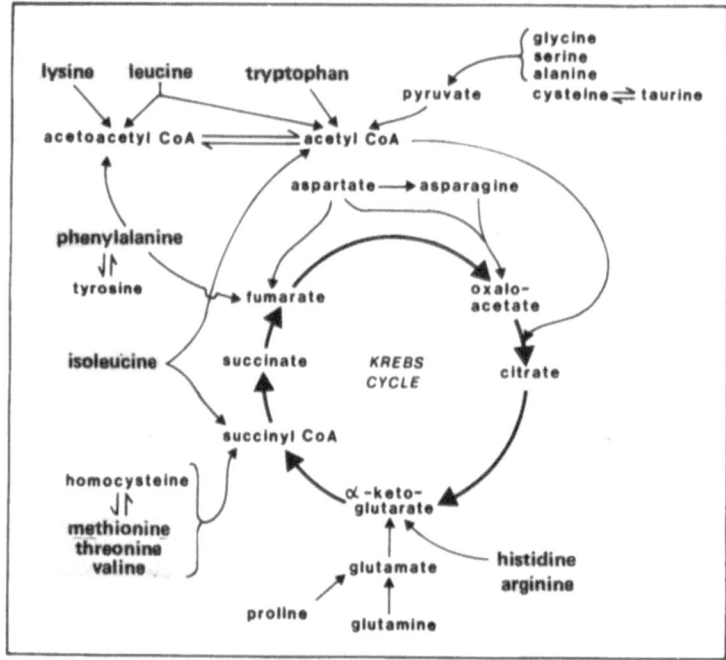

Figure 2.17 Sites of amino-acid entry into the energy pathways. Essential amino acids are shaded over

function will be discussed. Instances of disordered amino-acid metabolism have sometimes been discovered during studies screening the urinary excretions of amino acids and amino-acid metabolites[95] in populations of mentally retarded persons. In such cases it is sometimes unclear whether the metabolic disorders are causally or coincidentally related to the neurological manifestations. Amino-acid metabolism disorders have not been considered in this book unless it seems probable that the metabolic disturbances are the cause of neurological dysfunction. While the clinical manifestations of the disorders differ, the neuro-pathological changes occurring in disturbances of amino-acid metabolism have generally proved non-specific, and comprise delayed and defective myelination with a cystic degeneration (status spongiosus) of white and grey matter and fibrillary gliosis[96,97].

As well as the disorders of α-amino-acid metabolism, certain disorders of small (often N containing) molecules are considered in this section. Such disorders include porphyria, urea cycle disturbances, carnosine metabolism disturbance, aspartylglucosamine accumulation, and the effects of certain vitamin deficiencies which alter amino-acid metabolism.

2.5.1 Aromatic amino-acid disorders

The three aromatic α-amino acids are phenylalanine, tyrosine and tryptophan. Tyrosine is the m-hydroxy derivative of phenylalanine. It is convenient to discuss disorders of metabolism of phenylalanine and tyrosine together and to then deal with disorders of tryptophan metabolism. The most important disorder of tyrosine and phenylalanine metabolism is phenylketonuria. In addition three forms of tyrosinaemia occur.

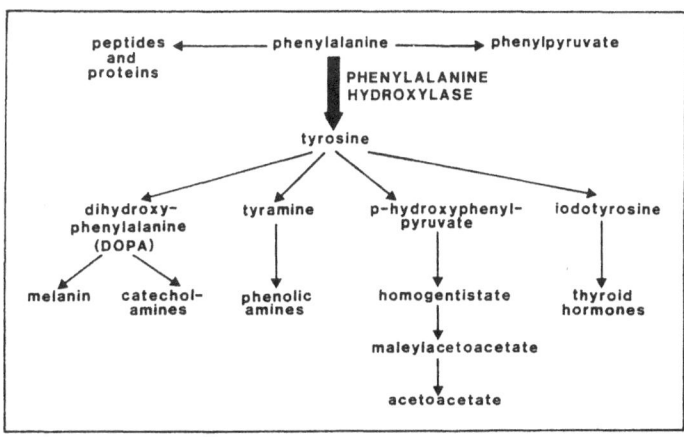

2.5.1.1 Phenylketonuria (hyperphenylalanaemia): phenylalanine hydroxylase deficiency

Hyperphenylalanaemia with phenylketonuria is due to deficient function of the enzyme phenylalanine hydroxylase. It is one of the more common and serious disorders of amino-acid metabolism. The homozygous state occurs about once in 12 000 Caucasian live births. Menkes and Koch[98], Tourian and Sidbury[99] and Scriver and Clow[100] have provided reviews of the disorder.

Biochemical abnormality – The pathways for phenylalanine metabolism are set out in Figure 2.18. The metabolic conversion of phenylalanine to tyrosine

Figure 2.18 Pathways of phenylalanine metabolism

is catalysed by the enzyme tyrosine hydroxylase which has tetrahydrobiopterin for a cofactor.

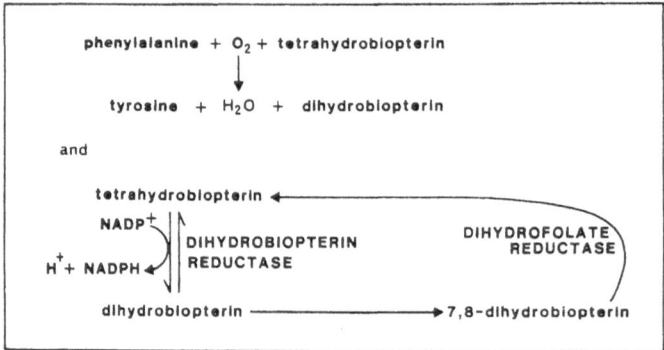

The hydroxylation of phenylalanine to tyrosine involves several stages (Figure 2.19). Phenylalanine hydroxylation does not occur in the brain. Tyrosine, preformed in the liver, is carried to the brain in the circulation, and is transported across the blood–brain barrier before being further metabolized in neural tissue.

Figure 2.19 Details of the phenylalanine hydroxylation reaction

Phenylketonuria (the classical variety) is due to deficient or absent tyrosine hydroxylase activity in extraneural tissues. At least two other lethal variants of phenylketonuria are known. In one the enzyme dihydrobiopterin reductase is absent or deficient[101]. Consequently tetrahydrobiopterin cannot be resynthesized once it is oxidized. Hence there is a functional defect in phenylalanine hydroxylase activity, though this enzyme is itself intact. There are associated defects in the brain's conversion of L-dopa to dopamine and tryptophan to 5-hydroxytryptophan[102]. Both these reactions also require continuing supplies of tetrahydrobiopterin. In the other lethal variant of phenylketonuria, liver tetrahydrobiopterin levels are greatly reduced, though phenylalanine hydroxylase and dihydrobiopterin reductase are intact. The cause of the co-factor deficiency in this variant is uncertain. It has been suggested[103] that there is a metabolic block between guanosine triphosphate and L-erythro-7,8-dihydrobiopterin so that 7,8-dihydroneopterin must be diverted to the synthesis of neopterin. Niederwieser et al.[104] proposed that, in this variant, the enzyme 7,8-dihydrobiopterin synthetase was defective.

When phenylalanine cannot be metabolized to tyrosine, deficient production of tyrosine metabolites (including catecholamines, phenolic amines and melanin) follows. Phenylalanine is diverted towards what are normally its minor routes of metabolism, in particular forming the phenylketone, phenyl-pyruvate (Figure 2.20).

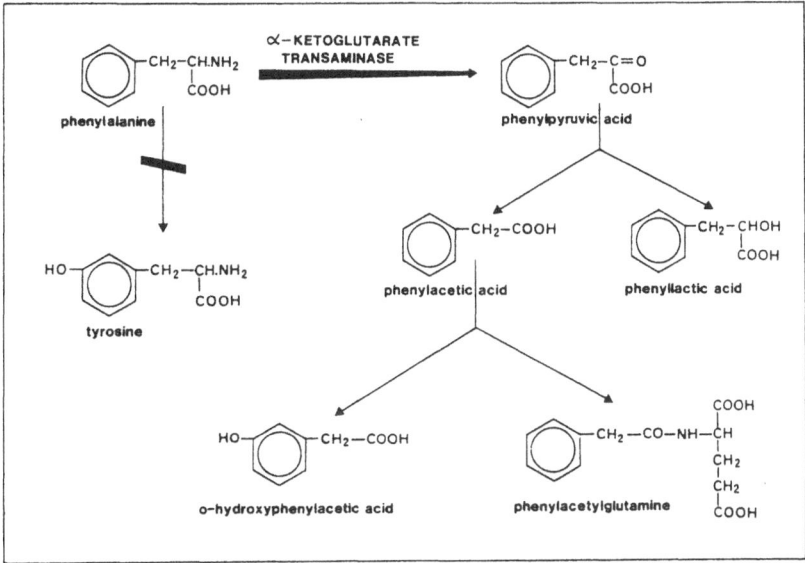

Figure 2.20   Pathways of altered phenylalanine metabolism in phenylketonuria, resulting from phenylalanine hydroxylase deficiency

In phenylketonuria, phenylalanine and its metabolites, other than tyrosine and its derivatives, occur at raised concentrations in blood, and appear in increased quantities in urine. The raised phenylalanine levels in blood are associated with altered levels of other amino acids. This occurs because active transport of other amino acids, notably the aromatic amino acids tryptophan and tyrosine, and also leucine[105], across the intestinal mucosa and the blood–brain barrier, is reduced. Glutamine uptake into the brain is also decreased[106]. As a result of decreased intestinal tryptophan transport, increased amounts of tryptophan metabolites are formed in the gut from the non-absorbed amino acid. These metabolites (indolyl-3-lactic acid, indolyl-3-acetic acid, indolyl-3-pyruvic acid, indican) absorb from the intestine and are excreted in urine. Decreased entry of tryptophan and tyrosine into the brain leads to deficiency in serotonin and catecholamine neurotransmitters, respectively, and also causes decreased protein synthesis. When phenylketonuria is due to tetrahydro-biopterin deficiency, there is additionally decreased hydroxylation of tyrosine and tryptophan, since these reactions are biopterin dependent. The possibility of inhibition of pyruvate kinase in these circumstances has also been raised and Land et al.[107] have shown that phenylpyruvate inhibits pyruvate transport across mitochondrial membranes. If this occurred to a significant extent there would be interference with acetate and ATP production.

It is not known which of the above biochemical changes is responsible for the neurological abnormalities of phenylketonuria, or whether some other, as yet not defined, metabolic derangement is culpable.

Aetiology – Classical phenylketonuria is inherited as an autosomal recessive trait. Heterozygotes, with 50% of normal phenylalanine hydroxylase activity, can metabolize phenylalanine normally.

Structural pathology – There is interference with brain maturation. Myelination is defective, and there are irregular areas of demyelination and status spongiosus of the cerebral white matter. The substantia nigra and locus coeruleus contain less melanin pigment than normal. More detailed accounts are available[97,108,109]

Clinical features – Affected infants tend to be blond haired and blue eyed, with rough dry eczematous skins. They vomit frequently and are irritable in the first 2 months of life. Mental retardation is obvious by the second 6 months of postnatal life. They often develop hypsarrhythmia and other patterns of epileptic seizure. Some cases are microcephalic and some have signs of pyramidal tract insufficiency. Older patients are restless, hyperactive, often autistic and commonly suffer from epilepsy. About 4% of cases with hyperphenylalanaemia are mentally normal.

Diagnosis
(1) Clinical: While the clinical picture described above should suggest the diagnosis, particularly if the family history is positive, the question of clinical diagnosis should not have to arise if neonatal urine screening for phenylketonuria is carried out effectively, and appropriate genetic counselling is given, and followed.

(2) Laboratory: Ferric chloride testing of urine for phenylpyruvate, and measurement of plasma phenylalanine content, will detect cases of phenylketonuria. Specific diagnosis of the type of phenylketonuria requires measurement of phenylalanine hydroxylase and dihydrobiopterin reductase activity in cultured skin fibroblasts. Exact diagnosis is of practical

importance, since the dietary measures that are effective in phenylalanine hydroxylase deficiency are not successful in dihydrobiopterin reductase deficiency.

Therapy – If classical phenylketonuria is diagnosed early in postnatal life, mental retardation can be prevented by a low phenylalanine diet (250–500 mg of the amino acid per day) which maintains plasma phenylalanine levels below 12 mg/100 ml. Such a diet probably should be continued through the first decade of life, and resumed during pregnancy in female homozygotes, in the hope of avoiding fetal damage associated with intrauterine hyperphenylalanaemia.

In dihydrobiopterin reductase deficiency, as well as dietary phenylalanine restriction being imposed, 5-hydroxytryptophan and L-dopa supplements are given to try to overcome the neurotransmitter shortage which results from defective decarboxylation of the corresponding amino acids. Tetrahydrobiopterin does not enter the brain readily from the circulation so that its direct replacement is not practicable. Although dietary phenylalanine restriction leads to good control of plasma phenylalanine levels, unless the supplements are given mental retardation will develop in children with dihydrobiopterin reductase deficiency.

2.5.1.2 Tyrosinaemia

There are at least three varieties of tyrosinaemia, (1) a benign neonatal variety, (2) a hereditary type with a good prognosis for life, and (3) a hereditary type with liver and renal involvement[110]. All three varieties are rare.

Biochemical abnormality – The metabolic pathways for tyrosine are shown in Figure 2.21. Tyrosine transaminase deficiency causes tyrosinaemia with a

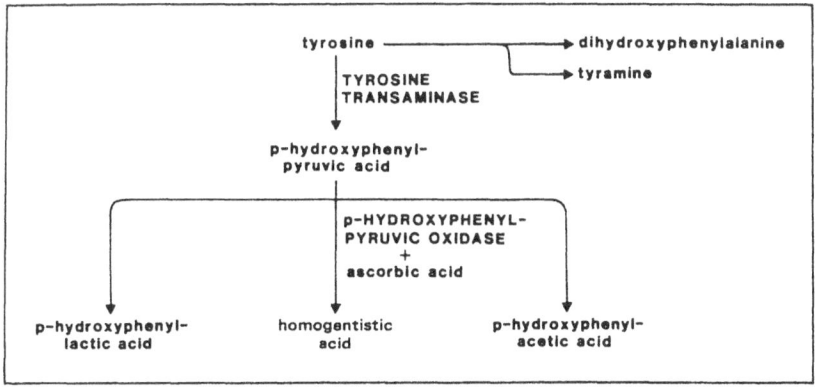

Figure 2.21 Pathways of tyrosine metabolism

55

benign prognosis for life, while p-hydroxyphenylpyruvic oxidase deficiency causes tyrosinaemia with hepato-renal involvement. The basis of the mental retardation in tyrosine transaminase deficiency is unclear, as is the reason for increased urine excretion of 5-aminolevulinic acid and porphobilinogen in p-hydroxyphenylpyruvic oxidase deficiency. In the latter disorder the accumulation of porphyrin precursors probably correlates with the neurological aspects of the condition, namely bouts of abdominal pain and polyneuropathy.

Aetiology – The mode of inheritance of tyrosine transaminase deficiency is uncertain. Deficiency of p-hydroxyphenylpyruvate oxidase is inherited as an autosomal recessive trait.

Structural pathology – Cases of p-hydroxyphenylpyruvate oxidase deficiency develop nodular cirrhosis of the liver, and tyrosine crystals are present in the bone marrow. Few details of the neuropathology of the various types of tyrosinaemia are available. One case showed slight delay in myelination with some spongy changes in cortical and subcortical regions[97].

Clinical features – Tyrosine transaminase deficiency is associated with mental retardation, microcephaly, e.e.g. abnormalities, corneal opacities and ulcers and punctate keratotic skin changes. There is a good prognosis for life.

p-Hydroxyphenylpyruvic oxidase deficiency is associated with vomiting, diarrhoea, hepatosplenomegaly, anaemia, a bleeding tendency, hypoglycaemia, hyperammonaemia and defective renal tubular function (which causes hypokalaemia, hypophosphataemic rickets, acidosis, glycosuria and generalized aminoaciduria). Porphyria-like attacks of abdominal pain and polyuria may occur.

Diagnosis
(1) Clinical: From the neurologist's point of view, p-hydroxyphenylpyruvic oxidase deficiency would enter the differential diagnosis of acute intermittent porphyria. However, the fully developed clinical picture might also raise the clinical question of leukaemia or related conditions. Mental retardation, in the presence of skin and corneal lesions, would permit suspicion of tyrosine transaminase deficiency.

(2) Laboratory: Raised blood tyrosine levels, with tyrosinuria, confirm the diagnosis of tyrosinaemia. Failure of the blood tyrosine level to fall after ascorbic acid intake suggests that the tyrosinaemia is probably due to tyrosine transaminase deficiency.

Therapy – A diet low in phenylalanine and tyrosine, i.e. a phenylketonuria diet, may benefit the corneal and skin lesions of tyrosine transaminase deficiency, but not the mental retardation. This diet is also beneficial in p-hydroxyphenylpyruvic oxidase deficiency.

2.5.1.3 *Hartnup disease*

Hartnup disease causes a pellagra-like syndrome which reflects the effects of tryptophan deficiency arising from the defective transport of certain neutral amino acids in the small intestine and kidney[111]. At least 53 cases were known up to 1978[112].

Biochemical abnormality – Impaired alimentary (mainly jejunal) absorption and impaired renal tubular resorption of neutral monoamino monocarboxylic acids lead to tryptophan deficiency. The fates of neutral amino acids other than tryptophan in the disorder are largely unknown, as is any contribution their altered metabolism may make to the clinical features of the disorder. Tryptophan undergoes a variety of metabolic reactions in man. These reactions involve:

(1)   Preservation of the indole nucleus (Figure 2.22),

(2)   Cleavage of the pyrrole ring (in the liver), and

(3)   Following the previous reaction, benzene ring cleavage (Figure 2.23).

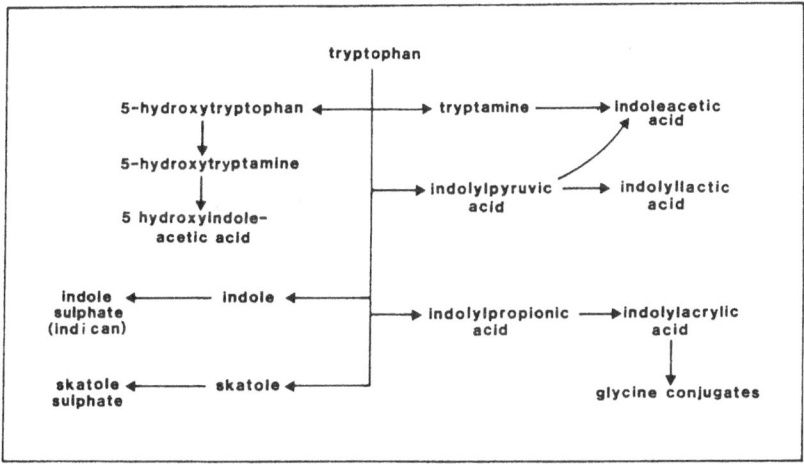

Figure 2.22   Tryptophan metabolic reactions involving preservation of the indole nucleus

Because of its pattern of metabolism, tryptophan deficiency ultimately leads to nicotinic acid deficiency (the basis of pellagra) and thus to a shortage of NAD^+, which is a cofactor in many reactions in the cellular energy yielding pathways. Hartnup disease includes the neurological and cutaneous manifestations of pellagra, though whether clinically significant brain serotonin depletion occurs is uncertain. In addition to the pellagrinous manifestations there is aminoaciduria due to failure of renal tubular amino-acid transport. This causes

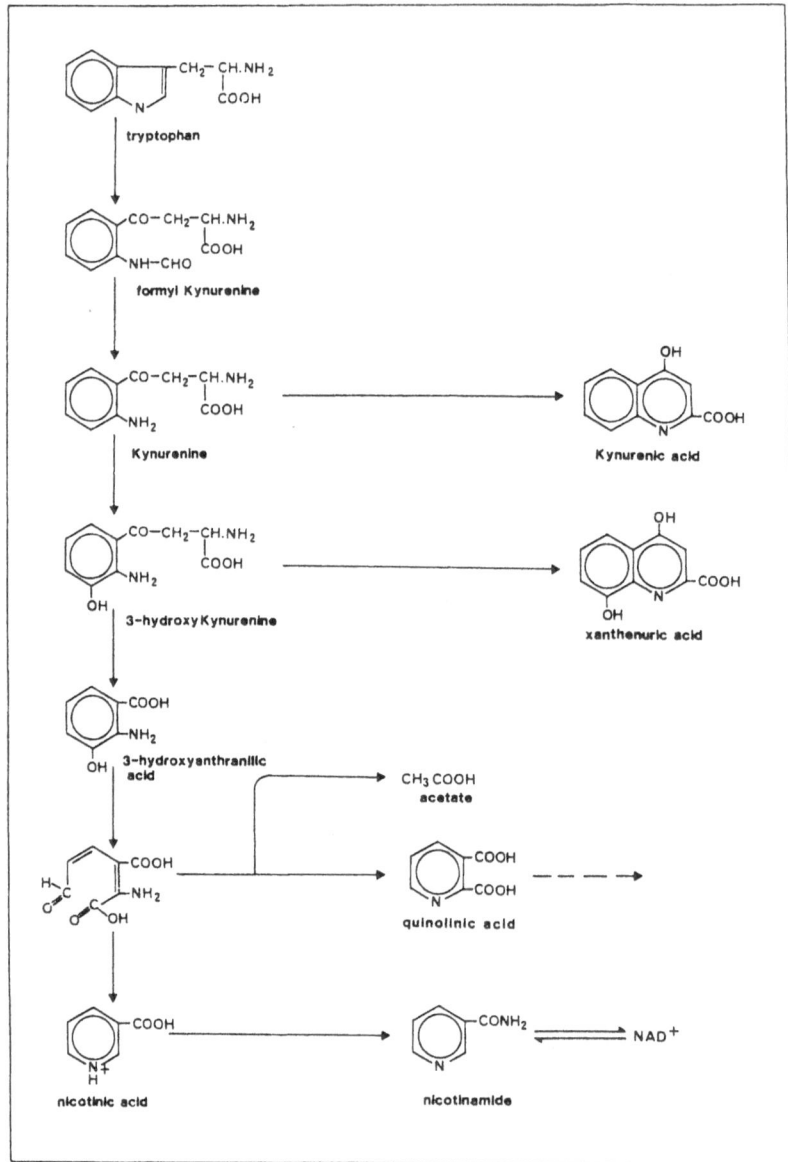

Figure 2.23 Tryptophan metabolic reactions involving pyrrole ring cleavage, and benzene ring scission

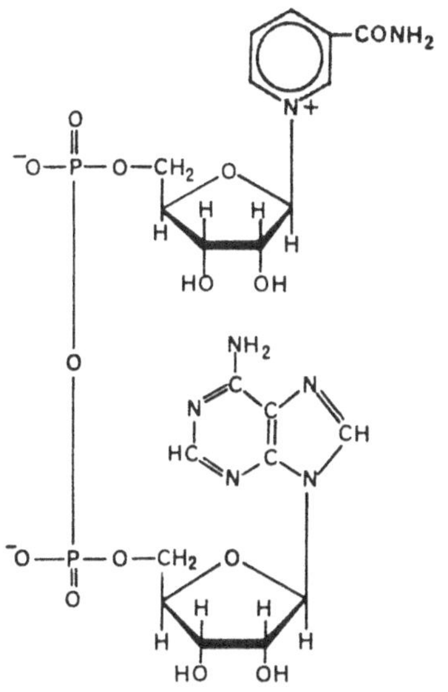

.nicotinamide adenine dinucleotide (NAD$^+$)

increased urinary loss of alanine, serine, threonine, asparagine, glutamine, valine, leucine, isoleucine, phenylalanine, tyrosine, tryptophan, histidine, glycine and citrulline. As well, any unabsorbed tryptophan remaining in the intestine is metabolized by intestinal bacteria; the indole and skatole so formed are absorbed, possibly to be further metabolized in the liver forming indican (which is subsequently excreted in urine) as well as other substances.

The neurological disturbance in Hartnup disease is episodic. Why this is so is not fully understood though there appears to be some association of the neurological manifestations with periods of poor diet.

Aetiology – Hartnup disease is inherited as an autosomal recessive trait.

Structural pathology – No morbid anatomical data appear to have been published.

Clinical features – Hartnup disease causes an intermittent, photosensitive, red scaly rash resembling the rash of pellagra, with episodic, severe but reversible cerebellar ataxia, faintness, headache and muscle discomfort. Occasionally psychological disturbance occurs. The e.e.g. is often severely but non-specifically abnormal. A generalized neutral aminoaciduria is present. The severity of symptoms tends to lessen in adult life. Some cases appear to remain virtually asymptomatic through life[113].

Diagnosis

(1) Clinical: Episodic cerebellar ataxia of itself raises the possibility of various types of structural lesion of the posterior cranial fossa, or of a neurometabolic disorder precipitated by dietary factors. The concomitant presence of an appropriate skin rash may suggest the possibility of nicotinic acid deficiency, making the diagnosis of pellagra very likely.

(2) Laboratory: An oral tryptophan load produces an abnormally flat plasma tryptophan level–time curve in cases of Hartnup disease. There is increased urinary excretion of indican and indolylacetic acid and its congeners. The pattern of aminoaciduria in Hartnup disease is diagnostic and does not occur in pellagra.

Therapy – Oral nicotinic acid supplementation will permit adequate synthesis of NAD^+ in patients with Hartnup disease. This corrects the nervous system and skin manifestations of the disorder. The aminoaciduria remains unaltered.

2.5.2 Early life ketoacidotic states

A number of disorders of amino-acid, particularly of branched chain amino-acid, metabolism produce severe progressive or intermittent ketoacidosis exaggerated by dietary protein uptake. Some are associated with disturbed glycine and NH_4^+ metabolism. Glycogen storage disease and pyruvate decarboxylase or dehydrogenase deficiency may produce similar ketoacidosis in the same age group.

2.5.2.1 Branched-chain amino-acid disorders (branched chain ketoaciduria: maple syrup urine disease)

The branched chain α-amino acids comprise leucine, isoleucine and valine. Disturbances of their metabolism produce one distinctive neurological disease (branched chain ketoaciduria), and also rare instances of other neurological disturbances, as outlined below[114].

Branched chain ketoaciduria is itself an uncommon biochemical syndrome comprising at least three genetically distinct disorders[115].

In these disorders there is impaired oxidative decarboxylation of the keto-acid products of branched chain amino-acid deamination. The syndrome is inherited, with over 80 cases reported up to 1977[115]. Nervous system function is affected in all variants of the syndrome. Defects in closely related metabolic reactions have been reported occasionally. These disorders are mentioned below in relation to the biochemical aspects of the syndrome.

Biochemical abnormality – Branched chain amino-acid metabolism involves several stages. It ultimately results in the formation of acetoacetate, acetyl CoA, propionyl CoA or methylmalonate semialdehyde. The latter two substances undergo further metabolic transformation before all four enter the energy yielding pathway (at the Krebs cycle). The metabolic reactions involved

60

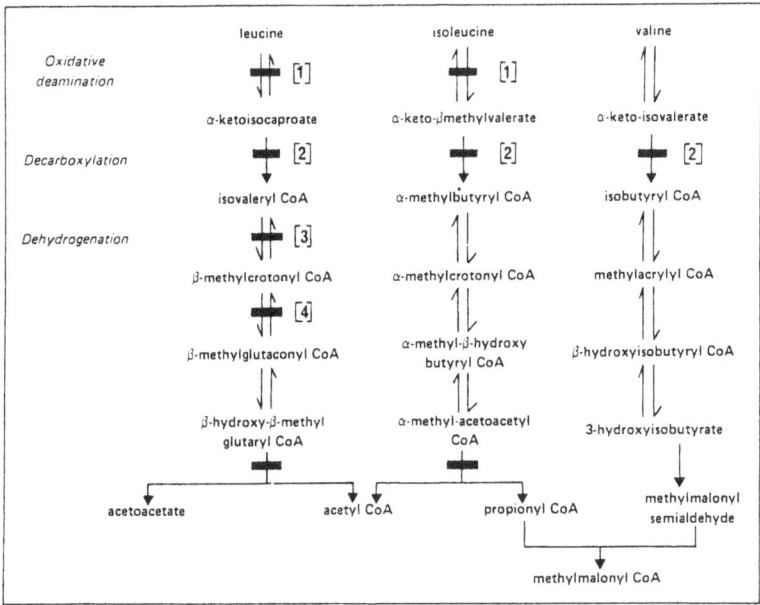

Figure 2.24 Branched chain amino-acid metabolic pathways, showing sites of metabolic blocks which produce human disease

in branched chain amino-acid metabolism are shown in Figure 2.24.

(1) *Transaminase deficiency* – The first stage of branched chain amino-acid biotransformation is a transamination, usually with α-ketoglutarate, the branched chain amino acid exchanging its —NH$_2$ group for a =O, thus forming the corresponding keto acid. Deficiency of branched chain amino-acid trans-aminase (the enzyme which catalyses the transaminations of leucine and iso-leucine) causes combined hyperleucinaemia and hyperisoleucinaemia. This is an exceedingly rare disorder, in which prolinaemia is also present. Mental retardation, seizures, retinal degeneration and deafness have been present in

```
CH₃   CH₃              COOH                    CH₃   CH₃            COOH
  \  /                 |                         \  /               |
   CH                  CH₂    TRANSAMINASE         CH                CH₂
   |                   |      ───────────▶         |                 |
   CH₂          +      CH₂                         CH₂        +       CH₂
   |                   |      (+ pyridoxal         |                 |
   CH.NH₂              C=O      phosphate)         C=O               CH.NH₂
   |                   |                           |                 |
   COOH                COOH                        COOH              COOH

leucine           α-ketoglutaric            α-ketoisocaproic       glutamic
                      acid                       acid               acid
```

the reported cases. Whether the metabolic abnormality caused the neurological features is uncertain.

At least one case of valine transaminase deficiency has been reported, associated with lethargy and mental retardation. Again the aetiological role of the metabolic defect is not established.

(2) *Decarboxylase deficiency* – Impairment of the next stage of the branched chain keto-acid metabolic pathway, the oxidative decarboxylation of all three branched chain keto acids, is the defect in branched chain ketoaciduria. The enzyme system involved in this stage is a multienzyme complex resembling the pyruvate dehydrogenase complex (Section 2.2.5). It uses thiamine pyrophosphate as a cofactor, and its activity results in the linkage of coenzyme A to the decarboxylated keto acids. The deficiency of the enzyme system may be total or partial: one variant is thiamine responsive. The relevant reactions are shown in Figure 2.25.

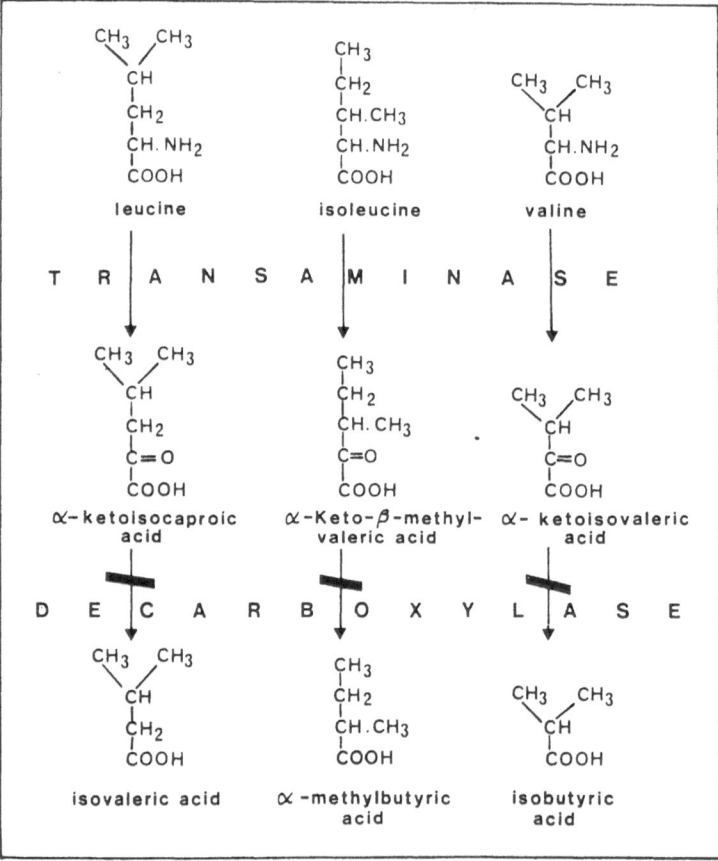

Figure 2.25 Details of the biochemical disturbance of branched chain ketoaciduria

In branched chain ketoaciduria the branched chain amino acids and their corresponding keto acids all accumulate. However, neurological dysfunction appears related in particular to high levels of α-ketoisocaproic acid[115]. α-Ketoisocaproate is a competitive inhibitor of pyruvate dehydrogenase in the liver and the brain, and of glutamate decarboxylase in the liver. Possibly α-ketoisocaproate also inhibits pyruvate transport across mitochondrial membranes[107], as well as acetoacetate transport[116]. Inhibition of pyruvate metabolism would lead to decreased production of acetyl CoA with consequent decreased metabolic flux through the Krebs cycle, and also decreased γ-amino-butyrate and glutamine formation. The ketoacidaemia in the decarboxylase deficiency also impairs isoleucine uptake into cells[117], inhibits alanine and aspartate aminotransferases[118] (thus causing reduced blood alanine levels) and may produce increased transamination between tryptophan and the keto acids, resulting in reduced concentrations of tryptophan and of its metabolite serotonin. High blood leucine levels may also interfere with amino-acid transport into the brain and with glutamine uptake[106] and thus may cause decreased cerebral protein synthesis. Acidosis with a decreased plasma pH is present in the syndrome. Some cases of branched chain ketoacidaemia have hypo-glycaemic episodes which are possibly due to impaired gluconeogenesis, itself resulting from high leucine levels. Hyperuricaemia may also occur, possibly because the keto-acid load interferes with renal tubular excretion of urate. The substances responsible for the characteristic urine odour of maple syrup in the syndrome are unknown.

(3) *Dehydrogenase deficiency* – It is known that metabolic defects can also occur in the subsequent stages of branched chain keto-acid metabolism. Iso-valeryl CoA dehydrogenase deficiency causes isovaleric acidaemia with increased urinary excretion of this substance and its conjugate isovalerylglycine. There is an associated severe ketoacidosis with a malodourous ('sweaty feet') urine. Affected persons have attacks of acidosis. Some are mentally retarded, but other are not. Intermittent cerebellar ataxia may be present. About 50% of such cases die in infancy.

(4) *β-Methylcrotonyl CoA-carboxylase deficiency* – The following stage in leucine metabolism may also be impaired, causing biotin-responsive β-methyl-crotonylglycinaemia. The disorder is very rare and neurological features are inconstant; there was no neurological abnormality in one reported case. The carboxylase deficiency may also occur as part of a more generalized biotin-dependent carboxylase deficiency.

$$\begin{array}{c} CH_3 \\ \diagdown \\ C=CH.COOH \\ \diagup \\ CH_3 \end{array}$$

(5) *β-Ketothiolase deficiency* – This disorder of isoleucine metabolism is discussed in relation to the ketotic hyperglycinaemic syndrome (Section 2.5.2.2.2).

(6) *Hydroxymethylglutaryl CoA lyase deficiency* – In one patient with this deficiency a viral illness caused malaise, drowsiness, hepatomegaly, hyperammonaemia, hypoglycaemia and mild acidosis. At liver biopsy, swollen hepatocytes were filled with lipid droplets. Leucine metabolites proximal to the metabolic block were excreted in excess in urine[119].

Aetiology – Branched chain ketoaciduria (maple syrup urine disease) comprises at least three genetically distinct disorders inherited as autosomal recessive traits. In the classical type, branched chain keto-acid decarboxylase activity is virtually absent. In intermittent branched chain ketoaciduria, decarboxylase activity in cultured skin fibroblasts is in the range of 2–15% of normal, while in the mild type activity of the enzyme is in the range 0–20%. In the two reported cases of the thiamine responsive variant, enzyme activity levels were 20% of normal.

Structural pathology – In fatal cases of branched chain ketoaciduria[96] the brain may be shrunken, or swollen, with the white matter spongy. Tracts which normally myelinate after birth are poorly myelinated, with areas of spongy change and gliosis. Axons appear relatively well preserved in hypomyelinated areas.

Clinical features – In the classical variety of branched chain ketoaciduria affected neonates feed poorly, cry weakly, and deteriorate rapidly with shallow respiration and intermittent cyanosis. The urine has a characteristic maple syrup-like odour. Tonic seizures and opisthotonus develop, and death (often from intercurrent infection) often supervenes in untreated cases within the first 4 months of life. Survivors may be mentally retarded. Cases of the mild variant of the disorder develop symptoms later and are moderately retarded mentally. Probably they can survive to adult life. Only a few such cases are known. Symptoms in the intermittent type begin during childhood. Affected children are of normal intellect, but in response to intercurrent illness or stress may develop anorexia, nausea, vomiting, lethargy and ataxia of gait. Recovery may occur within a few days, or the illness may progress to convulsions, stupor and coma. The thiamine-responsive form shows mild mental retardation.

Diagnosis
(1) Clinical: The clinical picture described above in association with the characteristic urine odour and occurring in a neonate with affected relatives should suggest the diagnosis in the classical type of branched chain keto-aciduria. It seems unlikely that a definitive diagnosis of the other variants could be made without laboratory investigations, unless the 'sweaty feet' odour of isovaleryl CoA dehydrogenase deficiency were recognized.

(2) Laboratory: In the classical variety, by the end of the first week of postnatal

64

life there are increased plasma and urinary levels of branched chain amino and keto acids, demonstrated by a variety of techniques ranging in complexity from thin layer chromatography to gas chromatography–mass spectrometry.

Decreased branched chain amino-acid decarboxylase activity can be measured in leukocytes, cultured skin fibroblasts, and cultured amniotic fluid cells (so that antenatal diagnosis is possible).

Therapy – The classical variety should be treated with a low protein diet containing the minimal amount of branched chain amino acid necessary for maintenance of normal plasma levels of branched chain amino acids. It may be necessary to supply much of the dietary nitrogen intake as purified amino acid. A low protein diet is needed in the intermittent form only during attacks.

2.5.2.2 *Hyperglycinaemia*

Several disorders of amino-acid metabolism can result in accumulation of glycine. This amino acid is involved in a variety of metabolic reactions (Figure 2.26).

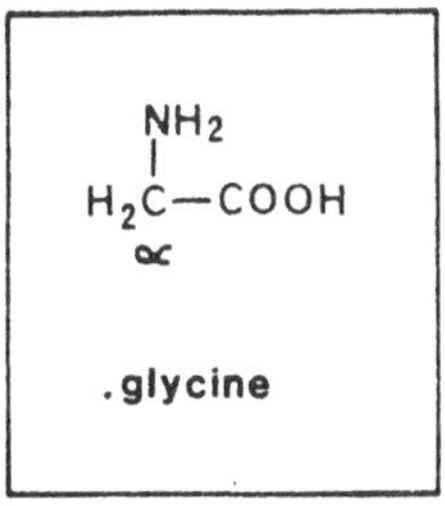

$$NH_2$$
$$H_2\overset{|}{C}-COOH$$
$$\alpha$$

.glycine

There are two types of hyperglycinaemia (1) non-ketotic, and (2) ketotic. In the non-ketotic variety the metabolic block appears to occur in the decarboxylation of glycine (Figure 2.26). The relevant biochemical mechanisms are not fully understood, but it is known that several acyl CoA compounds inhibit the reaction converting glycine to serine. In the ketotic type it is also thought that the causative metabolic block is in the conversion of glycine to serine (Figure 2.26). Such ketotic hyperglycinaemia occurs as a manifestation of certain amino-acid metabolism disorders, namely:

(1) Propionic acidaemia,

(2) Methylmalonic acidaemia,

(3) β-Ketothiolase deficiency. (β-Ketothiolase catalyses the reaction

β-ketoacyl CoA + HS.CoA $\rightleftharpoons$ acetyl CoA + acyl CoA),

(4) Isovaleric acidaemia,

65

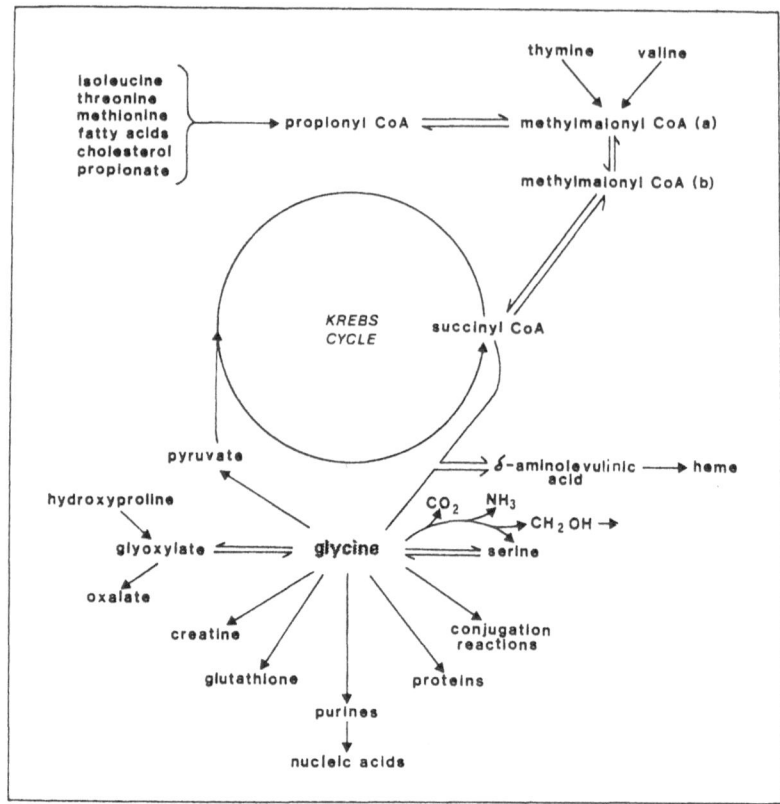

Figure 2.26 Metabolic pathways relevant to glycine

(5) Certain hyperammonaemic states (e.g. carbamyl phosphate synthetase deficiency).

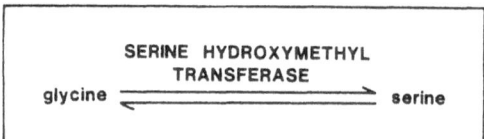

2.5.2.2.1 Non-ketotic hyperglycinaemia

This is a comparatively uncommon disorder, with about 20 cases reported up to 1977[120]. It appears to be due to a failure of glycine cleavage. The disorder causes severe neurological damage in early life[121]

66

Biochemical abnormality – The formation of serine from glycine involves the cleavage of one molecule of glycine to yield a methyl group which then methylates a second glycine molecule, thus forming serine (Figure 2.27). Folate coenzymes are involved in the $—CH_3$ group transfer.

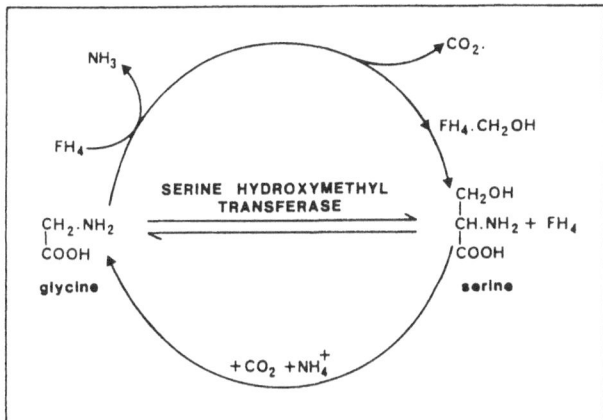

Figure 2.27   The formation of serine from glycine

Non-ketotic hyperglycinaemia is thought due to impairment of the glycine cleavage enzyme system with failure to produce hydroxymethyltetrahydro-folate. Unfortunately, details of the disordered biochemical mechanism are not adequately worked out. It is suspected that more than one biochemical defect may produce the raised glycine levels of the disorder[120]. Accumulation of glycine, an inhibitory neurotransmitter in the spinal cord, may interfere with neurological functions. However, it seems likely that other, as yet not defined, biochemical disturbances account for the neuropathological changes of the disorder. In this connection Dennis and Clarke[122] have shown that glyoxylate [CHO.COOH], a metabolite of glycine, is unlikely to be neurotoxic.

Aetiology – The disorder is probably inherited as an autosomal recessive trait.

Structural pathology – Neuropathological changes in the disorder are non-specific and resemble those of phenylketonuria and branched chain keto-aciduria[123]. There is delayed myelination and spongy degeneration of white matter and long tracts, with gliosis. Lipid inclusions occur in glial cells. The solid abdominal viscera are mildly enlarged.

Clinical features – Typically, non-ketotic hyperglycinaemia presents in neonates, who are lethargic, hypotonic, and inactive. Inadequate respiratory movements may cause respiratory acidosis. Hiccup, myoclonic jerks and convulsions develop and the e.e.g. often shows hypsarrhythmic changes. Survivors are mentally retarded.

Non-ketotic hyperglycaemia has also been reported in three adult siblings who had symptoms of spinal cord disorder[124].

Diagnosis
(1) Clinical: The clinical picture does not appear sufficiently distinctive to permit even a reasonably definite clinical diagnosis.

(2) Laboratory: Raised plasma glycine levels, with increased urine glycine excretion but no increase in excretion of proline and hydroxyproline, in the absence of organic acidaemia, suggest the diagnosis. (In man proline and hydroxyproline share a common renal tubular transport system with glycine.)

Therapy – Dietary measures do not appear to benefit the neurological disorder. There is no established effective treatment, but exchange transfusion may produce temporary benefit. It has been claimed that therapy with strychnine, a glycine antagonist, is of benefit[125].

2.5.2.2.2 Ketotic hyperglycinaemic syndrome

This syndrome is associated with recurrent episodes of vomiting, lethargy, coma and ketoacidosis. Unlike the non-ketotic variety, there is often neutropenia and thrombocytopenia. Osteoporosis may occur. Untreated cases develop epilepsy and mental retardation. As mentioned earlier, the hyperglycinaemia is probably a secondary phenomenon, and its biochemical mechanism of production is uncertain. Certain of the primary biochemical defects which lead to such ketotic hyperglycinaemia are considered below (namely β-ketothiolase deficiency, propionic acidaemia, methylmalonic acidaemia).

(1) *β-Ketothiolase deficiency* – This is a relatively recently described disorder of isoleucine metabolism. Only a few cases are known[126]. The enzyme deficiency causes an acidosis which is associated with episodic symptoms, and with glycine accumulation.

Biochemical abnormality – As illustrated in Figure 2.24, the metabolism of isoleucine occurs through several stages, the last of which, the conversion of α-methylacetoacetyl CoA to acetyl CoA and propionyl CoA, is catalysed by β-ketothiolase. When this enzyme is deficient α-methylacetoacetyl CoA, butanone and α-methyl-β-hydroxybutyryl CoA accumulate (Figure 2.28). The situation is likely to be made worse by a dietary protein load. It seems probable that one or more of the accumulated metabolites may interfere with glycine metabolism, since hyperglycinuria, with or without hyperglycinaemia, occurs in the disorder.

Aetiology – The enzyme defect appears to be inherited as an autosomal recessive trait.

68

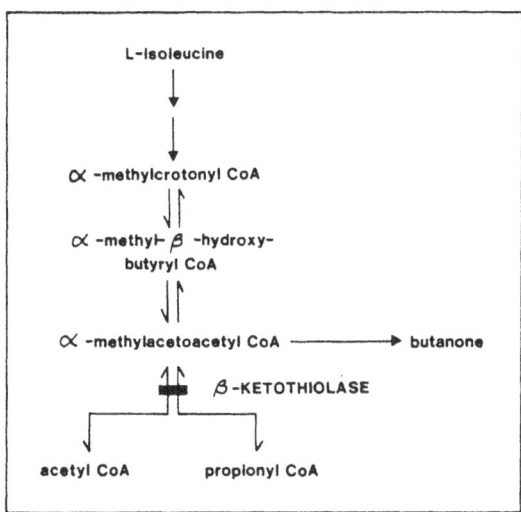

Figure 2.28 Metabolic pathways relevant to β-ketothiolase deficiency. Substances proximal to the metabolic block tend to accumulate

Structural pathology – No details are available.

Clinical features – The disorder appears to develop in infancy, with episodes of ketoacidosis. Some cases have had little or no mental or physical retardation but others have been retarded. At least one subject had neutropenia and thrombocytopenia with hyperammonaemia during an acidotic episode.

Diagnosis
(1) Clinical: On purely clinical grounds it is unlikely that the diagnosis could be taken past that of a (protein-dependent) ketoacidotic syndrome of early life.

(2) Laboratory: Diagnosis depends on finding increased quantities of iso-leucine metabolites in blood and urine. The demonstration of hyperglycinaemia or hyperglycinuria is not itself diagnostically specific.

Treatment – Dietary protein restriction, and correction of the secondary biochemical defects which occur during acute episodes, offers some prospect of avoiding significant mental and physical handicap in sufferers from the disorder.

(2) *Propionic acidaemia* – This uncommon disorder of the neonatal period (fewer than 20 cases described) also produces a ketotic hyperglycinaemic syndrome. Propionic acidaemia is caused by propionyl CoA carboxylase deficiency[126].

$$CH_2 \cdot CH_3$$
$$|$$
$$C-S.CoA$$
$$||$$
$$O$$

.propionyl CoA

Biochemical abnormality – The pathways of propionate metabolism are shown in Figure 2.29. Propionyl CoA carboxylase is a mitochondrial enzyme which utilizes biotin as its cofactor. Propionyl CoA carboxylase deficiency causes accumulation of propionic acid in blood and urine, with increased urinary excretion of 3-hydroxypropionic acid and methyl citrate. The synthesis of fatty acids may increase, possibly leading to fatty liver. There may also be some accumulation of intermediates of amino-acid metabolism, e.g. butanone, 2-methyl-3-oxovaleric acid, 3-oxovaleric acid, 3-hydroxyvaleric acid and 2-methyl-3-hydroxybutyric acid[127]. Increased dietary intake of the relevant amino acids (isoleucine, methionine, threonine) increases the degree of propionic acidaemia. It is not clear how the metabolic disturbance causes either the altered nervous system function, or the hyperglycinaemia and hyperammonaemia which often occur in the disorder.

Aetiology – The disorder is probably inherited as an autosomal recessive trait.

Structural pathology – One case which came to autopsy had a fatty liver and

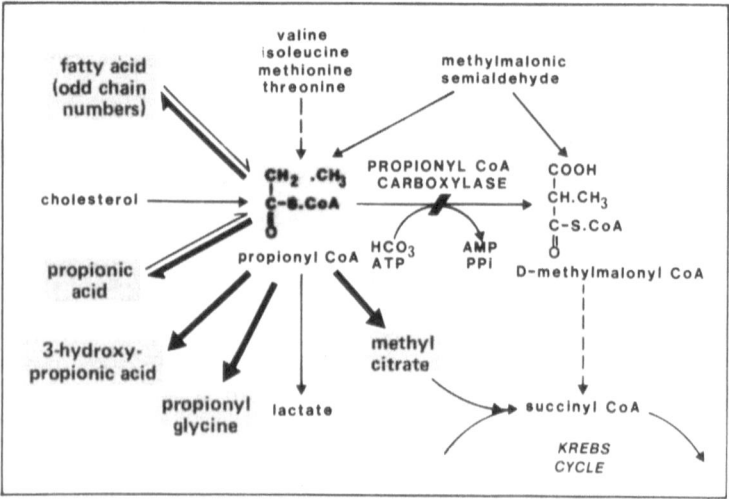

Figure 2.29 Propionate metabolism. Substances which accumulate in propionic acidaemia are shaded over

degeneration of the cerebellar cortex[128]. Another showed diffuse myelin pallor and vacuolation in the brain[123].

Clinical features – The clinical features of propionic acidaemia are those of a neonatal episodic ketotic hyperglycinaemic syndrome with exacerbation of ketoacidosis provoked by increased dietary protein intake, or by intercurrent infection. In the episodes dehydration, lethargy and coma may develop. Some (milder) cases present later in childhood, with mental retardation.

Diagnosis
(1) Clinical: There are no distinctive features which would permit diagnosis being taken beyond that of a ketotic syndrome of the neonatal period.

(2) Laboratory: Raised blood levels of glycine and propionic acid, with normal methylmalonic acid levels, are very suggestive. The enzyme defect, demonstrated in cultured fibroblasts or leukocytes, is diagnostic. Prenatal diagnosis of the enzyme deficiency in cultured amniotic fluid cells is possible.

Therapy – A low protein diet may be used to minimize the frequency and severity of attacks of ketoacidosis. Attacks of ketoacidosis are treated by withdrawal of all dietary protein and by parenteral administration of base. An instance of the syndrome which proved partly responsive to biotin (a cofactor for the carboxylase) has been described[129].

(3) *Methylmalonic acidaemia* – Up to 1977 only about 15 cases of this disorder had been reported[120]. Several different enzymatic defects may cause methylmalonic acid accumulation. The disorder presents as a ketotic hyperglycinaemic syndrome of early life[126].

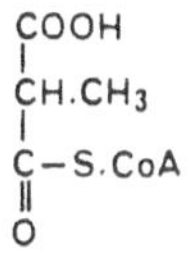

$$
\begin{array}{l}
\text{COOH} \\
|\\
\text{CH.CH}_3 \\
|\\
\text{C}-\text{S.CoA} \\
\|\\
\text{O}
\end{array}
$$

.methylmalonyl CoA

Biochemical abnormality – The relevant metabolic pathways are set out in Figure 2.30. Methylmalonic acidaemia may arise from defective function of either of the following enzymes, each of which catalyses a stage of methylmalonate metabolism:

(a) Methylmalonyl CoA racemase,
(b) L-Methylmalonyl CoA carbamyl mutase.

Decreased function of the latter enzyme can also be due to a deficiency of its cofactor (deoxyadenosyl vitamin B_{12}). In some cases, methyl B_{12} synthesis is

71

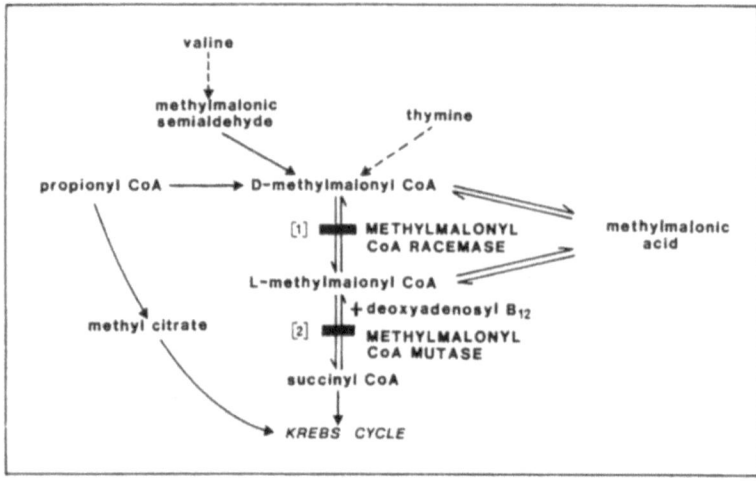

Figure 2.30 Metabolic pathways relevant to methylmalonic acidaemia

also defective. Thus four different biochemical mechanisms responsible for producing methylmalonic acidaemia are now recognized:

(a) Methylmalonyl CoA racemase deficiency,
(b) Methylmalonyl CoA mutase apoenzyme deficiency,
(c) Defective deoxyadenosylcobalamin synthesis, itself due to two different possible metabolic defects, and
(d) Defective deoxyadenosyl- and methylcobalamin synthesis.

When the disorder is due to deoxyadenosyl or methyl B_{12} deficiency, blood methionine level is low and homocystinuria and cystathionuria may also be present (for the explanation see Section 2.5.4). The clinical presentation of the disorder is then that of homocystinuria, and differs from the features of the other types of methylmalonic aciduria. In methylmalonic aciduria, as well as methylmalonic acid, propionic acid, 3-hydroxyvaleric acid, 3-oxovaleric acid, 2-methyl-3-oxovaleric acid and 2-methyl-3-hydroxybutyric acid accumulate in urine[127].

Why hyperglycinaemia, and occasionally hypoglycaemia and hyperammonaemia, should occur in methylmalonic aciduria remains obscure. The biochemical pathogenesis of the neurological disturbance and of the blood cellular changes is uncertain. The basis of the acidosis is also uncertain. It has been suggested that methylmalonic acid may inhibit the enzyme carbamyl phosphate synthetase. This inhibition would cause NH_4^+ to accumulate, since NH_4^+ could no longer enter the urea cycle.

Aetiology – Inheritance is through autosomal recessive mechanisms, though there has been difficulty in identifying heterozygous carriers biochemically.

Structural pathology – Neuropathological changes resembling those of subacute combined degeneration have occurred in cases in whom there was cobalamin deficiency[126].

Clinical features – The disorder usually appears in the first month of life, and nearly always within the first year. Affected infants have recurrent episodes of ketoacidosis and vomiting triggered by infection or increased dietary protein intake. Hepatomegaly occurs, with mild anaemia, intermittent neutropenia and thrombocytopenia, hypoglycaemia and osteoporosis. Irreversible metabolic acidosis may culminate in death.

Diagnosis
(1)   Clinical: Episodic ketoacidosis related to increased dietary protein intake may suggest disordered amino-acid metabolism, but laboratory studies are necessary for precise diagnosis unless manifestations of homocystinuria are present.

(2)   Laboratory: Methylmalonic acid levels (unmeasurable in normal blood) are raised in blood and urine. In the varieties related to defective methyl B_{12} activity there is associated homocystinuria and cystathionuria, and there is a high blood:CSF ratio of methylmalonic acid. Cobalamin levels can be measured in blood, and help identify these variants. Activity of the relevant enzymes of methylmalonate metabolism can be measured in leukocytes, cultured skin fibroblasts or cultured amniotic fluid cells (so that antenatal diagnosis is feasible).

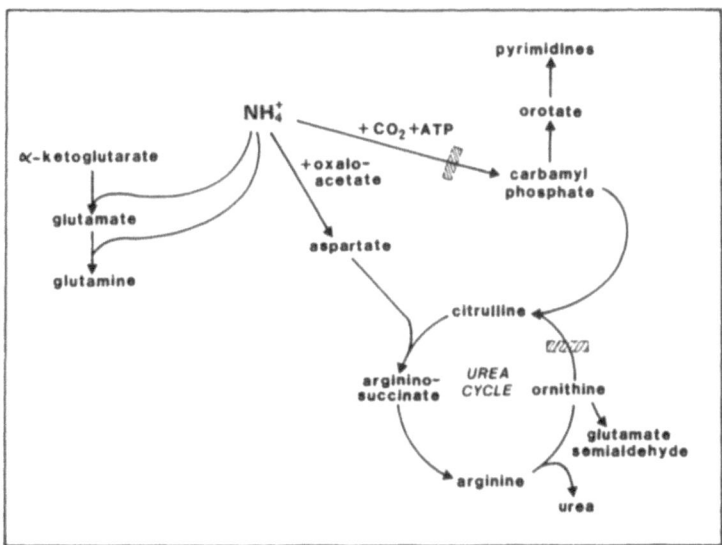

Figure 2.31   Pathways of ammonium metabolism

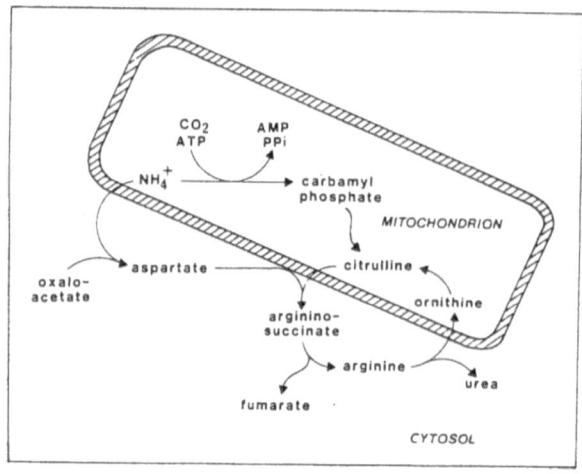

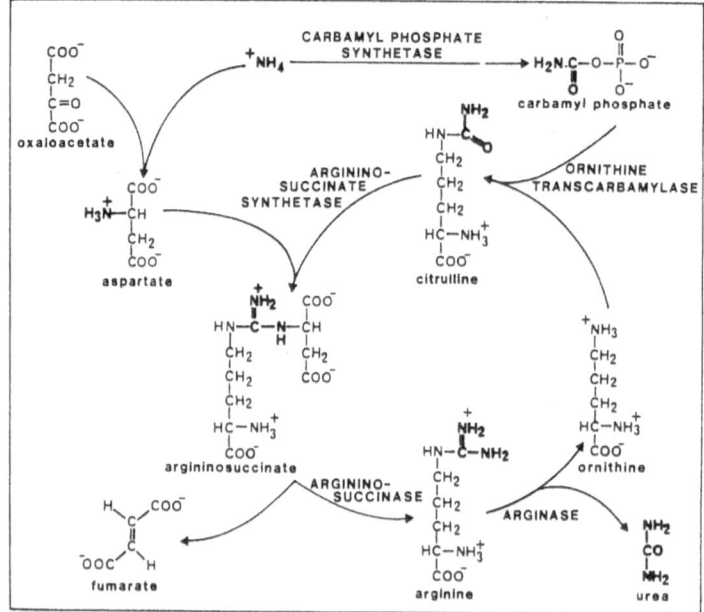

Figure 2.32   Details of the urea cycle

Therapy – The B_{12} deficiency varieties may be treated with high dosage vitamin B_{12}. When the racemase or the mutase is defective a diet low in amino acids that are metabolized to methylmalonic acid is prescribed, and acidosis is corrected as far as is possible.

2.5.3 Disorders of ammonium metabolism

Ammonium, produced from metabolic deamination of amino acids, from breakdown of purines and pyrimidines, or absorbed as the ion itself from the alimentary tract, undergoes the patterns of metabolism set out in Figure 2.31. The main pathway of NH_4^+ metabolism leads to urea which is subsequently excreted in urine. The urea cycle (Figure 2.32) in rat brain is incomplete, activities of the enzymes carbamyl phosphate synthetase and ornithine trans-carbamylase being virtually absent[130,131], though small amounts of these enzymes are said to be present in the human neonatal brain[132]. Consequently nearly all NH_4^+ formed in the brain must be transported to the liver to be there converted to urea. However, the brain can synthesize urea if supplied with citrulline, γ-aminobutyrate and aspartate[133]. The α-ketoglutarate–glutamate–glutamine mechanism has some capacity to 'mop-up' brain NH_4^+ production, though at the price of diverting metabolite flux from the Krebs cycle into the parallel GABA shunt (Figure 2.33 and, with structural details, Figure 5.11 where GABA is considered in relation to its neurotransmitter role).

A number of congenital defects of urea cycle enzymes exist. All the defects have in common the effect of decreasing the hepatic formation of urea. Consequently NH_4^+ accumulates in the liver, in blood, and in other tissues. Ammonium ion is neurotoxic and produces characteristic change in glial morphology. High NH_4^+ levels divert α-ketoglutarate from the Krebs cycle to glutamate formation and may thus reduce metabolic flux through the cycle. Therefore NH_4^+ accumulation may lead to decreased energy availability,

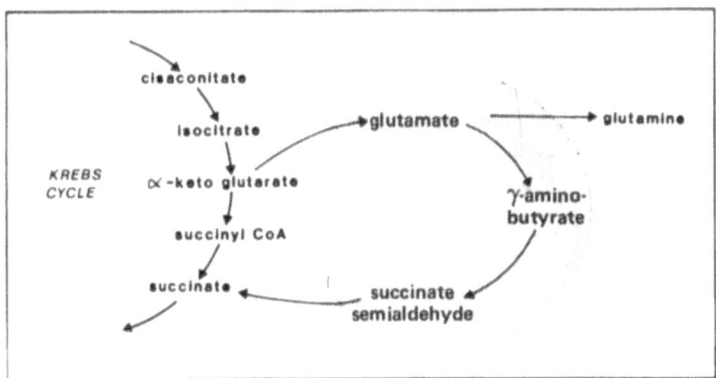

Figure 2.33 The γ-aminobutyrate (GABA) shunt

75

which may contribute to altered brain function. Whether this can cause the brain atrophy which often occurs in urea cycle disorders is uncertain.

2.5.3.1 *Urea cycle defects*

Defects in function of several urea cycle enzymes are known[134,135]. The laboratory detection of urea cycle disorders is discussed by Levin[136].

2.5.3.1.1 *Carbamyl phosphate synthetase deficiency*

Only a few cases of this defect have been recorded, and some are rather dubious. Hyperammonaemia and its consequences occur, but the clinical picture is heterogeneous. There are no specific chemical abnormalities in plasma. Some neuropathological data are available[137]. The defective enzyme may be measured in liver biopsies and leukocytes.

2.5.3.1.2 *Ornithine transcarbamylase (carbamyl transferase) deficiency*

This is the most common type of urea cycle disorder. Nearly 40 cases were reported up to 1978[135]. The transcarbamylase deficiency causes defective formation of citrulline from ornithine and carbamyl phosphate. The severity of the condition depends on whether the patient is homozygous or heterozygous for the defect.

Biochemical abnormality – Ornithine transcarbamylase deficiency blocks the urea cycle between ornithine and citrulline. NH_4^+ accumulates to high concentrations and blood glutamine level rises as the NH_4^+ is directed towards the glutamate–glutamine mechanism. Blood arginine level falls, due to failure of the urea cycle. Unconsumed carbamyl phosphate is directed toward pyrimidine synthesis, causing orotic acidaemia and aciduria (Figure 2.31).

Aetiology – Ornithine transcarbamylase deficiency is inherited as an X-linked dominant trait.

Structural pathology – In a fatal case with a severe form of the disorder the brain showed cerebral oedema with severe diffuse cerebral cortical atrophy, degenerated neurons, widespread demyelination and Alzheimer Type II astrocytes[97].

Clinical features – Hemizygous males die in the first week of life. They may show lethargy, dyskinesias, convulsions and respiratory distress and pass into coma. Heterozygous females have a partial deficiency of the enzyme. Their symptoms begin in infancy or childhood. They have episodes of anorexia, vomiting and ataxia, and develop hypertonus, irritability, screaming, inappropriate behaviour, confusion, lethargy and clouding of consciousness. Hepatomegaly is usually present. There is a variant type in males which begins in childhood with vomiting, followed by failure to thrive and ultimately coma.

Diagnosis
(1) Clinical: In the absence of a family history of the disorder, it is unlikely that purely clinical diagnosis could be taken beyond the suspicion of defective amino-acid metabolism, and then only if symptoms appeared to vary in parallel with dietary protein intake.

(2) Laboratory: The presence of raised blood NH_4^+ levels will narrow the differential diagnosis considerably. Orotic acidaemia may suggest the exact diagnosis. Precise diagnosis depends on the demonstration of absent ornithine transcarbamylase activity in leukocytes, liver or jejunal biopsies from hemizygous males, and a 65–95% reduction of activity of the enzyme in heterozygous females. It should be noted that there may be a temporary decrease in the activity of the enzyme in Reye's syndrome.

Therapy – Dietary protein restriction may be helpful in milder cases. Otherwise there is no useful treatment.

2.5.3.1.3 Citrullinaemia

This rare urea cycle defect (14 cases known to 1978[135]) is due to argininosuccinate synthetase deficiency.

Biochemical abnormality – Argininosuccinate synthetase deficiency blocks the formation of urea by impairing the conversion of citrulline to argininosuccinate. Blood citrulline, NH_4^+ and glutamine levels are consequently often considerably raised, and there is citrullinuria. A degree of orotic aciduria may also be present.

Aetiology – The enzyme deficiency is probably inherited as an autosomal recessive trait.

Structural pathology – The brain was oedematous, with neuronal and myelin degeneration, delayed myelination and glial cell enlargement, in a fatal case.

Clinical features – Neonatal cases usually die in the first few days of life. They suck poorly and are irritable and lethargic. They develop respiratory distress, convulsions and coma. A more chronic and benign variety of the disorder also occurs. In this variety affected patients are mentally retarded and suffer from convulsions, tremor, ataxia and episodes of vomiting. Some have enlarged livers.

Diagnosis
(1) Clinical: It seems unlikely that a purely clinical diagnosis could be taken beyond the suspicion of amino-acid metabolic disorder.

(2) Laboratory: Raised blood levels of NH_4^+ and glutamine suggest the category of the disorder, while raised levels of citrulline are diagnostic. The culprit enzyme defect can be demonstrated in liver biopsy specimens, circulating leukocytes or cultured skin fibroblasts.

Therapy – A low protein diet may be of some benefit.

2.5.3.1.4 Argininosuccinic acidaemia

This disorder, due to deficiency of argininosuccinate lyase, is the second most common type of urea cycle disturbance. Over 50 cases are known.

Biochemical abnormality – Argininosuccinate lyase deficiency impairs the conversion of argininosuccinate to arginine, and thus decreases urea formation. Argininosuccinate and citrulline accumulate in plasma, since their further metabolism is impeded. NH_4^+ is diverted to form glutamine. Hyperammonaemia occurs, particularly after meals.

Aetiology – Argininosuccinate lyase deficiency appears to be inherited as an autosomal recessive trait.

Structural pathology – In fatal cases[96,97] the brain has shown disturbed myelination with swollen Alzheimer Type II astrocytes. In some cases the brain has been swollen, and there have been areas of spongy degeneration. Changes in other organs are non-specific.

Clinical features – A fulminant neonatal type occurs with lethargy, respiratory distress, coma and death. There is also a more chronic variety of the syndrome with mental retardation, seizures, episodic lethargy and ataxia, hepatomegaly, coarse friable hair and a distaste for high protein foodstuffs.

Diagnosis
(1) Clinical: It does not seem possible to make an exact diagnosis on purely clinical grounds, though the diagnostic category of urea cycle defect might be suspected from the clinical features.

(2) Laboratory: Raised plasma NH_4^+ and glutamine levels indicate the diagnostic category, and raised argininosuccinate and citrulline levels in the blood and urine suggest the exact diagnosis. Reduced activity of the defective enzyme can be measured in red blood cells, amniotic fluid cells or liver biopsy specimens.

Therapy – A low protein diet may prevent exacerbation of symptoms in chronic cases.

2.5.3.1.5 Hyperargininaemia

Only four cases of this disorder are recorded. It is due to arginase deficiency and has the usual clinical pattern of a urea cycle disorder.

Biochemical abnormality – Arginase deficiency blocks the urea cycle between arginine and ornithine. Arginine and NH_4^+ accumulate in plasma. Arginine levels are increased in urine, together with increased amounts of cystine, lysine, ornithine and the disulphides of cysteine, citrulline and glutamine. Possibly arginine competitively blocks the renal tubular reabsorption of other amino acids, which consequently accumulate in plasma.

Aetiology – Arginase deficiency is probably inherited as an autosomal recessive trait.

Structural pathology – No details have been traced.

Clinical features – The disorder produces vomiting, seizures, spasticity and mental retardation.

Diagnosis – The diagnosis can be made only by laboratory methods. Raised blood levels of NH_4^+ and arginine suggest the diagnosis. The urine amino-acid excretion pattern can be confused with that of cystinuria. Reduced arginase levels can be detected in red blood cells.

Therapy – Dietary protein restriction may be of some use.

2.5.3.1.6 Hyperornithinaemia

This rare disorder is associated with mental retardation, and raised plasma levels of ornithine and raised urine levels of homocitrulline. The causative metabolic defect is uncertain. Reduced ornithine transport into mitochondria and ornithine decarboxylase deficiency have both been suggested. There may be more than one type of defect responsible for the syndrome.

2.5.3.2 Familial lysinuric protein intolerance

This is an intermittent hyperammonaemic syndrome occurring at times of increased dietary protein intake. Neurological features are not severe and the biochemical mechanisms involved are not completely clear. Over 20 cases are known, mostly of Finnish origin[138].

Biochemical abnormality – It is suspected that the disorder is due to impaired uptake of dibasic amino acids (lysine and arginine) into liver cells. Consequently there is an intracellular deficiency of ornithine, and this retards metabolic flux through the urea cycle. Therefore NH_4^+ tends to accumulate after protein intake, yet the blood urea level is low. A similar defect in renal tubular transport of dibasic amino acids may account for the increased urine loss of lysine, ornithine and arginine which occurs in the syndrome and the low levels of these substances which are found in plasma. Raised blood glutamine levels occur as a consequence of diversion of NH_4^+ towards the glutamate–glutamine mechanism. Presumably the neurotoxicity of the disorder is related to the hyperammonaemia and altered amino-acid availability.

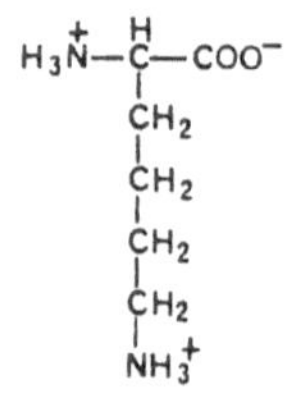

$$H_3\overset{+}{N}-\overset{\overset{\displaystyle H}{|}}{C}-COO^-$$
$$|$$
$$CH_2$$
$$|$$
$$CH_2$$
$$|$$
$$CH_2$$
$$|$$
$$CH_2$$
$$|$$
$$\overset{+}{N}H_3$$

.lysine

Aetiology – The disorder is inherited as an autosomal recessive trait.

Structural pathology – Liver cells contain increased smooth endoplasmic reticulum and some cells contain large vesicles. No information is available regarding neuropathological changes in the disorder.

Clinical features – Once affected infants begin to take cow's milk they develop vomiting and diarrhoea and a dislike of protein-rich food appears. Physical growth is poor and the liver becomes enlarged with fatty change. Splenomegaly often occurs. Cases become hypotonic and osteoporosis, neutropenia and thrombocytopenia often develop. A minority appears mentally retarded. Most cases have increased thyroxine and thyroxine-binding globulin levels in plasma.

Diagnosis
(1) Clinical: The story of dietary protein intolerance with the other clinical features should raise suspicion of an amino-acid metabolism disorder.

(2) Laboratory: Postprandial hyperammonaemia, with low blood levels of dibasic amino acids and high urinary levels of lysine, make the diagnosis probable.

Therapy – Supplying additional dietary arginine or citrulline together with lysine can go some way to correcting the deficiency of function of the urea cycle.

2.5.3.3 Congenital lysine intolerance with periodic NH_4^+ intoxication

This disorder is due to deficient lysine NAD^+-oxidoreductase. Lysine accumulation in blood and tissues inhibits the enzyme arginase (Section 2.5.3.1.5) therefore causing raised arginine and NH_4^+ levels. One well documented case of the disorder had vomiting and periodic lethargy, dependent on dietary protein intake. Episodes progressed to dehydration, spasticity, seizures and coma[139,140].

2.5.3.4 Other disorders of lysine metabolism

Familial hyperlysinaemia is due to deficiency of lysine-α-ketoglutarate reductase[141] and is associated with failure to form saccharopine (ε-N-glutaryl-2-L-lysine). Saccharopinuria is due to failure to degrade saccharopine, probably because of deficiency in aminoadipic semialdehyde glutamate reductase (saccharopine dehydrogenase). Neither disorder causes any definite neurological disturbance, though a case of saccharopinuria associated with spastic diplegia has been reported[142].

As well as being metabolized through saccharopine and glutamic acid to α-aminoadipic acid, lysine may be metabolized to this substance via pipecolic acid. Hyperpipecolaemia, associated with progressive neurological demyelinating disease and hepatomegaly, has been recorded[143].

2.5.3.5 Further causes of hyperammonaemia

Raised blood NH_4^+ levels also occur in hepatocellular failure and portosystemic blood shunting (Section 2.6.4), in ketotic hyperglycinaemic syndromes (Section 2.5.2.2.2), in non-ketotic hyperglycinaemia (Section 2.5.2.2.1), in various conditions where there is mitochondrial injury[144] and in certain other rare metabolic disorders.

2.5.4 Disorders of sulphur-containing amino acids

The pattern of metabolism of the sulphur-containing amino acids is set out in Figure 2.34.

Blocking the metabolism of cysteine and cystathione produces disorders (cystinosis and cystathionuria respectively) which are inocuous so far as the nervous system is concerned[145]. A single instance of myopathy and low intelligence associated with methioninaemia has been reported, though it was not established that the biochemical and neurological abnormalities were causally related[146]. These disorders will not be considered further. However, impaired metabolism of homocysteine produces a neurologically important disorder, homocystinuria.

2.5.4.1 Homocystinuria

Apart from phenylketonuria, homocystinuria is the most frequent disorder of amino-acid metabolism[147]. Mudd and Levy[148] were able to trace almost 300

81

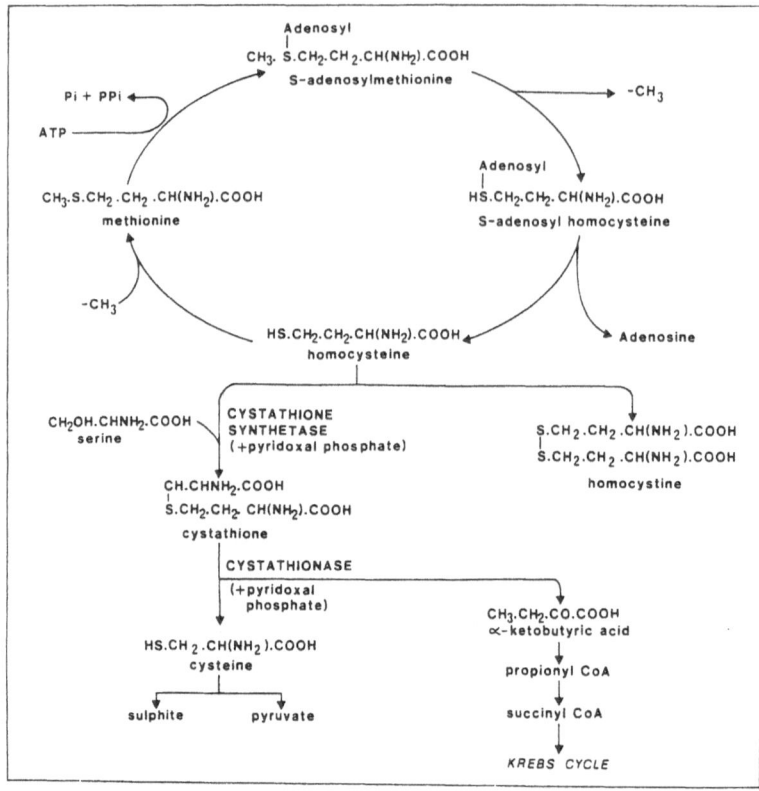

Figure 2.34 Metabolic pathways of sulphur-containing amino acids

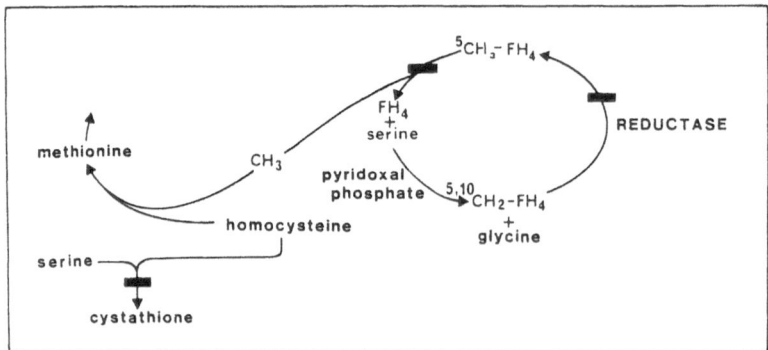

Figure 2.35 Reactions involved in homocysteine metabolism

cases of one aetiological type of the disorder in the literature. Skeletal deformity occurs in the condition. Some cases are mentally retarded, and in addition the nervous system may be damaged by thromboembolism. Homocystinuria is a syndrome due to several different biochemical disturbances which interfere with homocysteine metabolism.

Biochemical abnormality – The metabolic fate of homocysteine is as shown in Figure 2.34 and, in more detail, in Figure 2.35.
Homocystinuria may be due to

(1) Failure of the cystathione synthetase reaction, because of
   (a) Cystathione synthetase deficiency (the most common cause of homocystinuria),
   (b) Pyridoxal phosphate deficiency.

(2) Failure of methionine formation, because of lack of available $-CH_3$ groups, due to
   (a) 5,10-Methylene tetrahydrofolate reductase deficiency,
   (b) Methyltetrahydrofolate methyltransferase deficiency (from failure of activation of its coenzyme, deoxyadenosyl vitamin B_{12}).

Homocysteine itself may damage blood vessel endothelium, but the biochemical pathogenesis of the skeletal abnormalities in homocystinuria is unclear.

Aetiology – Cystathione synthetase deficiency appears to be inherited as an autosomal recessive trait.

Structural pathology – Neuropathological changes are consequences of occlusion of brain arteries. Skeletal and ocular abnormalities also occur[96,149].

Clinical features – The manifestations of the disorder include lens dislocations (causing glaucoma, myopia, retinal detachment and cataracts), fair skin, light hair, osteoporosis, genu valgum, and often scoliosis, pectus excavatum and vertebral abnormalities. High arched palate may be present. Platelet stickiness is increased and thromboembolism may occur and affect various organs, including the brain. Some patients are mentally retarded, probably from causes other than cerebrovascular disorder in the majority of instances.

Diagnosis
(1) Clinical: The presence of the clinical features described above should suggest the diagnosis.

(2) Laboratory: The diagnosis of homocystinuria is made by finding increased levels of the amino acid in blood and urine. The diagnosis of the causative biochemical defect requires additional laboratory study. Plasma methionine levels are raised in cystathione synthetase deficiency. Activity of the synthetase can be measured in liver biopsy specimens. In deoxyaden-

osyl vitamin B_{12} deficiency methylmalonic aciduria is also present (Section 2.5.2.2.2).

Therapy – The biochemical defect in cystathione synthetase insufficiency can be corrected in some cases by providing pyridoxine either in low, or in high, dose. However, it is not clear that correction of this defect will produce clinical recovery. Dipyridamole can be used to try to remedy the platelet disturbance. Vitamin B_{12} may be given in instances of deoxyadenosyl B_{12} deficiency.

2.5.4.2 Sulphite oxidase deficiency

Sulphite oxidase deficiency is a very rare disorder of sulphur metabolism. Descriptions of three cases are available[150]. A neurological disorder was present from the time of birth in one case, and in the other two cases acute neurological deficits developed in infancy, with progressive choreoathetosis and seizures. Lens dislocations were present.

The relevant biochemical defect appears to have been a deficiency in the oxidation of sulphite, formed from metabolism of *S*-containing amino acids (Figure 2.34), to sulphate. This deficit leads to increased excretion of sulphite, thiosulphate and *S*-sulphocysteine in urine, but there is no homocystinuria. Neuropathological changes in one instance were recorded by Rosenblum[151]. Sulphite oxidase measurements in the parents of one case produced findings consistent with an autosomal recessive inheritance for the disorder[150].

2.5.5 Miscellaneous rare amino-acid disorders

2.5.5.1 Disorders of proline metabolism

The amino acid proline is derived from pyroglutamate, itself formed from the cyclization of glutamate. The pathway for proline catabolism is set out in Figure 2.36.

Hydroxyproline undergoes an analogous series of reactions, again culminating in the production of glutamate. Metabolic disorders of these amino acids

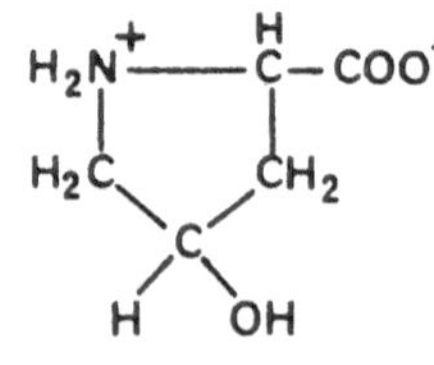

.hydroxyproline

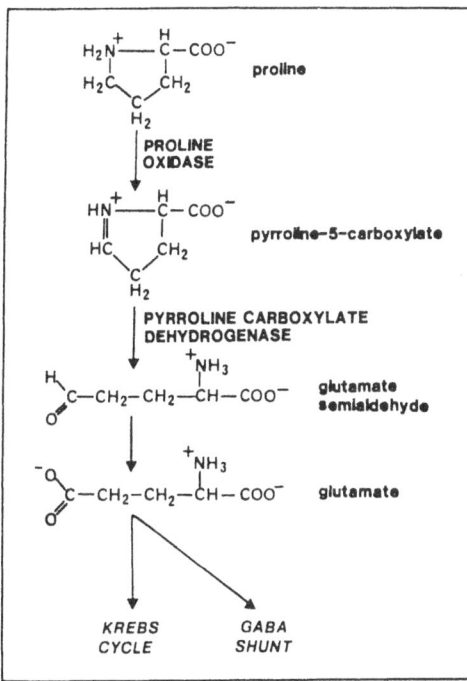

Figure 2.36 Pathways of proline metabolism

are discussed by Kivirikko and Simila[152] and Scriver[153], and the disorders are considered in relation to glutamate metabolism by Prusiner[154].

Proline oxidase deficiency, which causes hyperprolinaemia, is sometimes but not always associated with mental retardation[153]. There does not seem to be a definite causal relation between the disorder and the neurological disturbance.

Pyrroline carboxylate dehydrogenase deficiency, causing increased plasma levels of both proline and pyrroline-5-carboxylate, is a less common disorder than proline oxidase deficiency. It is sometimes associated with mental retardation, convulsions and e.e.g. abnormalities, but it is uncertain whether the relation is causal or coincidental.

2.5.5.2 Oxoprolinuria

Meister[155] reviewed three cases of 5-oxoprolinuria, one of whom showed neurological abnormalities (oligophrenia, bilateral spasticity and a cerebellar deficit) as well as episodic jaundice. It is not certain whether the neurological and biochemical defects were causally related.

2.5.5.3 Histidinaemia

This disorder, due to decreased histidase activity, is not consistently associated with nervous system abnormality, though most cases reported have been mentally retarded[156,157].

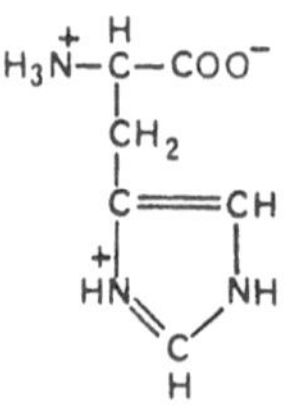

.histidine

2.5.5.4 Carnosinase deficiency

Only seven cases of this disorder were known prior to 1978[158]. Carnosinase deficiency is associated with carnosine accumulation and neurological symptoms.

Biochemical abnormality – Carnosinase catalyses the cleavage of the dipeptide carnosine (β-alanyl histidine) into its component amino acids. The biological function of carnosine is not understood. Therefore it is impossible to determine how the enzyme defect affects nervous system function. It is not certain that neurological disorder necessarily accompanies the biochemical disturbance.

Aetiology – The disorder is probably inherited as an autosomal recessive trait, though there is a possibility of X-linked inheritance.

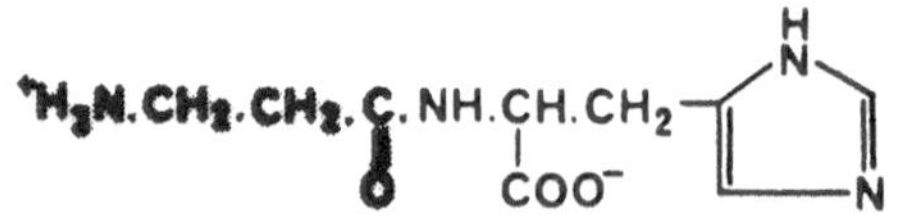

.carnosine (β-alanyl histidine)

Structural pathology – The brain in affected cases has shown loss of neurons in the cerebral isocortex, and loss of Purkinje cells, with cerebellar gliosis. The pyramidal and spinocerebellar traits were oedematous. Peripheral nerves showed widespread severe axonal degeneration, with demyelination[159].

86

Clinical features – Affected infants show slow development in the first 6.months of life, and may suffer from myoclonic seizures. As they grow older they prove to be severely retarded mentally, with deafness, blindness, nystagmus, spasticity, and autism.

Diagnosis
(1) Clinical: The clinical features of the disorder do not appear sufficiently distinctive to permit diagnosis. The differential diagnosis would include a range of metabolic disorders as well as mechanical lesions such as obstructive hydrocephalus.

(2) Laboratory: Raised plasma carnosine levels may be present even when the patient is taking a meat-free diet (considerable quantities of carnosine are present in muscle). The serum carnosinase level has been decreased in affected persons.

Therapy – No known therapy is of benefit.

2.5.5.5   Homocarnosinase deficiency

A few cases of familial progressive spastic paraplegia associated with mental deterioration and retinal degeneration have been reported[160,161], in whom increased CSF homocarnosine (γ-amino-n-butyryl histidine) levels were present However, the neurologically normal mother of one of the affected families also had raised CSF homocarnosine levels. One affected case lacked homocarnosinase activity in a brain biopsy, though homocarnosine–carnosine synthetase activity was normal.

2.5.5.6   Glutaric acidaemia

At least eight instances of neurological disease associated with glutaric acidaemia are known[162].

Biochemical abnormality – Deficiency of the enzyme glutaryl CoA dehydrogenase is believed to impair the oxidation of glutaryl CoA to glutaconyl CoA, one of the stages in the catabolism of lysine and tryptophan to acetoacetyl CoA (Figure 2.37). The accumulated glutaryl CoA is hydrolysed to glutaric acid and reduced coenzyme A. The detailed biochemical pathogenesis of the neurological manifestation is not known, though it is possible that the glutarate interferes with glutamate at its receptors, or inhibits glutamate decarboxylase. Reduced activity of the latter enzyme in the striatum occurs in association with Huntington's chorea (Section 5.3.1.2), which features a pattern of involuntary movement disorder similar to that recorded in glutaric acidaemia.

Aetiology – The cause of the enzyme deficiency is uncertain.

87

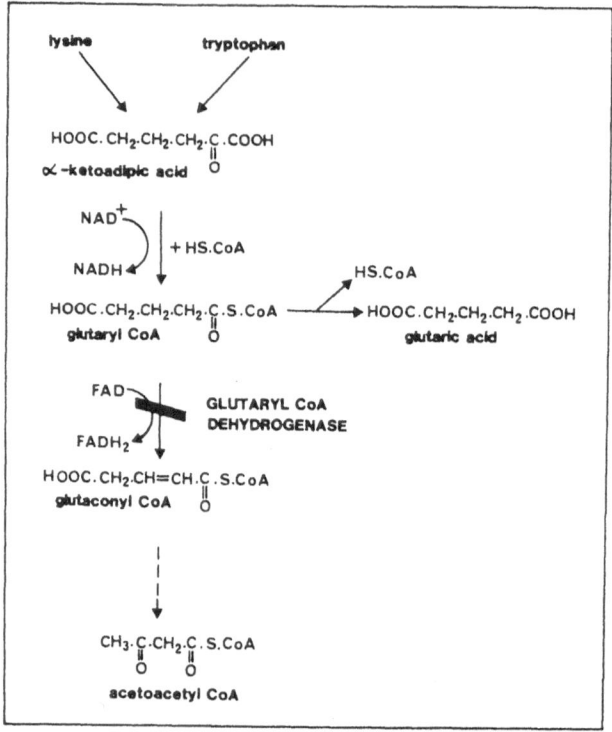

Figure 2.37 Metabolic pathways relevant to glutaric acidaemia

Structural pathology – In one case who died at the age of 3.5 years the striatum at autopsy was shrunken, with neuronal loss, but the pallidum and thalamus were normal[162].

Clinical features – The disorder presents in early life as a progressive choreo-athetosis and dystonia, with spastic quadriparesis. Mental retardation occurred in one case only. Four cases had recurrent acidotic episodes, with increased blood and urine levels of glutaric acid.

Diagnosis
(1) Clinical: The differential diagnosis of early life progressive choreoathetosis includes cerebral birth palsy, D-glyceric acidaemia, sulphite oxidase deficiency, Lesch–Nyhan syndrome, certain sphingolipidoses, Wilson's disease, early onset Huntington's disease and torsion dystonia. Associated clinical features in some of these conditions may help narrow the range of diagnostic possibilities.

(2) Laboratory: The diagnosis is established by the presence of excess glutaric

acid in blood and urine, and by a deficiency of glutaryl CoA dehydrogenase in cultured skin fibroblasts.

Treatment – No treatment is of proven value. Restriction of dietary lysine and tryptophan intakes does not appear to be useful.

2.6 DISORDERS OF OTHER N-CONTAINING SMALL MOLECULES

2.6.1 Lesch–Nyhan syndrome

This inherited disorder of urate metabolism, caused by deficiency of the enzyme hypoxanthine-guanine phosphoribosyl transferase, produces mental retardation, spasticity and self-mutilatory behaviour. At least 100 cases are known[163]. The subject has been reviewed by Nyhan[164] and Kelley and Wyngaarten[165].

Biochemical abnormality – The relevant biochemical pathway is shown in Figure 2.38. Hypoxanthine-guanine phosphoribosyl transferase (HGPT) catalyses the salvage synthesis of nucleotides (inosine and guanosine monophosphates) from the purine bases hypoxanthine and guanine. These purine bases have been formed by nucleotide and nucleic acid catabolism.

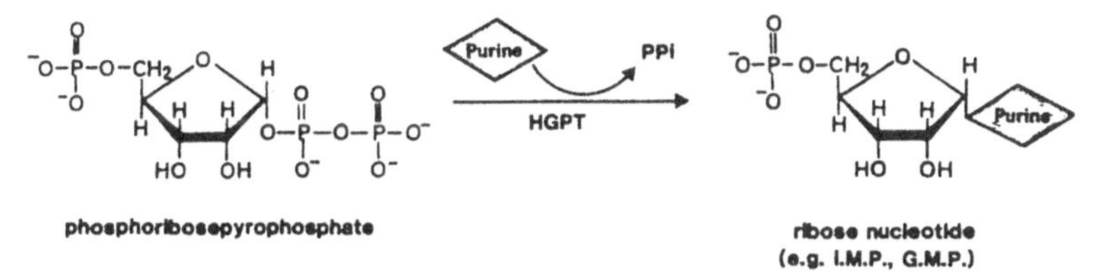

In deficiency of this enzyme, reduced salvage synthesis of adenine and guanine monophosphates leads to decreased feedback inhibition of the *de novo* pathway for purine synthesis. Further, the accumulation of phosphoribose pyrophosphate proximal to the metabolic block at the HGPT stage provides increased substate for the *de novo* purine biosynthesis pathway. Consequently there is overproduction of purines, ultimately resulting in excess formation of urate. The consequent hyperuricaemia may lead to gout. The chemical pathogenesis of the cerebral disturbance in the Lesch–Nyhan syndrome is uncertain. Possibly altered purine nucleotide concentrations at critical phases of brain maturation are injurious to neural tissue[166]. Plasma levels of dopamine-β-hydroxylase, the enzyme catalysing the formation of noradrenaline from dopamine, are very high in cases of Lesch–Nyhan syndrome with self-mutilating tendencies[167]. The interpretation of this association is unclear. The hyper-

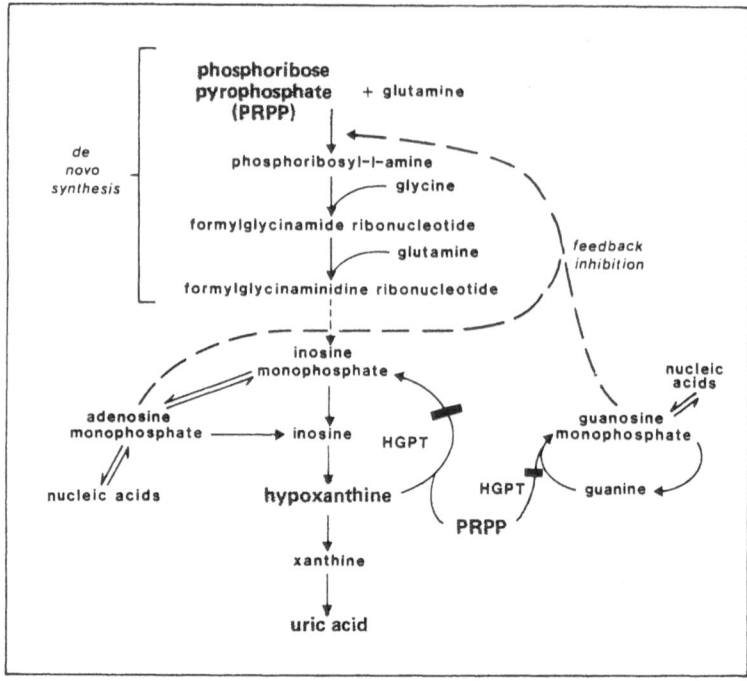

Figure 2.38 Pathways of purine metabolism. In HGPT deficiency there is not only decreased feedback inhibition of the *de novo* purine synthesis pathway, but the accumulated PRPP proximal to the metabolic block also serves to drive the *de novo* pathway

uricaemia that develops in the syndrome may cause renal damage as well as gout.

Aetiology – Hypoxanthine guanine phosphoribosyl transferase deficiency is inherited as a sex-linked recessive trait. Rare instances of partial deficiency of the enzyme are known[168].

Structural pathology – Autopsy findings in cases of the syndrome have included growth retardation, evidences of self-mutilation and shrunken kidneys with deposits of monosodium urate and uric acid. There are no consistently present gross or microscopic neuropathological changes[165].

Clinical features – The disorder[168] occurs in males, who though apparently normal at birth show delayed motor development in the first 6 months of life. Thereafter extrapyramidal signs appear (choreoathetosis and dystonia). Evidence of severe pyramidal tract insufficiency and often seizures develop. Compulsive severe self-destructive behaviour occurs with biting of fingers, lips and buccal mucosa. Most sufferers are mentally retarded and often aggressive.

Uric acid crystalluria and urinary calculi may develop, and ultimately renal insufficiency. Occasionally clinical gout occurs.

Diagnosis
(1) Clinical: The fully developed clinical picture is very suggestive of the diagnosis, though occasional dyskinetic children with mental retardation from other causes may appear to injure themselves deliberately and compulsively.

(2) Laboratory: The presence of hyperuricaemia, with virtually absent red cell hypoxanthine-guanine phosphoribosyl transferase activity, will establish the diagnosis. Heterozygotes can be detected by the presence of low levels of hypoxanthine-guanine phosphoribosyl transferase in cultured skin fibroblasts (though there is no deficiency in red cells). Prenatal diagnosis of the disease may be made on the basis of enzyme deficiency in cultured amniotic fluid cells.

Therapy – Allopurinol administration may prevent the hyperuricaemic complications of the Lesch–Nyhan syndrome, but no therapy is available to remedy the neurological manifestations.

2.6.2 Porphyrin metabolism disturbances

At least three disturbances of porphyrin metabolism are known in which nervous system function is disturbed[169]. Neurological disturbance appears somehow related to retention of the porphyrin precursors δ-aminolevulinate

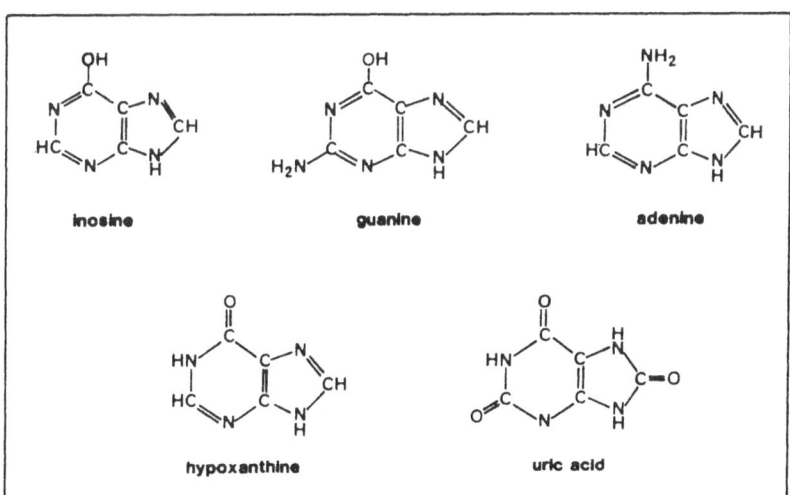

Figure 2.39 Structures of certain purine derivatives

and porphobilinogen. The occurrence of cutaneous photosensitivity appears related to protoporphyrin accumulation. The best understood and most important of these three disorders is acute intermittent porphyria.

2.6.2.1 Acute intermittent porphyria

Acute intermittent porphyria is an inherited disorder of pyrrole metabolism which produces episodic neuropsychiatric and abdominal disturbances in adult life. The cause of the disorder is a partial deficiency of the enzyme uroporphyrinogen synthetase. This deficiency indirectly permits increased activity of δ-aminolevulinate synthetase. Roughly one case occurs per 100 000 persons.

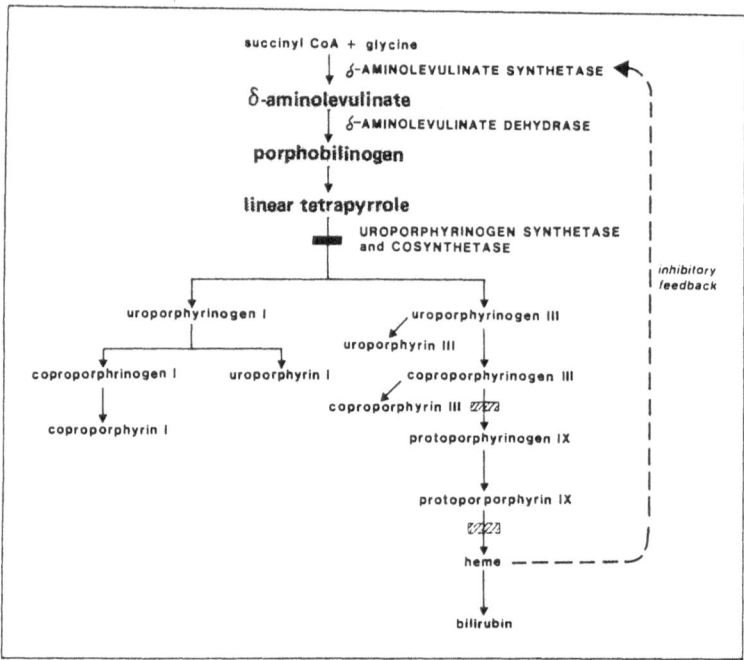

Figure 2.40   Porphyrin metabolism

Biochemical abnormality – The relevant biochemical pathway is shown in Figure 2.40. Deficiency of hepatic uroporphyrinogen synthetase ultimately leads to deficient formation of heme. Heme normally exerts a feedback inhibition on δ-aminolevulinate synthetase, the enzyme that catalyses the committed step of the porphyrin synthesis pathway. With decreased heme formation, this feedback inhibition lessens. Therefore more δ-aminolevulinate and porphobilinogen form. However, porphobilinogen molecules cannot condense

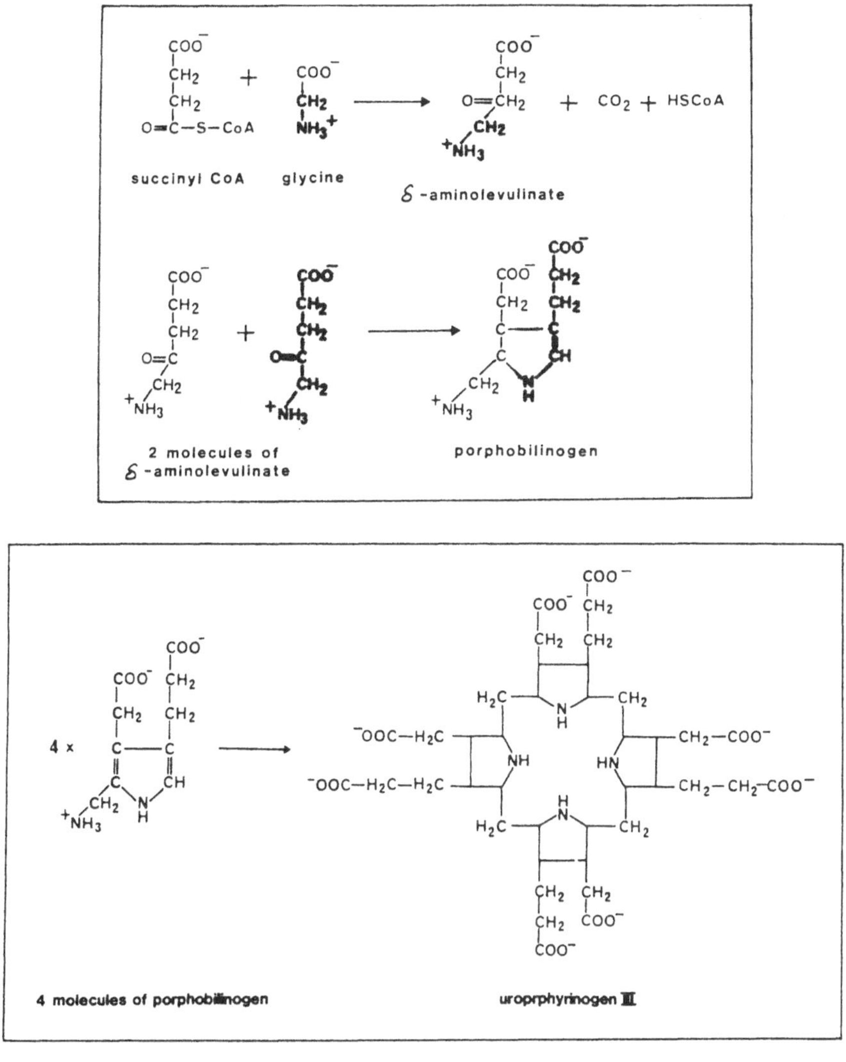

Figure 2.41 Structural details of the porphyrin biosynthetic pathway

to form uroporphyrinogens as in the normal because uroporphyrinogen syn-thetase is deficient (Figure 2.41). Hence δ-aminolevulinate and porphobilinogen accumulate in plasma and are excreted in urine. Since δ-aminolevulinate syn-thetase can be induced by drugs such as sulphonamides, oestrogens, bar-biturates and phenytoin, exposure to these drugs can increase the severity of the biochemical disorder in acute intermittent porphyria. Intake of these drugs can also precipitate attacks of clinical porphyria. The intimate biochemical mechanisms responsible for the neurological manifestations of acute inter-

93

mittent porphyria are obscure. Neither δ-aminolevulinate nor porphobilinogen is directly neurotoxic, though the former inhibits Na^+,K^{+-} adenosine triphosphatase, the 'sodium pump' and, at substantially higher concentrations than those encountered in human porphyria, impairs acetylcholine release from nerve endings[170]. δ-Aminolevulinate is a structural analogue of γ-aminobutyrate, and inhibits potassium stimulated γ-aminobutyrate release from rat brain synaptosomes[171]. It is not known if this effect occurs in acute intermittent porphyria in humans.

Aetiology – Acute intermittent porphyria is inherited as an autosomal dominant trait.

Structural pathology – In fatal cases of acute intermittent porphyria[75] the nervous system shows perivascular demyelination, with loss of neurons in the hypothalamus. There is a demyelinative neuropathy of peripheral somatic and autonomic nerves.

Clinical features – The disorder tends to appear in adult life and is more severe in women. The condition may be silent throughout life in some carriers of the trait, or else may produce only mild and vague symptoms. Typically, acute attacks cause abdominal and other bodily pains, nausea, vomiting, dark urine and autonomic disturbances (blood pressure rise or fall, tachycardia, bowel and bladder distension and fever). Mental changes may develop, with altered fluid and electrolyte metabolism (e.g. hyponatraemia) and epileptic seizures. Occasionally hypertensive encephalopathy and rarely coma, focal cranial nerve or peripheral nerve signs, polyneuritis and long tract signs appear. Episodes tend to become more frequent with advancing age. Many episodes appear to be precipitated by drug intake. Sufferers often have a continuing mild psychiatric disturbance of neurotic type before the onset of acute episodes, and also between episodes.

Diagnosis
(1) Clinical: The fully developed clinical picture of acute intermittent porphyria is characteristic. However, the early manifestations, and incomplete variants of the full picture, can be misleading. Porphyria is one of the conditions which should be born in mind in any patient with obscure neurological symptoms.

(2) Laboratory: Diagnosis depends on the detection of increased urinary excretion of δ-aminolevulinate and porphobilinogen, particularly during acute attacks. Porphobilinogen may be estimated by semiquantitative ward tests, but more exact measurement requires chromatographic assay. Lead intoxication can produce abdominal pain and polyneuritis, with increased urine δ-aminolevulinic acid excretion, but porphobilinogen level usually is not greatly increased in lead intoxication. Increased porphyrin precursors may not be present in latent cases of acute intermittent porphyria, but defective uroporphyrinogen synthetase activity can be

94

measured in red blood cells or cultured fibroblasts. Tyrosinaemia, due to
p-hydroxyphenylpyruvic oxidase deficiency (Section 3.5.1.2), can also cause
raised blood levels of δ-aminolevulinate and porphobilinogen, with
porphyria-like symptoms.

Therapy – Avoidance of drugs which induce δ-aminolevulinate synthetase is
essential. These drugs include many of the anticonvulsants, which might otherwise be given to treat epileptic seizures caused by the disorder[172]. Attacks of
acute porphyria may be managed by general supportive and symptomatic
measures, and correction of individual defects, e.g. fluid depletion. Intravenous
glucose infusion appears beneficial in attacks and intravenous haematin can
be infused to produce increased feedback inhibition of δ-aminolevulinate and
porphobilinogen synthesis[173]. This treatment consistently improves the biochemical abnormalities, but is less often of clinical benefit[174].

2.6.2.2 Hereditary coproporphyrinuria

This disorder produces episodic neurological and abdominal symptoms similar
to those of acute intermittent porphyria. It is believed that the condition results
from a failure to metabolize coproporphyrinogen III to protoporphyrinogen
III. The defective enzyme appears to be coproporphyrinogen oxidase, and the
defect is inherited as an autosomal dominant trait. Many carriers of the defect
remain asymptomatic. As in acute intermittent porphyria decreased heme
formation leads to derepression of δ-aminolevulinate synthetase (Figure 2.40).

The condition is distinguished from acute intermittent porphyria by an
increased excretion of coproporphyrin III in urine and particularly in faeces,
and by reduced coproporphyrinogen oxidase activity in fibroblasts and leukocytes. In acute attacks δ-aminolevulinate and porphobilinogen excretion in
the urine is increased.

2.6.2.3 Variegate porphyria

In addition to abdominal and neuropsychiatric manifestations similar to those
of acute intermittent porphyria, photodermatitis occurs in this porphyria
variant. The presence of large amounts of protoporphyrin and coproporphyrin
in faeces, the provocation of attacks by drugs, and knowledge that heme
represses δ-aminolevulinate synthetase in the condition, suggest that, by
analogy with acute intermittent porphyria, variegate porphyria is due to a
metabolic block between protoporphyrinogen and heme. However, the postulated enzyme defect, though inherited as an autosomal dominant trait, does
not appear to have yet been defined.

2.6.3 Vitamin deficiencies

The effects of deficiencies of particular vitamins on certain biochemical
reactions have already been discussed at several points in this text. These

matters have arisen in relation to individual genetic metabolic defects. Thiamine deficiency has also been considered as an entity in its own right. The consequences of vitamin deficiency in man often result from simple failure of vitamin intake, as well as from failure of specific steps in metabolic pathways. It therefore seems desirable to discuss various vitamin deficiencies as entities in their own rights. To some extent this can be done by cross-reference to material presented earlier in this book.

2.6.3.1 Nicotinic acid deficiency

Nicotinic acid deficiency causes pellagra. This disorder occurs in the malnourished, in whom the effects of deficiencies of other vitamins and nutrients often coexist.

Biochemical abnormality – Nicotinamide nucleotides (NAD^+ and $NADP^+$) are involved as cofactors in many H^+ transfer reactions in the body. The formula of nicotinamide adenine dinucleotide (NAD^+) was shown in Section 2.5.1.3. That of nicotinamide adenine dinucleotide phosphate ($NADP^+$) is shown below. In general, NAD^+ is involved in reactions of the energy yielding pathway and $NADP^+$ in biosynthetic reactions. Nicotinamide, as well as being absorbed from the alimentary tract preformed or as its precursor nicotinic

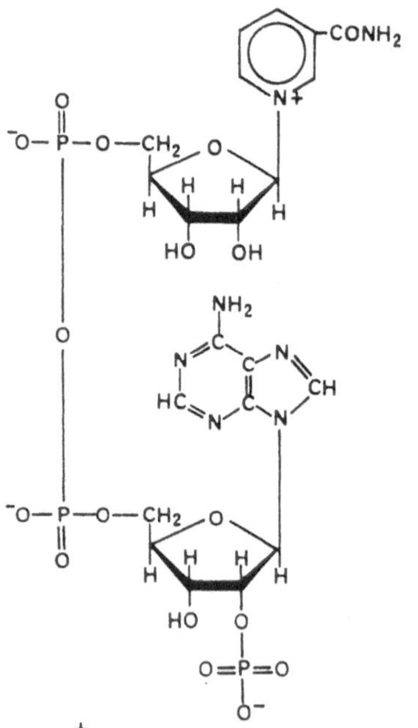

96

acid, may be synthesized within cells from tryptophan (see Hartnup disease, Section 2.5.1.3). Whether clinical manifestations of nicotinamide deficiency occur is determined by the intake of both the vitamin itself and of tryptophan. It is easy to see that NAD^+ and $NADP^+$ deficiencies could disturb a variety of biochemical reactions, but it is not known which of the altered reactions is responsible for the clinical manifestations of pellagra.

The possibility of coincidental nutritional deficiencies arises when interpreting some of the manifestations that have been ascribed to pellagra.

Aetiology – Nicotinamide deficiency is almost always due to a poor diet resulting from social and economic factors or chronic disease. Malignant carcinoid, which may divert a substantial proportion of dietary tryptophan intake to serotonin formation, is a very rare cause of nicotinamide deficiency, as is Hartnup disease.

Structural pathology – In the nervous system encephalopathic, myelopathic and peripheral neuropathic changes have been reported in fatal cases[75].

Clinical features – The disorder produces a chronic photosensitive skin rash, glossitis, diarrhoea, psychic and emotional changes with disorientation, delirium, hallucinations and dementia. There may be manifestations of peripheral neuropathy and myelopathy.

Diagnosis
(1) Clinical: The clinical picture and the circumstantial evidence often permit the diagnosis of nicotinic acid deficiency.

(2) Laboratory: The question of laboratory diagnosis is largely academic, though nicotinic acid and nicotinamide levels can be measured in blood, if desired. Once clinical suspicion of the diagnosis arises, the vitamin should be given in adequate dosage as a confirmatory therapeutic test.

Therapy – Treatment comprises high dosage nicotinic acid, with other vitamins and an adequate diet if the cause of the deficiency appears to be insufficient dietary intake.

2.6.3.2 *Pyridoxine deficiency*

Pyridoxal phosphate deficiency, of dietary origin or induced by drugs, is a comparatively uncommon event. It can lead to several different patterns of neurological disturbance. The subject is discussed by Weiner and Klawans[175] and Weiner[176].

Biochemical abnormality – Pyridoxine undergoes a number of metabolic interconversions in forming the biologically active pyridoxal-5'-phosphate and pyridoxamine phosphates (Figure 2.42).

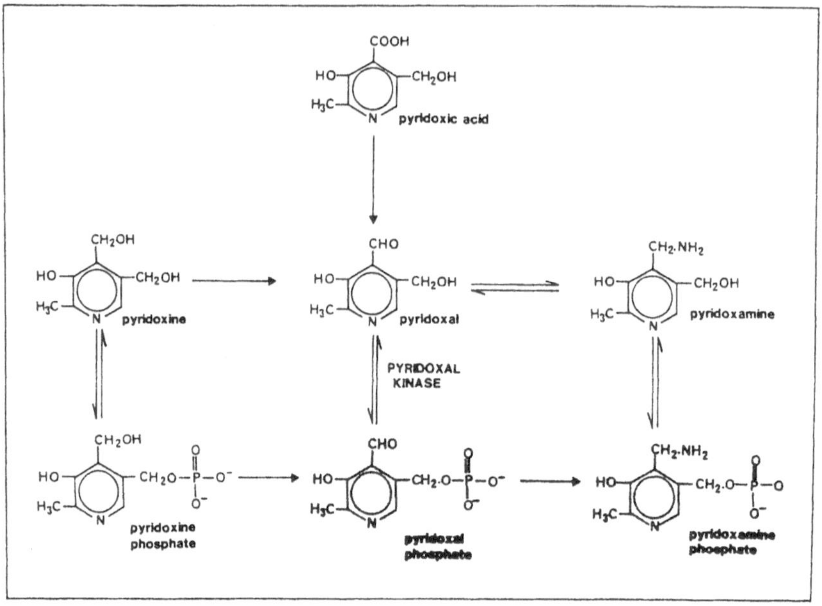

Figure 2.42   Pyridoxine metabolism

Pyridoxal phosphate is a cofactor to the following enzymes:

(1)   Aromatic amino-acid decarboxylase, which catalyses the formation of dopamine and serotonin from L-dopa (L-dihydroxyphenylalanine) and 5-hydroxytryptophan, respectively (Sections 5.3.1 and 5.4.1).

(2)   Kynureninase, which catalyses the formation of 3-hydroxyanthranilic acid (the precursor of nicotinic acid) from 3-hydroxykynurenine (Section 2.5.1.3).

(3)   Glutamate decarboxylase, which catalyses the conversion of glutamate to the inhibitory neurotransmitter γ-aminobutyrate (Section 5.5.1).

(4)   γ-Aminobutyrate transaminase, which catalyses the deamination of γ-aminobutyrate to succinic semialdehyde (Section 5.5.1).

(5)   Cystathioninase and cystathionine synthetase, which catalyse the formations of cysteine from cystathionine, and cystathionine from homocysteine, respectively (Section 2.5.4,1).

Because of the role of pyridoxine in these reactions, deficiency of the vitamin may cause decreased formation of several neurotransmitters and also of NAD^+ and $NADP^+$. Further, it may alter metabolism of sulphur-containing amino acids.

Aetiology – Pyridoxine deficiency may be of dietary origin, occurring as part

98

of a more general malnutritional situation or as a selective deficiency. Such a specific deficiency occurred when infants were fed a food preparation lacking the vitamin but otherwise nutritionally adequate. Certain drugs produce the equivalent of a pyridoxine deficiency state. Isoniazid may form hydrazones with pyridoxal, thus inactivating it, and also may possibly inhibit pyridoxal kinase, the enzyme which catalyses the phosphorylation of pyridoxal. Penicillamine also has an antipyridoxine effect, possibly by binding to the carbonyl group of pyridoxal phosphate.

Structural pathology – The neuropathological changes of homocystinuria are mentioned in Section 2.5.4.1. Lott *et al.*[177] described the neuropathological changes in a child with pyridoxine-dependent epilepsy.

Clinical features – Pyridoxal phosphate deficiency is reported as producing the following neurological disturbances:

(1) Generalized convulsive epilepsy in infants aged 1–4 months fed a pyridoxine deficient diet. Pyridoxine therapy fully reverses this disorder. The clinical and e.e.g. picture of hypsarrhythmia may occur. Possibly this pattern of epilepsy arises mainly from consequences of γ-aminobutyrate depletion. In one variant of this syndrome much higher pyridoxine dosage may be required to control the epilepsy than in the dietary deficiency cases, and there is risk of mental retardation in affected persons.

(2) Peripheral sensori-motor polyneuropathy, as occurs with isoniazid therapy, particularly in slow acetylators of the drug (who tend to accumulate unmetabolized isoniazid).

Cystathioninuria may be present in instances of pyridoxine deficiency. The clinical features of pyridoxine responsive homocystinuria are described in Section 2.5.4.1.

Diagnosis
(1) Clinical: The clinical circumstances may suggest the possibility of pyridoxine deficiency, but it seems unlikely that a definite clinical diagnosis would be justified without the therapeutic test of giving the vitamin.

(2) Laboratory: Increased production of xanthenuric acid in response to a tryptophan load is said to be suggestive of pyridoxine deficiency, though there is some doubt over the interpretation of tryptophan loading tests.

Therapy – Pyridoxine is often given on suspicion of the possibility of pyridoxine deficiency (e.g. in hypsarrhythmia, isoniazid neuropathy) without formal confirmation of the diagnosis. For practical purposes pyridoxine is virtually non-toxic.

2.6.3.3 Folate deficiency

Biochemical folate deficiency, due to a variety of causes, is sometimes encountered in neurological practice. However, the extent to which the deficiency
disturbs nervous system function is uncertain, and is still the subject of
argument. The subject is reviewed in Botez and Reynolds[178].

Biochemical abnormality – The term 'folate' is used collectively for a number
of folic acid derivatives. The structural formula of tetrahydrofolate (FH_4) is
shown (Figure 2.43).

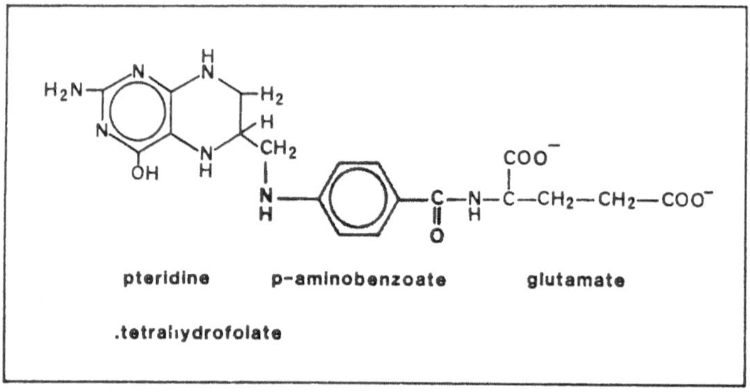

pteridine   p-aminobenzoate   glutamate

.tetrahydrofolate

Figure 2.43   Tetrahydrofolate

The reactive portion of the folate molecule is shown below. The various
tetrahydrofolates serve as one-carbon group donors in a number of biosyntheses, such as:

N^5-methyl FH_4 : homocysteine → methionine

(See Section 2.5.4.1), a reaction which is vitamin B_{12} dependent.

$$\left. \begin{array}{l} N^5,N^{10}\text{-methenyl } FH_4: \\ N^5\text{-formyl } FH_4 \qquad : \end{array} \right\} \rightarrow \text{purines}$$

N^5,N^{10}-methylene FH_4 : → thymine

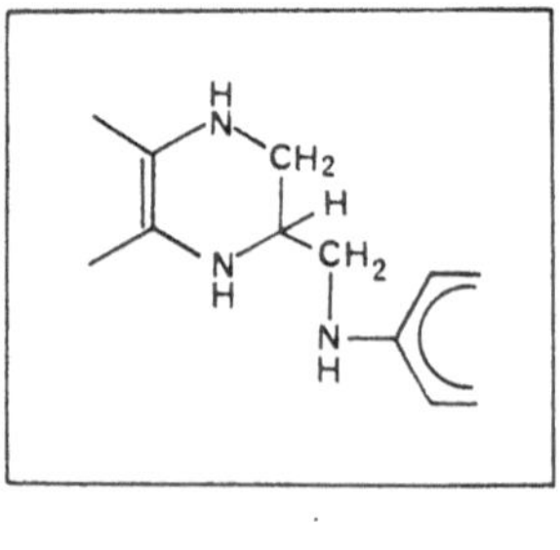

100

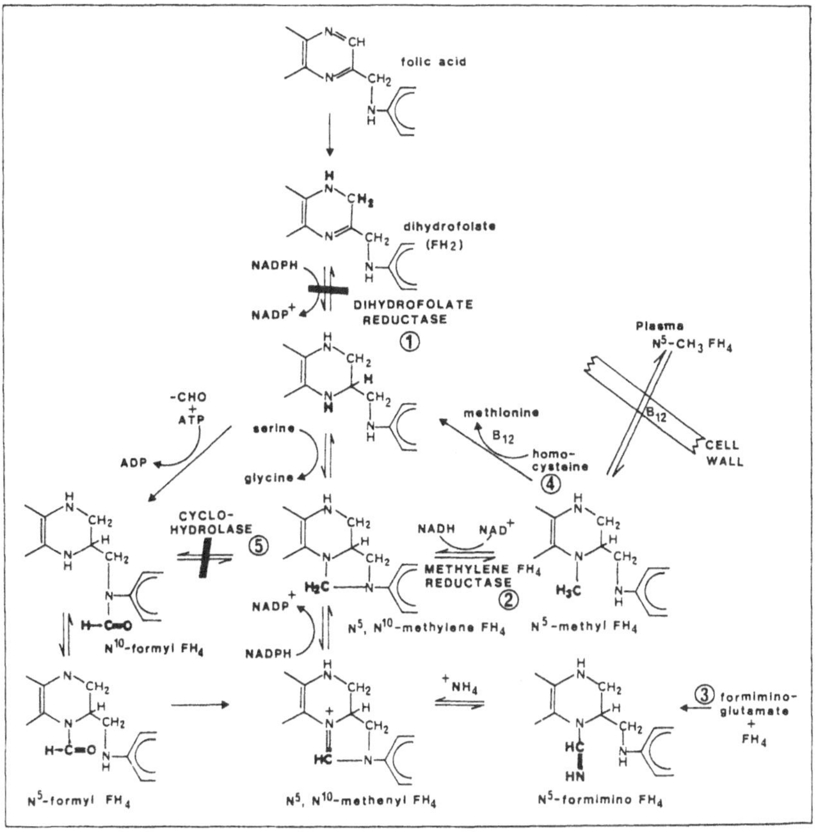

Figure 2.44 Interconversions of various folate derivatives. Sites of enzyme deficiency which interfere with folate interconversions are indicated by numbers which refer to enzymes mentioned in the text

Tetrahydrofolates also accept one-carbon groups from degradative reactions, such as:

the demethylation of serine to glycine $\rightarrow N^5,N^{10}$-methylene FH_4

(Section 2.5.2.2.1)

formiminoglutamate (formed from histidine breakdown)$\rightarrow N^5$-formimino FH_4.

Folic acid itself is derived from dietary pteroylpolyglutamates, mainly methyl- and formyl-FH_4. It is transported in blood mainly as N^5-methyl FH_4. Folic acid deficiency, or enzymatic defects causing defective folate interconversion, might be expected to disturb the cellular biochemical economy. It is thought that deficient thymine synthesis, and therefore DNA synthesis, explains the megaloblastic anaemia and other manifestations of folate deficiency. However, it is still uncertain whether folate deficiency does disturb the functions of the central and peripheral nervous system. Because of this uncertainty there may be little point in considering in greater detail the possible biochemical mechanisms involved in the postulated neurological disturbance.

Aetiology – Folate deficiency may be due to dietary deficiency or to folate malabsorption, as in certain intestinal disorders, e.g. sprue, coeliac disease and the rare congenital specific defect in intestinal folate absorption. Enzymatic defects, which disturb various stages of cellular folate metabolism (Figure 2.44), include:

(1)   Dihydrofolate reductase deficiency,
(2) N^5,N^{10}-Methylene FH_4 deficiency,
(3)   Formiminotransferase deficiency,
(4)   Homocysteine–methionine methyltransferase deficiency,
(5)   Cyclohydrolase deficiency.

All these defects are rare.

Folate depletion may develop after prolonged exposure to anticonvulsant drugs (phenytoin, phenobarbitone, carbamazepine), possibly due to impaired alimentary absorption of dietary pteroylglutamates though this question is not settled. Effectively, cellular folate deficiency also occurs in vitamin B_{12} deficiency. As mentioned above, folate is carried in plasma mainly as N^5-methyl FH_4. B_{12} is required for N^5-methyl FH_4 entry into cells, or retention within cells; the demethylation of N^5-methyl FH_4 to FH_4 is linked to the B_{12} dependent conversion of homocysteine to methionine, and FH_4 formation is required before biochemical activity of folate is possible.

Structural pathology – The only proven structural pathology due to folate depletion is megaloblastic haematopoiesis, with similar megaloblastic changes in other cells, e.g. those of the tongue and alimentary mucosae.

Clinical features – Folate deficiency produces megaloblastic anaemia and glossitis. Nervous system disturbances attributed to folate deficiency (though the relation appears unproven) include intellectual deterioration and peripheral neuropathy.

Diagnosis
(1)   Clinical: There are no neurological clinical features which could permit a justified diagnosis of folate deficiency, though the presence of such deficiency might be suspected from circumstantial evidence, e.g. a history of long-term anticonvulsant intake.

(2)   Laboratory: Reduced serum folate levels (< 7 ng/ml) are diagnostic, though they may anticipate reduced intracellular folate levels by many months, or longer. Low folate levels in red cells occur in both folate and B_{12} deficiency, since B_{12} is required to retain N^5-methyl FH_4 within cells. The morpholog-

102

ical changes in formed elements of peripheral blood, and bone marrow, will not differentiate between folate and B_{12} deficiency.

Therapy – Oral folic acid will correct folate deficiency of dietary or absorptive origin, and may help overcome the consequences of defective activity of enzymes involved in folate interconversion. In the latter syndromes specific folate derivatives may be required, e.g. formyl FH_4 in dihydrofolate reductase deficiency.

2.6.3.4 Cobalamin (vitamin B₁₂) deficiency

Vitamin B_{12} deficiency is the cause of subacute combined degeneration of the spinal cord. This disorder is occasionally encountered in contemporary neurological practice, and enters into the differential diagnosis of a number of neurological syndromes. The topic of B_{12} deficiency has been reviewed by Kunze and Leitenmaier[179].

Biochemical abnormality – Cobalamin (vitamin B_{12}) is a comparatively large molecule with a planar corrin core, comprising four pyrrole units (Figure 2.45).

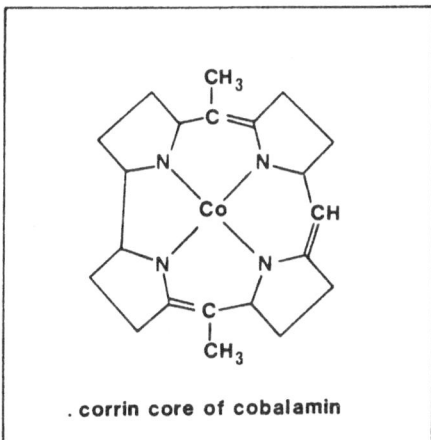

. corrin core of cobalamin

Figure 2.45   Corrin core of cobalamin

The core is similar to the structure of a porphyrin, though not identical with it. The corrin portion of the cobalamin molecule contains a central cobalt atom, linked to each of the four pyrrole rings, to a 5,6-dimethylbenzimidazole moiety and to a sixth substituent, which may be $-CH_3$, $-OH$ or deoxyadenosyl in its nature, or $-CN$ (in the form in which cobalamin is usually available commercially). The Co atom may exist in an oxidized (trivalent) or reduced (divalent) state.

Cobalamin is absorbed from the ileum after forming a complex with a glyco-protein (intrinsic factor) secreted by the gastric mucosa. After absorption, the complex dissociates and the released methyl or hydroxy cobalamin is trans-ported in the blood, bound to one or other of two globulins. Deoxyadenosyl cobalamin is formed intracellularly from hydroxycobalamin.

Cobalamin is a known cofactor in the following two reactions which occur in mammals:

(1) Methionine–homocysteine interconversion (Sections 2.5.2.2.2 and 2.5.4.1),

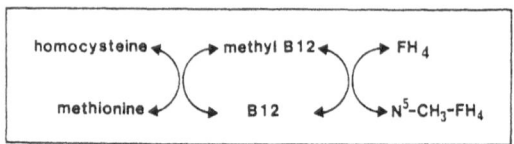

where methylcobalamin is the cofactor for methionine synthetase. The N^5-methyl FH_4, formed in this reaction, is converted to N^5,N^{10}-methylene FH_4, which is required for thymidilate synthesis, which leads to DNA formation. Thus B_{12} deficiency alters DNA formation by causing func-tional folate deficiency.

(2) L-Methylmalonyl CoA $\rightleftharpoons$ methylmalonyl CoB_{12} interconversion, cataly-sed by methylmalonyl CoA mutase (Section 2.5.2.2). Deoxyadenosyl cobalamin is required as a cofactor for this reason. Failure of this reaction causes methylmalonic aciduria (Figure 2.46).

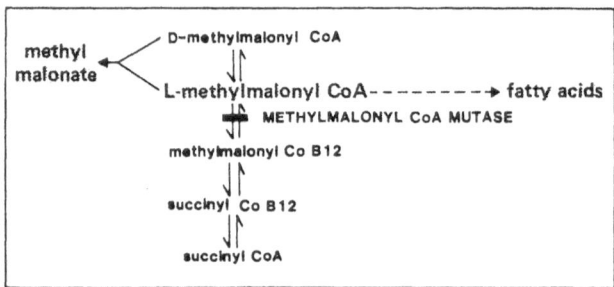

Figure 2.46 Methylmalonate metabolism

The L-methylmalonyl CoA which then accumulates may be diverted to abnormal fatty acid synthesis, which might lead to abnormalities of myelin and other nervous system lipids.

The biochemical consequences of B_{12} deficiency are thus:

(1) Defective DNA synthesis, leading to megaloblastic haematopoiesis and nuclear abnormalities in rapidly dividing cells,

(2) Defective formation of nervous system lipids, including myelin, and

(3) Methylmalonic aciduria.

The neurological defects in cobalamin deficiency do not appear to be mediated through the various folate mechanisms. Folate therapy will not correct the neurological manifestations of subacute combined degeneration, though it will correct the blood changes of B_{12} deficiency. Nor does B_{12} deficiency appear to exert its effects through decreased energy availability, resulting from decreased succinyl CoA entry into the Krebs cycle[180].

Aetiology – Cobalamin deficiency can be due to:

(1) Inability to form intrinsic factor, due to, for example, pernicious anaemia or gastrectomy.

(2) Excess cobalamin consumption by intestinal parasites.

(3) Failure to absorb the B_{12}-intrinsic factor complex, as in ileal disease.

(4) Defective biosynthesis of deoxyadenosyl-, or deoxyadenosyl- and methyl-cobalamins (Section 2.5.2.2.2).

Structural pathology – Cobalamin deficiency[75] causes peripheral neuropathy, with loss of axons in peripheral nerve and demyelination of the posterior and lateral columns of the spinal cord. This demyelination begins in the thoracic region (where the whole of the cord white matter may be affected), and extends into the lumbar and cervical regions. The affected spinal cord tends to have a loose vacuolated appearance. Axons break down in the thoracic region, with resultant Wallerian degeneration above and below this level. Some reactive gliosis occurs. Perivascular foci of demyelination may occur in the brain. Megaloblastic haemopoiesis develops.

Clinical features – The clinical manifestations of the disorder are due to a combination of peripheral neuropathy and dysfunction of the posterior and lateral columns of the spinal cord. At an early stage there is tingling and sensory blunting in the legs, with a mild ataxic paraparesis. This progresses, if untreated, to severe ataxia and paraplegia with extensive sensory impairment, mainly in the lower limbs. Mental changes may occur, ranging from mild neurotic symptoms to apparent dementia and paranoid psychosis. Optic atrophy occurs rarely. Pernicious anaemia is often present, and glossitis may develop.

Diagnosis
(1) Clinical: The combination of progressive peripheral polyneuropathy with bilateral pyramidal tract insufficiency, mainly affecting the legs, and anaemia (particularly with atrophic glossitis), is highly suggestive of subacute combined degeneration. In the absence of haematological change, as may happen, the possibility of B_{12} deficiency should still be considered.

(2) Laboratory: The diagnosis is made by the presence of a low serum cobalamin level. The presence of methylmalonic aciduria also raises the

possibility of the diagnosis. Schilling's test, measuring the urinary excretion of a dose of radioactive B_{12} before and after oral administration of intrinsic factor, detects cases due to pernicious anaemia (intrinsic factor lack).

Therapy – Once the diagnosis is made, parenteral cobalamin therapy is necessary for life. Such treatment halts the neurological deterioration and may lead to improvement in signs and symptoms which have not been present for too long a time.

2.6.3.5 Biotin

There have been recent reports of a few instances of neurological and other disturbances associated with an apparent deficiency of biotin in tissues[52,53].

Biochemical abnormality – Biotin is a prosthetic group on several carboxylase enzymes, including pyruvate carboxylase, acetyl CoA carboxylase, propionyl CoA carboxylase and methylcrotonyl CoA carboxylase. The biotin molecule serves as a temporary carrier of CO_2, and is attached via an amide bond at the end of its long aliphatic tail to an amino-acid residue (often lysine) on the carboxylase apoenzyme.

   While isolated deficiencies of pyruvate, propionyl CoA and methylcrotonyl CoA carboxylases are known (Sections 2.2.5.1, 2.5.2.2.2, 2.5.2.1 respectively), in tissue biotin deficiency there are biochemical changes consistent with simultaneously reduced activity of all these three enzymes. The clinical picture comprises a combination of the deficiencies of the individual enzymes plus defective immunological functioning. Thus there is intermittent lactic acidosis and increased urinary excretion of β-hydroxypropionate, methylcitrate, β-methylcrotonylglycine and β-hydroxyisovalerate. One patient had a selective IgA deficiency and another had decreased numbers of circulating T lymphocytes.

Aetiology – The reported cases of the disorder have been familial, but the

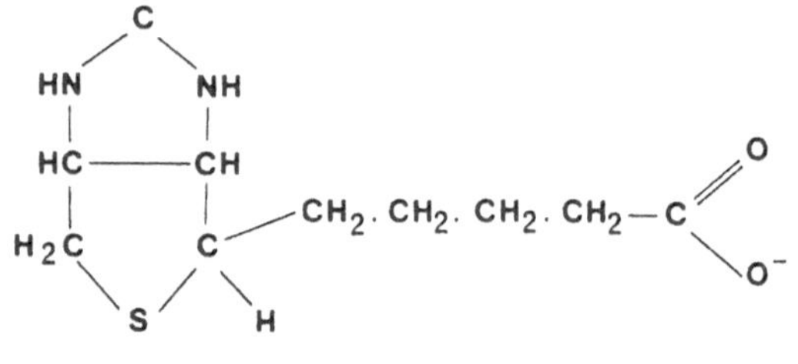

biotin

biochemical basis of the condition is not adequately understood. There was no evidence of deficient dietary biotin intake in the cases recorded.

Structural pathology – In the one autopsy reported there was atrophy of the cerebellar folia with virtually total disappearance of Purkinje cells, loss of granule cells and proliferation of the Bergmann glia. The cerebellar white matter was gliotic. The spinal cord showed a subacute necrotic myelopathy affecting the posterior, lateral and anterior columns and there was a widespread acute but low grade meningoencephalitis.

Clinical features – Patients suffering from the disorder presented in infancy or early childhood with a progressive neurological disorder comprising intermittent ataxia, opisthotonus, myoclonus and convulsive epilepsy. In two cases, death resulted from overwhelming infection. The patients had candida dermatitis, kerato-conjunctivitis and alopecia.

Diagnosis
(1) Clinical: The occurrence of intermittent ataxia in early life would raise the possibility of a defect in pyruvate oxidation, a disturbance of amino-acid metabolism or a mechanical lesion in the posterior fossa. The concurrence of clinical features of lactic acidosis or other ketoacidosis would favour a biochemical basis, and the presence of immune deficiency manifestations might suggest the exact diagnosis.

(2) Laboratory: The presence of ketoacidosis and the pattern of organic acid excretion in urine would suggest the diagnosis, which could be confirmed by finding decreased activity of the three biotin-dependent enzymes pyruvate carboxylase, propionyl CoA carboxylase and methylcrotonyl CoA carboxylase in circulating leukocytes.

Therapy – In one case oral biotin therapy led to return towards normal of the urinary organic acid excretory pattern. The ataxia episodes subsided and the cutaneous candidiasis cleared.

2.6.4 Hepatic encephalopathy

Hepatic cellular insufficiency and/or porto-systemic shunting of blood, are reasonably common causes of diffusely disturbed cerebral function.

Biochemical abnormality – Liver cell metabolic failure leads to the following:

(1) Altered amino-acid and fatty acid metabolism and, if the failure is sufficiently severe, impaired gluconeogenesis, with hypoglycaemia.

(2) Altered protein synthesis which causes changed plasma protein composition, with hypoalbuminaemia and hypoprothrombinaemia. The latter

may cause a bleeding tendency which may lead to intracranial haemorrhage and bleeding into other sites.

(3)   Failure of the hepatic urea cycle leads to hyperammonaemia.

Portal-systemic blood shunting allows substances absorbed from the alimentary tract and normally degraded by the liver to enter the general circulation
in increased amounts and subsequently to gain access to the brain. Such
substances include NH_4^+ and polyamines.

Several of the above biochemical disturbances have been invoked as possible
causes of altered brain function in hepatic encephalopathy.

(1)   Raised NH_4^+ levels may overload the brain's capacity to 'mop up' ammonium ion by forming glutamate and glutamine. There appears to be a
correlation between blood NH_4^+ levels and the glial cell structural changes
of hepatic encephalopathy. However, it is thought that NH_4^+ effects alone
are not sufficient to explain the brain disorder in liver failure. NH_4^+
accumulation does not itself interfere with brain energy metabolism[181].
There are suggestions, but not proof[182], that increased formation of α-
ketoglutaramate, which occurs as a result of NH_4^+ overload, may lead to
neurotoxicity. There is a correlation between CSF α-ketoglutaramate levels
and the severity of the encephalopathy[182]. The reactions involved in α-
ketoglutaramate formation are shown in Figure 2.47.

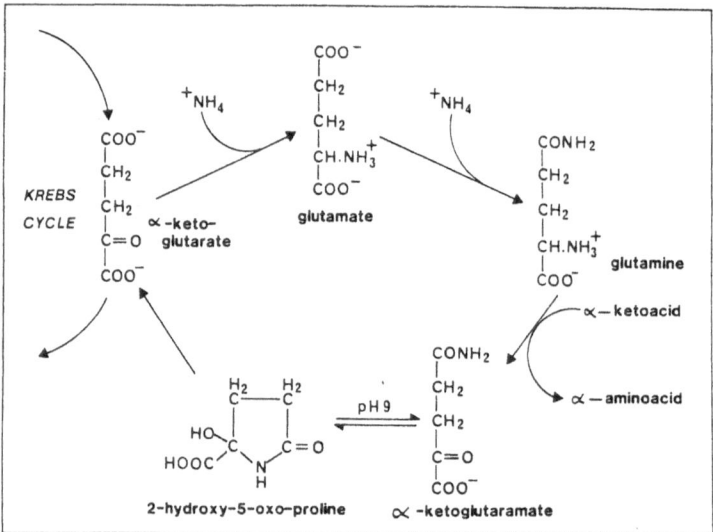

Figure 2.47   Reactions involved in α-ketoglutaramate metabolism

(2)   Transport of the aromatic amino acids tyrosine and tryptophan into the
brain is increased in hepatic failure, possibly because

(a)   Plasma levels of other neutral amino acids which normally compete for a shared amino-acid transport mechanism into brain are reduced, thus favouring brain entry of tyrosine and tryptophan, and

(b)   Plasma levels of tryptophan and tyrosine are increased, probably due to decreased metabolism of these amino acids in the liver[184]. These raised concentrations also favour the entry of these amino acids into the brain along concentration gradients.

In rats with experimental portacaval shunts the raised levels of these aromatic amino acids lead, in the case of tryptophan, to raised brain levels of the established neurotransmitter 5-hydroxytryptamine (serotonin) and its metabolite 5-hydroxy-indoleacetic acid and, in the case of tyrosine, to increased levels of the possible neurotransmitter octopamine[185]. Similar alterations to serotonin and its metabolite occur in the human brain in hepatic coma[186]. In the brain, tyrosine may be hydroxylated to L-dopa, and thence metabolized to catecholamine neurotransmitters. Alternatively, tyrosine may be decarboxylated to tyramine, and thence oxidized to octopamine. The hydroxylation of tyrosine has a substantially lower K_m than the decarboxylation reaction. Therefore, as tyrosine concentration in brain rises, relatively more is diverted towards octopamine formation than towards catecholamine formation[187].

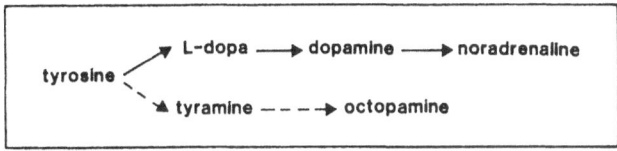

Octopamine levels become raised in brain, urine[188] and plasma in instances of hepatic encephalopathy in man[187]. The octopamine may displace catecholamine neurotransmitters from storage sites and thus lead to dopamine and noradrenaline deficiency. (Oral L-dopa therapy helps reverse manifestations of hepatic encephalopathy temporarily[189].) There have also been suggestions that dietary phenolamines may escape hepatic degradation when portal-systemic shunting occurs. If these amines can enter the brain it is possible that they may act as false neurotransmitters and also competitively displace catecholamine neurotransmitters.

(3)   There may also be decreased entry of short chain fatty acids into the brain in hepatic encephalopathy.

The exact biochemical changes disturbing brain function in hepatic encephalopathy are not yet determined, but it seems likely that altered NH_4^+ and amino-acid metabolism contribute to the disorder.

Aetiology – Various forms of hepatocellular disease may produce both hepatic cell failure and intra- and extrahepatic portal-systemic shunting of blood.

Shunting, not necessarily associated with liver cell failure, may occur after portacaval anastomosis for portal hypertension or after portal vein thrombosis (with development of a collateral circulation).

Structural pathology – In acute hepatic coma, neurons show little change but protoplasmic astrocytes are increased in size and number in the deep layers of the neocortex, and in the lenticular nuclei, thalamus, substantia nigra, red nucleus, dentate and pontine nuclei[190]. In chronic hepatocerebral disease cortical laminar necrosis may occur, with microcavities in the putamen and cerebellar white matter and with some neuronal degeneration in the cerebral and cerebellar cortices, and in the dentate and lenticular nuclei. Alzheimer II cells are present, and glycogen accumulates in astrocyte nuclei[191].

Clinical features – Acute hepatic encephalopathy presents as confusion, seizures, delirium, stupor and finally coma. There may be a flapping tremor (asterixis) and diffuse e.e.g. changes[190]. Chronic (non-Wilson's) hepatocerebral disease causes persisting dementia, dysarthria, intention tremor, ataxia and choreo-athetosis with signs of mild pyramidal tract insufficiency and diffuse e.e.g. abnormalities.

Non-neurological manifestations of liver cell failure are likely to be present also.

Diagnosis
(1) Clinical: The occurrence of manifestations of diffuse brain dysfunction, in the presence of known liver disease, is very suggestive of hepatocerebral disease.

(2) Laboratory: Biochemical tests of liver function may confirm the presence of liver cell disease, and raised blood levels of NH_4^+, glutamine and octopamine (when the latter can be measured) reinforce the diagnostic probability of encephalopathy. However hepatocerebral dysfunction is basically a clinical diagnosis which may be supported by laboratory evidence.

Therapy – Management comprises general supportive and symptomatic treatment, and supplying calories mainly as glucose. Dietary protein intake is substantially restricted, any blood coagulation defect or hypoproteinaemia is corrected (partly to prevent gastrointestinal bleeding which could produce a protein load for absorption from the alimentary tract) while non-absorbable antibiotics (e.g. neomycin) are given orally to inactivate intestinal bacteria (thus preventing intestinal bacterial production of unwanted N-containing metabolites). The latter measures have the aim of decreasing direct or indirect NH_4^+ intake. Lactulose, which alters bacterial metabolism in the large intestine, decreases colonic pH and in consequence reduces blood NH_4^+ levels, may be helpful. The use of dietary L-dopa therapy might be considered.

2.6.5 Unconjugated bilirubin encephalopathy

Unconjugated bilirubin may be neurotoxic for the premature infant or neonate[192]. Prevention of such neurotoxicity is a frequent consideration in neonatal paediatric practice, though the hazard has lessened with modern management of Rh and other blood group incompatibilities between mother and offspring.

Biochemical abnormality – Heme catabolism yields bilirubin, which binds competitively to circulating albumin, and in this form is transported to the liver. Here it is conjugated to glucuronic acid and then excreted in bile as a glucuronide. If the liver cannot handle a bilirubin load, increased amounts of unconjugated bilirubin occur in blood and tissues. Unconjugated bilirubin is lipid soluble and can enter the brain. There it appears to damage the parenchymal elements, possibly by uncoupling oxidative phosphorylation in mitochondria and thus leading to inappropriate and wasteful ATP production. The unconjugated bilirubin may also alter protein and DNA synthesis.

Aetiology – Increased concentrations of unconjugated bilirubin occur when there is

(1) Increased haemoglobin breakdown, as in haemolytic disease of the newborn, or

(2) A decreased capacity to glucuronidate bilirubin, as in the neonate (particularly the premature neonate) and rarely in congenital deficiency of the glucuronyl transferase enzyme system.

Structural pathology – In fatal cases of kernicterus[193] there is bilirubin staining of the leptomeninges, pallidum, Ammon's horn, the subthalamic nuclei and the dentate nuclei. Staining is less severe in the spinal cord, the neocortex and the brain stem nuclei in proximity to the ventricular system. Pigment granules occur in both neurons and glia. Neurons in pigmented regions often appear shrunken or chromatolytic. In infants who survive the acute encephalopathy, gliosis and poverty of myelination occur.

Clinical features – Typically, jaundiced neonates with hepatosplenomegaly on the second to the fifth postnatal day begin to refuse to feed, become quiet and develop a 'cerebral' cry. The disturbance progresses to opisthotonus, limb rigidity, cyanosis, respiratory irregularity and cardiovascular collapse. Survivors are left mentally retarded, often with choreoathetoid dyskinesias and perceptive deafness.

Diagnosis
(1) Clinical: The development of any evidence of brain dysfunction in a

jaundiced neonate makes the possibility of unconjugated bilirubin enceph-alopathy very likely, though there is the possibility that associated hypo-prothrombinaemia may have caused intracranial bleeding.

(2) Laboratory: No laboratory investigation is itself diagnostic, though levels of unconjugated bilirubin in plasma should be high to sustain the diag-nosis.

Therapy – As far as possible the effects of blood group incompatibility should be prevented, or at least detected prenatally. In premature infants, and full term neonates known to be at risk, plasma levels of unconjugated bilirubin should be measured sequentially at short intervals over the first few postnatal days. Exchange transfusion to remove unconjugated bilirubin should be carried out before the unconjugated bilirubin levels reach values at which there is significant danger of neurotoxicity.

2.7 DISORDERS OF THE INTERMEDIARY METABOLISM OF FATTY ACIDS

The normal pattern of fatty acid catabolism involves the transport of the longer chain fatty acids across mitochondrial membranes, followed by intra-mitochondrial serial β-oxidations of the fatty acid chains to yield molecules of acetyl CoA which then enter the energy yielding pathway at the Krebs cycle stage. Medium chain fatty acids $[C_{4-8}]$ do not cross mitochondrial membranes easily. Neural tissue is unable to obtain useful amounts of energy from fatty acid metabolism, but skeletal muscle can. Disorders of fatty acid metabolism therefore tend to affect skeletal muscle function rather than nervous system function. The known disorders of fatty acid catabolism involve the mito-chondrial membrane transport of fatty acids rather than their subsequent β-oxidations. However, there is at least one exception to this generalization. Angelini et al.[194] have reported a disorder in which proximal muscle wasting with elevated serum levels of creatine kinase, glutamate-oxaloacetate and glutamate-pyruvate transaminases, hepatosplenomegaly and ichythosis were associated with triglyceride accumulation in muscle, liver, granulocytes and bone marrow. Fatty acid transport into mitochondria appeared normal, but the oxidation of oleic acid was abnormally slow within mitochondria.

Fatty acid transport across mitochondrial membranes begins with an initial activation of the fatty acid, forming a fatty acid CoA thioester, on the outer surface membranes of mitochondria. The acyl portion of the fatty acid CoA ester is then transported across the mitochondrial membrane united with carnitine. At the inner side of the mitochondrial membrane the acyl carnitine compound dissociates, reforming carnitine and the fatty acid CoA ester. The ester then undergoes sequential β-oxidation, yielding acetyl CoA residues, each oxidation shortening the fatty acid ester chain by two carbon residues.

These transacylations between acyl CoA and carnitine are catalysed by fatty

112

acyl CoA: carnitine fatty acid transferase, deficiency of which enzyme can cause muscle disease.

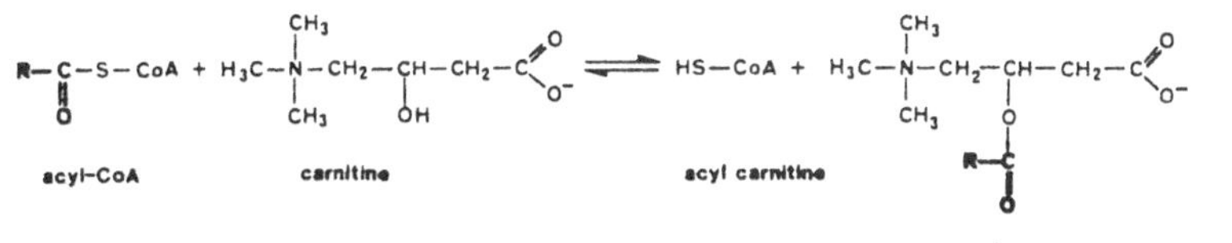

2.7.1 Carnitine deficiency

Carnitine deficiency is a rare cause of myopathy.

Biochemical abnormality – Carnitine deficiency limits the transfer of longer chain fatty acids to the interiors of mitochondria, where in muscle the fatty acids provide a major source of energy. The fatty acids that cannot be transported accumulate within muscle cell cytosol. As a result, muscle capacity for sustained or repeated contraction is reduced, since an increased proportion of muscle energy must then be derived from glucose, and the supply of this may become insufficient for continued contraction.

Aetiology – It has been suggested that there are two forms of carnitine deficiency, both of unknown cause[195]. In one type, carnitine levels in blood are normal, but carnitine levels are low in muscle, suggesting failure of carnitine absorption into muscle. The entry of carnitine into muscle involves active transport[196]. The second type is systemic carnitine deficiency, with a reduced carnitine content of multiple organs[197,198]. In this variant of carnitine deficiency there can be episodic liver dysfunction with lactic acidosis and ketoacidosis, as well as a myopathy.

Structural pathology – The histological picture is that of a lipid-storage myopathy[199] with, in systemic carnitine deficiency, lipid accumulation in other organs also[195].

Clinical features – There is a widespread myopathy with weakness, particularly during exercise and after fasting. The disorder may be fatal in infancy. There may be an associated peripheral neuropathic element[199]. In the systemic variety of carnitine deficiency there may be associated hepatomegaly, episodic liver insufficiency with hypoglycaemia, with or without hepatic encephalopathy[198], and also cardiomyopathy[200].

Diagnosis
(1) Clinical: Myopathic wasting with increasing weakness on exercise could raise suspicion of a failure of muscle energy metabolism, as from insufficient intramitochondrial availability of metabolites of glucose or fatty acids.

113

(2) Laboratory: The condition may need to be distinguished from McArdle's syndrome (Section 2.1.2.2), from phosphofructokinase deficiency (Section 2.2.1.1), and from carnitine-palmityl transferase deficiency (Section 2.7.2). The diagnosis depends on measurement of a reduced muscle carnitine content. Blood carnitine levels may be reduced in the systemic variety. Muscle biopsy shows lipid storage (rather than glycogen storage, as in McArdle's syndrome). In the systemic variety of the disorder, the absence of ketosis during episodes of hypoglycaemia may suggest the presence of defective fatty acid metabolism.

Therapy – Oral carnitine therapy may help correct the biochemical abnormalities in carnitine deficiency. The effect of carnitine therapy on the clinical condition is not clear but the treatment has been reported as correcting the muscle weakness and cardiomegaly of systemic carnitine deficiency[200].

2.7.2 Carnitine-palmityl transferase deficiency

Carnitine-palmityl transferase (fatty acyl CoA carnitine : fatty acid transferase) deficiency is the cause of a rare muscle disorder which has features of an energy insufficiency type myopathy.

Biochemical abnormality – Carnitine-palmityl transferase deficiency (which may be partial) limits the availability of longer chain fatty acids as an energy source for skeletal muscle and makes the tissue abnormally dependent on carbohydrate for its energy requirement[201]. When the enzyme is deficient fatty acids accumulate in muscle cells. Attempted exercise in the presence of energy limitation leads to cramp, and sometimes to myoglobin release into the blood stream with subsequent myoglobinuria. Ionasescu *et al.*[202] have described combined partial deficiency of muscle carnitine itself and of carnitine-palmityl transferase, while Di Donato *et al.*[203] have suggested that there are two types of carnitine-palmityl transferase, and that deficiency of each type can occur separately.

Aetiology – The deficiency is sometimes hereditary.

Structural pathology – Muscle cells have an increased lipid content.

Clinical features – The myopathy severely limits exercise capacity. Attempted exercise is painful and may cause myoglobinuria.

Diagnosis
(1) Clinical: The presence of a myopathy, with limited capacity for exercise and the occurrence of exercise-associated pain, suggests a metabolic myopathy due to deficient energy production. The differential diagnosis is as for carnitine deficiency myopathy. More exact diagnosis depends on laboratory studies.

(2) Laboratory: Unlike myophosphorylase deficiency (McArdle's syndrome), ischaemic exercise of a limb in carnitine-palmityl transferase deficiency results in a rise in venous blood lactate[204]. Storage of lipid, rather than glycogen, demonstrated histochemically in a muscle biopsy, will suggest the diagnostic category. Definitive diagnosis depends on the demonstration of deficient transferase activity in a muscle biopsy specimen.

Therapy – The tendency to muscle pain on exercise in the condition is decreased by a diet with a high carbohydrate and low fat content.

2.7.3 Other disorders of fatty acid metabolism

It would be logical to consider both Refsum's syndrome, due to the defective degradation of the fatty acid phytanic acid, and adrenoleukodystrophy at this point. However, because of their clinical features and the molecular sizes of the substances involved in these disorders, it has been decided to deal with them in relation to large molecule disorders (Chapter 3).

2.8 DISORDERS OF GENERAL MITOCHONDRIAL FUNCTION

At the time of writing, a concept of mitochondrial disease is being developed[205]. This rubric embraces disorders affecting several different mitochondrial biochemical functions (energy metabolism, fatty acid degradation, parts of the urea cycle and aspects of amino-acid degradation). As well as including disorders already discussed (e.g. Leigh's syndrome, porphyria) the concept of mitochondrial disorder takes in a variety of conditions in which there is abnormal mitochondrial morphology. Such disorders may involve skeletal muscle and other tissues[206] as well as the nervous system, e.g. the Kearns Sayre syndrome[207]. In addition, the concept of mitochondrial disease may embrace relatively selective biochemical disorders such as the malignant hyperthermia syndrome, an autosomal dominantly inherited condition. In this syndrome certain general anaesthetic agents seem able to uncouple mitochondrial oxidative phosphorylation in patients who have a number of different forms of myopathy, e.g. central core disease[208]. At present it may seem premature to develop the idea of mitochondrial disease too extensively, but it seems reasonable to discuss Reye's syndrome in relation to this concept.

2.8.1 Reye's syndrome of toxic encephalopathy with fatty infiltration of the viscera

Reye's syndrome is one of the most common potentially fatal, virus-associated diseases of the nervous system in childhood. More than 1000 cases have been described[209]. In the light of contemporary knowledge Reye's syndrome is probably best regarded as an acute generalized failure of mitochondrial function[210].

Biochemical abnormality – Although there is still some doubt, the primary biochemical disturbance in Reye's syndrome appears to be a general failure of mitochondrial enzyme activity. This failure probably occurs in many tissues, though the relevant enzymes have been measured mainly in the liver. All liver cytosol enzymes appear to be normal. However, activities of mitochondrial enzymes including those of the Krebs cycle, cytochrome oxidase, enzymes of pyruvate metabolism and of the intramitochondrial component of the urea cycle, are lowered. Consequences of failure of pyruvate metabolism occur, namely accumulation of pyruvate, lactate and alanine in blood and tissues. Reduced gluconeogenesis may cause hypoglycaemia. There is failure of adequate oxaloacetate production to prime the Krebs cycle and keep it turning. Decreased activity of the urea cycle enzymes carbamyl phosphate synthetase and ornithine transcarbamylase causes reduced conversion of NH_4^+ to urea. This leads to hyperammonaemia. Decreased ability to transport longer chain fatty acids into mitochondria, and to metabolize them within these organelles, causes fatty acidaemia and also ketonaemia. Short chain fatty acids also accumulate in plasma. Metabolic diversion of accumulating fatty acids to sterification in the cytosol may occur and could explain the microvesicular hepatic steatosis which occurs in, and is characteristic of, Reye's syndrome. Mitochondrial failure in cerebral capillary endothelial cells may cause a reduction in the blood–brain barrier effect. This increase in cerebral capillary endothelial permeability, plus biochemical disturbance in neural elements, may explain the development of cerebral oedema in Reye's syndrome. Decreased energy (ATP) availability could explain reduced protein synthesis in the liver, which leads to a decrease in circulating clotting factors and lipoproteins.

There is some evidence that serum from patients with Reye's syndrome during their actual illness contains a factor, probably a short or medium chain fatty acid, which decreases respiratory control and mitochondrial phosphorylation in rat brain[211,212]. A number of C_4–C_8 fatty acids, e.g. caprylic acid $(CH_3.(CH_2)_6.COOH)$, and the anticonvulsant valproic acid (*n*-propylpentanoic acid)[213], are known to be toxic to mitochondria[205]. Short and medium chain fatty acids cause an encephalopathy when infused into experimental animals[214]. Biochemical disturbances in many respects similar to those of Reye's syndrome occur in human diseases in which short branched-chain keto acids (as in β-ketothiolase deficiency, fatty acids (e.g. methylmalonic acid or the straight chain 3-carbon acid propionic acid (Section 2.5.2.2.2), accumulate to excess. Thus altered fatty acid metabolism may be a causative rather than a secondary event in the biochemical pathogenesis of Reye's syndrome.

Aetiology – The disorder usually follows several days after an acute viral illness, often an infection with influenza virus. It is not known how this infection sets in train the subsequent biochemical events of the syndrome.

Structural pathology – In fatal cases the brain shows widespread oedema, with loss of neurons in the cerebral cortex and cerebellum, in a pattern suggesting the consequences of hypoxia[215]. Cerebral oedema alone may occur, without inflammatory change. The liver is enlarged and the hepatocytes are packed with small lipid-containing vesicles (steatosis). The hepatic glycogen content is reduced. No inflammatory changes are seen. A similar but less severe fatty infiltration of the kidneys occurs. At electron microscopy, mitochondria from many tissues show structural abnormalities.

Clinical features – The disorder nearly always occurs in children, though rarely in adults[216], and usually follows a non-specific febrile illness which often involves the upper respiratory tract. Vomiting, stupor and delirium develop. Later coma and hyperventilation may occur. About one in three cases die. Survivors often recover quickly. At the stage when the encephalopathy becomes clinically obvious the liver is usually enlarged, though there is little clinical evidence of liver cell dysfunction. Jaundice is uncommon. Either respiratory alkalosis or metabolic acidosis may occur. Hypoglycaemia and raised blood urea levels are often present and hyperammonaemia and an increased prothrombin time are almost invariable. Liver enzyme levels in blood are increased and plasma levels of pyruvate, lactate, alanine, glutamine, lysine, propionate, butyrate, isobutyrate, isovalerate and octanoate (caprylate) may be raised.

Diagnosis
(1) Clinical: The clinical evidence of encephalopathy in a child with an enlarged liver following a presumed viral infection is strongly suggestive of Reye's syndrome, though there are other diagnostic possibilities.

(2) Laboratory: The various biochemical abnormalities described above collectively add considerable weight to the diagnostic probability of Reye's syndrome. The liver biopsy finding of microvesicular steatosis is characteristic of the syndrome.

Therapy – Treatment of the condition is discussed by Trauner[217]. Careful monitoring of the clinical state, with parenteral glucose intake, correction of disturbances of fluid, H^+ and electrolyte balance, maintenance of adequate ventilation, and administration of vitamin K and fresh frozen plasma, appear helpful. Intracranial pressure monitoring may be useful, raised intracranial pressure being controlled with hypertonic mannitol or glycerol, high dose glucocorticoids or induced barbiturate coma. Exchange transfusion has been attempted[209]. The optimal therapy for Reye's syndrome is still being worked out, while knowledge of the biochemical disturbance unfolds.

REFERENCES

1 Sacks, W. (1969). Cerebral metabolism *in vivo*. In Lajtha, A. (ed.) *Handbook of Neurochemistry*. Vol. 1, pp. 301–324. (New York: Plenum Press)

2 Austin, J. H. (1972). Disorders of glycogen and related macromolecules in the nervous system. In Lajtha, A. (ed.) *Handbook of Neurochemistry*. Vol. 7, pp. 1–15. (New York: Plenum Press)

3 Austin, J. and Sakai, M. (1976). Disorders of glycogen and related macromolecules in the nervous system. In Vinken, P. J. and Bruyn, G. W. (eds.) *Handbook of Clinical Neurology*. Vol. 27, pp. 169–219. (Amsterdam: North Holland)

4 Robitaille, Y., Carpenter, S., Karpati, G. and Dimauro, S. (1980). A distinct form of adult polyglucosan body disease with massive involvement of central and peripheral processes and astrocytes. *Brain*, **103**, 315–336

5 Seitelberger, F. (1968). Myoclonus body disease. In Minckler, J. (ed.) *Pathology of the Nervous System*. Vol. 1, pp. 1121–1134. (New York: McGraw-Hill)

6 Peress, N. S., DiMauro, S. and Roxburgh, V. A. (1979). Adult polysaccharidosis. Clinicopathological, ultrastructural, and biochemical features. *Arch. Neurol.* **36**, 840–845

7 Nishimura, R. N., Ishak, K. G., Reddick, R., Porter, R., James, S. and Barranger, J. A. (1980). Lafora disease: diagnosis by liver biopsy. *Ann. Neurol.*, **8**, 409–415

8 Howell, R. R. (1978). The glycogen storage disease. In Stanbury, J. B., Wyngaarten, J. B. and Frederickson, D. S. (eds.) *The Metabolic Basis of Inherited Disease*. 4th Edn. pp. 137–159. (New York: McGraw-Hill)

9 Hug, G. (1979). Five lysosomal disorders. *Pharmacol. Rev.*, **30**, 565–591

10 Hers, H. G. and De Barsy, T. (1973). Type II glycogenosis (acid maltase deficiency). In Hers, H. G. and Van Hoof, F. (eds.) *Lysosomes and Storage Diseases*. pp. 197–216. (New York: Academic Press)

11 Danon, M. J., Shin, O. J., DiMauro, S., Manaligod, J. R., Estwood, A., Naidu, S. and Schliselfeld, L. H. (1981). Lysosomal glycogen storage disease with normal acid maltase. *Neurology*, **31**, 51–57

12 Engel, A. G. (1970). Acid maltase deficiency in adults: studies in four cases of a syndrome which may mimic muscular dystrophy or other myopathies. *Brain*, **93**, 599–616

13 Sivak, E. D., Salanga, V. D., Wilbourn, A. J., Mitsumoto, H. and Golish, J. (1981). Adult-onset acid maltase deficiency presenting as diaphragmatic paralysis. *Ann. Neurol.*, **9**, 613–615

14 O'Brien, J. S., Bernett, J., Veath, M. L. and Paa, D. (1975). Lysosomal storage disorders. Diagnosis by ultrastructural examination of skin biopsy specimens. *Arch. Neurol.*, **32**, 592–599

15 Shanske, S. and Di Mauro, S. (1981). Late-onset acid maltase deficiency. Biochemical studies of leukocytes. *J. Neurol. Sci.*, **50**, 57–62

16 Schotland, D. L., Spiro, D., Rowland, L. P. and Carmel, P. (1965). Ultrastructural studies of muscle in McArdle's disease. *J. Neuropathol. Exp. Neurol.*, **24**, 629–644

17 Di Mauro, S. and Hartlage, P. L. (1978). Fatal infantile form of muscle phosphorylase deficiency. *Neurology*, **28**, 1124–1129

18 Rowland, L. P., Lovelace, R. E., Schotland, D. L., Araki, S. and Carmel, P. (1966). The clinical diagnosis of the McArdle's disease. *Neurology*, **16**, 93–100

19 Murase, I., Ikeda, H., Muro, I., Nakao, K. and Subita, H. (1973). Myopathy associated with Type III glycogenosis. *J. Neurol. Sci.*, **20**, 287–295

20 Brooke, M. (1977). *A Clinician's View of Neuromuscular Diseases*. (Baltimore: Williams and Wilkins)

21 Di Mauro, S., Hartwig, G. B., Hays, A., Eastwood, A. B., Franco, R., Olarte, M., Chang, M., Roses, A. D., Fetell, M., Schoenfeldt, R. S. and Stern, L. Z. (1979). Debrancher deficiency: neuromuscular disorder in 5 adults. *Ann. Neurol.*, **5**, 422–436

22 Zellweger, H., Mueller, S., Ionasescu, V., Schochet, S. S. and McCormick, W. F. (1972). Glycogenosis IV. A new cause of infantile hypotonia. *J. Pediatr.*, **80**, 842–844

23 Menkes, J. H. (1974). *Textbook of Child Neurology*. (New York: Lea and Febiger)

24 Layzer, R. B., Rowland, L. P. and Ranney, H. M. (1967). Muscle phosphofructokinase deficiency. *Arch. Neurol.*, **17**, 512–523

25 Prockop, L. D. (1976). Hyperglycaemia: effects on the nervous system. In Vinken, P. J. and Bruyn, G. W. (eds.) *Handbook of Clinical Neurology*. Vol. 27, pp. 79–98. (Amsterdam: North Holland)

26 Partridge, W. M. and Oldendorf, W. H. (1977). Transport of metabolic substrates through the blood–brain barrier. *J. Neurochem.*, **28**, 5–12

27 Dyck, P. J., Sherman, W. R., Hallcher, L. M., Service, F. J., O'Brien, P. C., Grina, L. A., Palumbo, P. J. and Swanson, C. J. (1980). Human diabetic endonurial sorbitol, fructose, and myoinositol related to sural nerve morphometry. *Ann. Neurol.*, **8**, 590–596

28 Natarajan, V., Dyck, P. J. and Schmid, H. H. O. (1981). Alterations of inositol lipid metabolism of rat sciatic nerve in streptozotocin-induced diabetes. *J. Neurochem.*, **36**, 413–419

29 Satran, R. and Griggs, R. C. (1979). Metabolic encephalopathy. In Tyler, H. R. and Dawson, D. M. (eds.) *Current Neurology*. Vol. 2, pp. 474–505. (Boston: Houghton Mifflin)

30 Wilkinson, D. S. and Prockop, L. D. (1976). Hypoglycaemia: effects on the central nervous system. In Vinken, P. J. and Bruyn, G. W. (eds.) *Handbook of Clinical Neurology*. Vol. 27, pp. 53–78. (Amsterdam: North Holland)

31 Lewis, L. D., Ljunggre, B., Ratcheson, R. A. and Siesjö, B. K. (1974). Cerebral energy state in insulin-induced hypoglycemia, related to blood glucose and to EEG. *J. Neurochem.*, **23**, 673–679

32 Agardh, C.-D., Chapman, A. G., Nilsson, B. and Siesjö, B. K. (1981). Endogenous substances utilized by rat brain in severe insulin-induced hypoglycemia. *J. Neurochem.*, **36**, 490–500

33 Hernandez, M. J., Vannucci, R. C., Salcedo, A. and Brennan, R. W. (1980). Cerebral blood flow and metabolism during hypoglycemia in newborn dogs. *J. Neurochem.*, **35**, 622–628

34 Norberg, K. and Siesjö, B. K. (1976). Oxidative metabolism of the cerebral cortex of the rat in severe insulin-induced hypoglycaemia *J. Neurochem.*, **26**, 345–352.

35 Gorell, J. M., Dockart, P. H. and Ferrendelli, J. A. (1976). Regional levels of glucose, amino acids, high energy phosphates, and cyclic nucleotides in the central nervous system during hypoglycemic stupor and behavioural recovery. *J. Neurochem.*, **27**, 1043–1049

36 Gorell, J. M., Law, M. M., Lowry, O. H. and Ferrendelli, J. A. (1977). Levels of cerebral cortical glycolytic and citric acid cycle metabolites during hypoglycemic stupor and its reversal. *J. Neurochem.*, **29**, 187–191

37 Dirks, B., Hanke, J., Krieglstein, J., Stock, R. and Wickop, G. (1980). Studies on the linkage of energy metabolism and neuronal activity in the isolated perfused rat brain. *J. Neurochem.*, **35**, 311–317

38 Gibson, G. E. and Blass, J. P. (1976). Impaired synthesis of acetylcholine in brain accompanying mild hypoxia and hypoglycemia. *J. Neurochem.*, **27**, 37–42

39 Agardh, C.-D., Folbergrova, J. and Siesjö, B. K. (1978). Cerebral metabolic changes in profound, insulin-induced hypoglycemia, and in the recovery period following glucose administration. *J. Neurochem.*, **31**. 1135–1142

40 Brierley, J. B. (1976). Cerebral hypoxia. In Blackwood, W. and Corsellis, J. A. N. (eds.) *Greenfield's Neuropathology*. 3rd Edn., pp. 43–85. (London: Arnold)

41 Coffey, G. L., O'Sullivan, D. J. and Burke, W. J. (1979). Hypoglycaemia secondary to pancreatic islet cell adenoma. *Clin. Exp. Neurol.*, **16**, 149–165

42 Froesch, E. R. (1978). Essential fructosuria, hereditary fructose intolerance, and fructose-1,6-diphosphatase deficiency. In Stanbury, J. B., Wyngaarden, J. B. and Fredrickson, D. S. (eds.) *The Metabolic Basis of Inherited Disease*. 4th Edn., pp. 121–136. (New York: McGraw-Hill)

43 Kaz, N. C., Pearson, C. M. and Verity, M. A. (1980). Muscle fructose-1,6-diphosphatase deficiency associated with an atypical central core disease. *J. Neurol. Sci.*, **48**, 243–256

44 Gitzelmann, R. and Baerlocher, K. (1977). Hereditary disorders of fructose and galactose metabolism. In Vinken, P. J. and Bruyn, G. W. (eds.) *Handbook of Clinical Neurology*. Vol. 29, pp. 255–262. (Amsterdam: North Holland Publishing)

45 Segal, S. (1978). Disorders of galactose metabolism. In Stanbury, J. B., Wyngaarten, J. B. and Fredrickson, D. S. (eds.) *The Metabolic Basis of Inherited Disease*. 4th Edn., pp. 160–181. (New York: McGraw-Hill)

46 Yandrasitz, J., Hwang, S. M., Cohn, R. and Segal, S. (1979). On the involvement of serotonin in galactose brain toxicity. *J. Neurochem.*, **33**, 1321–1323

47 Crome, L. (1962). A case of galactosemia with the pathological and neuropathological findings. *Arch. Dis. Child.*, **37**, 415–421

48 Haberland, C., Perou, M., Brunngraber, E. G. and Hof, H. (1971). The neuropathology of galactosemia. A histopathological and biochemical study. *J. Neuropathol. Exp. Neurol.*, **30**, 431–447

49 Guggenheim, M. A., McCabe, E. R. B., Roig, M., Goodman, S. I., Lum, G. M., Bullen, W. W. and Ringel, S. P. (1980). Glycerol kinase deficiency with neuromuscular, skeletal and adrenal abnormalities. *Ann. Neurol.*, **7**, 441–449

50 Stumpf, D. A. (1978). Friedreich's ataxia and other hereditary ataxias. In Tyler, H. R. and Dawson, D. M. (eds.) *Current Neurology.* Vol. 1, pp. 86–111. (Boston: Houghton Mifflin)

51 Blass, J. P. (1979). Disorders of pyruvate metabolism. *Neurology*, **29**, 280–286

52 Cowan, M. J., Wara, D. W., Packman, S., Ammann, A. J., Yoshino, M., Sweetman, L. and Nyhan, W. (1979). Multiple biotin-dependent carboxylase deficiencies associated with defects in T-cell and B-cell immunity. *Lancet*, **2**, 115–118

53 Sander, J. E., Malamud, N., Cowan, M. J., Packman, S., Amman, A. J. and Wara, D. W. (1980). Intermittent ataxia and immunodeficiency with multiple carboxylase deficiencies: a biotin responsive disorder. *Ann. Neurol.*, **8**, 544–547

54 Kark, R. A. P. and Rodriguez-Budelli, M. (1979). Pyruvate dehydrogenase deficiency in spinocerebellar degenerations. *Neurology*, **29**, 126–131

55 Kark, R. A. P., Rodriguez-Budelli, M. and Blass, J. P. (1978). Evidence for a primary defect of lipoamide dehydrogenase in Friedreich's ataxia. In Kark, R. A. P., Rosenberg, R. N. and Schut, L. J. (eds.) *Advances in Neurology.* Vol. 21, pp. 163–180. (New York: Raven Press)

56 Kark, R. A. P., Budelli, M. M. R., Becker, D. M., Weiner, L. P. and Forsythe, A. B. (1981). Lipoamide dehydrogenase: rapid heat inactivation in platelets of patients with recessively inherited ataxia. *Neurology*, **31**, 199–202

57 Barbeau, A., Melancon, S., Butterworth, R. F., Filla, A., Izumi, K. and Ngo, T. T. (1978). Pyruvate dehydrogenase complex in Friedreich's ataxia. In Kark, R. A. P., Rosenberg, R. N. and Schut, L. J. (eds.) *Advances in Neurology.* Vol. 21, pp. 203–217. (New York: Raven Press)

58 Constantopoulos, G., Chang, C. S. C. and Barranger, J. A. (1980). Normal pyruvate dehydrogenase complex activity in patients with Friedreich's ataxia. *Ann. Neurol.*, **8**, 636–639

59 Stumpf, D. A. and Parks, J. K. (1978). Friedreich's ataxia. I. Normal pyruvate dehydrogenase complex activity in platelets. *Ann. Neurol.*, **4**, 366–368

60 Stumpf, D. A. and Parks, J. K. (1979). Friedreich's ataxia. II. Normal kinetics of lipoamide dehydrogenase. *Neurology*, **29**, 820–826

61 Plaitakis, A., Nicklas, W. J. and Desnick, R. J. (1980). Glutamate dehydrogenase deficiency in three patients with spinocerebellar syndrome. *Ann. Neurol.*, **7**, 297–303

62 Livingstone, I. R., Mastaglia, F. L. and Pennington, R. J. T. (1980). An investigation of pyruvate metabolism in patients with cerebellar and spinocerebellar degeneration. *J. Neurol. Sci.*, **48**, 123–132

63 Williams, L. L. (1979). Pyruvate oxidation in Charcot–Marie–Tooth disease. *Neurology*, **29**, 1492–1498

64 Blass, J. P., Kark, R. A. P. and Engel, W. K. (1971). Clinical studies of a patient with pyruvate-decarboxylase deficiency. *Arch. Neurol.*, **25**, 449–461

65 Kark, R. A. P., Blass, J. P. and Spence, A. (1975). Physostigmine in patients with familial ataxias. *Neurology*, **27**, 70–72

66 Lawrence, C. M., Millac, P., Stout, G. S. and Ward, J. W. (1980). The use of choline chloride in ataxic disorders. *J. Neurol. Neurosurg. Psychiatry*, **43**, 452–454

67 Livingstone, I. R., Mastaglia, F. L., Pennington, R. J. T. and Skilbeck, C. (1981). Choline chloride in the treatment of cerebellar and spinocerebellar ataxia. *J. Neurol. Sci.*, **50**, 161–174

68 Chamberlain, S., Robinson, N., Walker, J., Smith, C., Benton, S., Kennard, C., Swash, M., Kilkenny, B. and Bradbury, S. (1980). Effect of lecithin on disability and plasma free-choline levels in Friedreich's ataxia. *J. Neurol. Neurosurg. Psychiatry*, **43**, 843–845

69 Di Donato, S., Rimoldi, M., Moise, A., Bertagnolio, B. and Uziel, G. (1979). Fatal ataxic encephalopathy and carnitine acetyltransferase deficiency: a functional defect of pyruvate oxidation. *Neurology*, **29**, 1578–1583

70 Hamel, E., Butterworth, R. F. and Barbeau, A. (1979). Effect of thiamine deficiency on levels of putative amino acid transmitters in affected regions of rat brain. *J. Neurochem.*, **33**, 575–577

71 Gubler, C. J., Adams, B. L., Hammond, B., Yuan, E. C., Guo, S. M. and Bennion, M. (1974). Effect of thiamine deprivation and thiamine antagonists on the level of γ-aminobutyrate acid and on 2-oxoglutarate metabolism in rat brain. *J. Neurochem.*, **22**, 831–836

72 McEntee, W. J. and Mair, R. G. (1980). Memory enhancement in Korsakoff's psychosis by clonidine: further evidence for a noradrenergic deficit. *Ann. Neurol.*, **7**, 466–470

73 Plaitakis, A., Van Woert, M. H., Hwang, E. C. and Berl, S. (1978). The effect of acute thiamine deficiency on brain tryptophan, serotonin and 5-hydroxyindoleacetic acid. *J. Neurochem.*, **31**, 1087–1089

74 Loken, A. C. (1971). Vitamin deficiencies. In Minckler, J. (ed.) *Pathology of the Nervous System.* Vol. 2, pp. 1568–1575. (New York: McGraw-Hill)

75 Smith, W. T. (1976). Nutritional deficiencies and disorders. In Blackwood, W. and Corsellis, J. A. N. (eds.) *Greenfield's Neuropathology.* 3rd Edn., pp. 194–237. (London: Arnold)

76 David, R. B., Mamunes, P. and Rosenblum, W. I. (1976). Necrotizing encephalopathy (Leigh). In Vinken, P. J. and Bruyn, G. W. (eds.) *Handbook of Clinical Neurology.* Vol. 28, pp. 349–363. (Amsterdam: North Holland)

77 De Vivo, D. C., Hammond, M. W., Obert, K. A., Nelson, J. S. and Pagliara, A. S. (1979). Defective activation of the pyruvate dehydrogenase complex in subacute necrotizing encephalomyelopathy (Leigh disease). *Ann. Neurol.*, **6**, 483–494

78 Pincus, J. H., Solitaire, G. B. and Cooper, J. R. (1976). Thiamine triphosphate levels and histopathology. Correlation in Leigh disease. *Arch. Neurol.*, **33**, 759–763

79 Pincus, J. H., Cooper, H. R., Piros, K. and Turner, V. (1974). Specificity of the urine inhibitor test for Leigh's disease. *Neurology*, **24**, 885–890

80 Murphy, J. V., Craig, L. J. and Glew, R. H. (1974). Leigh disease. Biochemical characteristics of the inhibitor. *Arch. Neurol.*, **31**, 220–227

81 Kalimo, H., Lundberg, P. O. and Olsson, Y. (1979). Familial subacute necrotizing encephalomyelopathy of the adult form (adult Leigh syndrome). *Ann. Neurol.*, **6**, 200–206

82 Dayan, A. D., Oegenden, B. G. and Crome, L. (1970). Necrotizing encephalomyelopathy of Leigh. Neuropathological findings in 8 cases. *Arch. Dis. Child.*, **45**, 39–48

83 Plaitakis, A., Whetsell, W. O. Jr., Cooper, J. R. and Yahr, M. D. (1980). Chronic Leigh disease: a genetic and biochemical study. *Ann. Neurol.*, **7**, 304–310

84 Feigin, I. and Budzilovich, G. N. (1977). Further observations on subacute necrotizing encephalomyelopathy in adults. *J. Neuropathol. Exp. Neurol.*, **36**, 128–139

85 Di Mauro, S., Mendell, J. R., Sahenk, Z., Bachman, D., Scarpa, A., Schofield, R. M. and Reiner, C. (1980). Fatal infantile mitochondrial myopathy and renal dysfunction due to cytochrome-c-oxidase myopathy. *Neurology*, **30**, 795–804

86 Land, J. M., Morgan-Hughes, J. A. and Clark, J. B. (1981). Mitochondrial myopathy. Biochemical studies revealing a deficiency of NADH-cytochrome b reductase activity. *J. Neurol. Sci.*, **50**, 1–13

87 Bachelard, H. S., Lewis, L. D., Pontén, U. and Siesjö, B. K. (1974). Mechanisms activating glycolysis in the brain in arterial hypoxia. *J. Neurochem.*, **22**, 395–401

88 Gibson, G. E., Shimada, M. and Blass, J. P. (1978). Alterations in acetylcholine synthesis and cyclic nucleotides in mild cerebral hypoxia. *J. Neurochem.*, **31**, 757–760

89 Hicks, S. P. (1968). Vascular pathophysiology and acute and chronic oxygen deprivation. In Minckler, J. (ed.) *Pathology of the Nervous System.* Vol. 1, pp. 341–350. (New York: McGraw-Hill)

90 Lindenberg, R. (1971). Systemic oxygen deficiencies. In Minckler, J. (ed.) *Pathology of the Nervous System.* Vol. 2, pp. 1583–1617. (New York: McGraw-Hill)

91 Yatsu, F. M. (1976). Biochemical mechanisms of ischemic brain infarction. In Vinken, P. J. and Bruyn, G. W. (eds.) *Handbook of Clinical Neurology.* Vol. 27, pp. 27–37. (Amsterdam: North Holland)

92 Welsh, F. A., Durity, F. and Langfitt, T. W. (1977). The appearance of regional variations in metabolism at a critical level of diffuse cerebral oligemia. *J. Neurochem.*, **28**, 71–79

93 Kobayashi, M., Lust, W. D. and Passonneau, J. V. (1977). Concentrations of energy metabolites and cyclic nucleotides during and after bilateral ischemia in the gerbil cerebral cortex. *J. Neurochem.*, **29**, 53–59

94 Levy, D. E. and Duffy, T. E. (1977). Cerebral energy metabolism during transient ischemia and recovery in the gerbil. *J. Neurochem.*, **28**, 63–70

95 Saifer, A. (1971). Rapid screening methods for the detection of inherited and acquired aminoacidopathies. In Bodansky, O. and Latner, A. L. (eds.) *Advances in Clinical Chemistry.* Vol. 14, pp. 145–218. (New York: Academic Press)

96 Menkes, J. H. (1971). Disturbances of amino acid metabolism. In Minckler, J. (ed.) *Pathology of the Nervous System.* Vol. 2, pp. 1273–1280. (New York: McGraw-Hill)

97 Martin, J. J. and Scholte, W. (1972). Central nervous system lesions in disorders of amino acid metabolism: a neuropathological study. *J. Neurol. Sci.*, **15**, 49–76

98 Menkes, J. H. and Koch, R. (1977). Phenylketonuria. In Vinken, P. J. and Bruyn, G. W. (eds.) *Handbook of Clinical Neurology.* Vol. 29, pp, 29–51. (Amsterdam: North Holland)

99 Tourian, A. Y. and Sidbury, J. B. (1978). Phenylketonuria. In Stanbury, J. B., Wyngaarten, J. B. and Fredrickson, D. S. (eds.) *The Metabolic Basis of Inherited Disease.* 4th Edn., pp. 240–255. (New York: McGraw-Hill)

100 Scriver, C. R. and Clow, C. L. (1980). Phenylketonuria: epitome of human biochemical genetics. *N. Engl. J. Med.*, **303**, 1336–1342; 1394–1400

101 Danks, D. M., Cotton, R. G. H. and Schlesinger, P. (1976). Variant forms of phenylketonuria. *Lancet*, **1**, 1236–1237

102 Butler, I. J., Koslow, S. H., Grumholz, A., Holtzman, N. A. and Kaufman, S. (1978). A disorder of biogenic amines in dihydropteridine reductase deficiency. *Ann. Neurol.*, **3**, 224–230

103 Nixon, J. C., Lee, C.-L., Milstien, S., Kaufman, S. and Bartholomé, K. (1980). Neopterin and biopterin levels in patients with atypical forms of phenylketonuria. *J. Neurochem.*, **35**, 898–904

104 Niederwieser, A., Curtius, H.-Ch., Bettoni, O., Bieri, J., Schircks, B., Visconti, M. and Schaub, J. (1979). Atypical phenylketonuria caused by 7,8-dihydrobiopterin synthetase deficiency. *Lancet*, **1**, 131–133

105 Antonas, K. N. and Coulson, W. F. (1975). Brain uptake and protein incorporation of amino acids studied in rats subjected to prolonged hyperphenylalaninaemia. *J. Neurochem.*, **24**, 309–314

106 Benjamin, A. M., Verjee, Z. H. and Quastel, J. H. (1980). Effects of branched-chain L-amino acids, L-phenylalanine, and L-methionine on the transport of L-glutamine in rat brain cortex *in vitro.* Influence of cations. *J. Neurochem.*, **35**, 78–87

107 Land, J. M., Mowbray, J. and Clark, J. B. (1976). Control of pyruvate and β-hydroxybutyrate utilization in rat brain mitochondria and its revelance to phenylketonuria and maple syrup urine disease. *J. Neurochem.*, **26**, 823–830

108 Malamud, N. (1966). Neuropathology of phenylketonuria. *J. Neuropathol. Exp. Neurol.*, **25**, 254–268

109 Jervis, G. A. (1971). Phenylketonuria. In Minckler, J. (ed.) *Pathology of the Nervous System.* Vol. 2, pp. 1280–1284. (New York: McGraw-Hill)

110 La Du, B. N. and Gjessing, L. R. (1978). Tyrosinosis and tyrosinemia. In Stanbury, J. B., Wyngaarten, J. B. and Fredrickson, D. S. (eds.) *The Metabolic Basis of Inherited Disease.* 4th Edn., pp. 256–267. (New York: McGraw-Hill)

111 Seakins, J. W. T. (1977). Hartnup disease. In Vinken, P. J. and Bruyn, G. W. (eds.) *Handbook of Clinical Neurology.* Vol. 29, pp. 149–170. (Amsterdam: North Holland)

112 Jepson, J. B. (1978). Hartnup disease. In Stanbury, J. B., Wyngaarten, J. B. and Fredrickson, D. S. (eds.) *The Metabolic Basis of Inherited Disease.* 4th Edn., pp. 1563–1577. (New York: McGraw-Hill)

113 Wicklen, B., Yu, T. S. and Brown, D. R. (1977). Natural history of Hartnup disease. *Arch. Dis. Child.*, **52**, 38–40

114 Dancis, J. and Levitz, M. (1978). Abnormalities of branched chain amino acid metabolism. In

Stanbury, J. B., Wyngaarten, J. B. and Fredrickson, D. S. (eds.) *The Metabolic Basis of Inherited Disease*. 4th Edn., pp. 397–410. (New York: McGraw-Hill)

115 Moser, H. W. (1977). Maple syrup urine disease (branched chain ketonuria). In Vinken, P. J. and Bruyn, G. W. (eds.) *Handbook of Clinical Neurology*. Vol. 29, pp. 53–55. (Amsterdam: North Holland)

116 Patel, T. B., Booth, R. F. G. and Clark, J. B. (1977). Inhibition of acetoacetate oxidation by brain mitochondria from the suckling rat by phenylpyruvate and α-ketoisocaproate. *J. Neurochem.*, **29**, 1151–1153

117 Bissell, M. G., Bensch, K. G. and Herman, M. M. (1974). Effects of maple syrup urine disease metabolites on mouse L-fibroblasts *in vitro*: a fine structural and biochemical study. *J. Neurochem.*, **22**, 957–964

118 Lysiak, W., Pienkowska-Vogel, M., Szutowicz, A. and Angielski, S. (1974). Inhibition of alanine and aspartate aminotransferases by α-oxoderivatives of the branched-chain amino acids. *J. Neurochem.*, **22**, 75–83

119 Robinson, B. H., Oei, J., Sherwood, W. G., Slyper, A. H., Heininger, J. and Mamer, D. A. (1980). Hydroxymethylglutaryl CoA lyase deficiency: features resembling Reye syndrome. *Neurology*, **30**, 714–718

120 Shih, V. E. (1977). Miscellaneous metabolic disorders involving aminoacids and organic acids. In Vinken, P. J. and Bruyn, G. W. (eds.) *Handbook of Clinical Neurology*. Vol. 29, pp. 195–243. (Amsterdam: North Holland)

121 Nyhan, W. L. (1978). Nonketotic hyperglycinemia. In Stanbury, J. B., Wyngaarten, J. B. and Fredrickson, D. S. (eds.) *The Metabolic Basis of Inherited Disease*. 4th Edn., pp. 518–527. (New York: McGraw-Hill)

122 Dennis, M. J. and Clarke, J. T. R. (1979). Effects of glycine and glyoxylate on cerebral glucose oxidation *in vitro*. *J. Neurochem.*, **33**, 383–385

123 Shuman, R. M., Leech, R. W. and Scott, C. R. (1978). The neuropathology of the nonketotic and ketotic hyperglycinemias: three cases. *Neurology*, **28**, 139–146

124 Bank, W. J., Pizer, L. and Pfender, W. (1978). Glycine metabolism and spinal cord disorder. In Kark, R. A. P., Rosenberg, R. N. and Schut, L. J. (eds.) *Advances in Neurology*. Vol. 21, pp. 267–278. (New York: Raven Press)

125 Gitzelmann, R., Steinmann, B., Otten, A., Dumermuth, G., Herdon, M., Reubi, J. C. and Geunod, M. (1978). Nonketotic hyperglycinemia treated with strychnine, a glycine receptor antagonist. *Helv. Paediatr.*, **32**, 517–525

126 Rosenberg, L. E. (1978). Disorders of propionate, methylmalonate, and cobalamin metabolism. In Stanbury, J. B., Wyngaarten, J. B. and Fredrickson, D. S. (eds.) *The Metabolic Basis of Inherited Disease*. 4th Edn., pp. 411–429. (New York: McGraw-Hill)

127 Kuhara, T. and Matsumoto, I. (1980). Studies on the urinary acidic metabolites from three patients with methylmalonic aciduria. *Biomed. Mass Spectr.*, **7**, 424–428

128 Hommes, F. A., Kuipers, J. R., Elema, J. D., Jansen, J. F. and Jonxis, J. J. P. (1968), Proprionic-acidemia, a new inborn error of metabolism. *Paediatr. Res.*, **2**, 519–524

129 Barnes, N. D., Hull, D., Baigobin, L. and Gompertz, D. (1970). Biotin-responsive propionicacidaemia. *Lancet*, **2**, 244–245

130 Buniatian, H. C. (1971). The urea cycle. In Lajtha, A. (ed.) *Handbook of Neurochemistry*. Vol. 5(A), pp. 235–247. (New York: Plenum Press)

131 Sadasivudu, B. and Hanumantharao, T. I. (1974). Studies on the distribution of urea cycle enzymes in different regions of rat brain. *J. Neurochem.*, **23**, 267–269

132 Glick, N. R., Snodgrass, P. J. and Schafer, I. A. (1976). Neonatal arginino-succinic aciduria with normal brain and kidney but absent liver arginino-succinate lyase activity. *Am. J. Hum. Genet.*, **28**, 22–30

133 Sadasnudu, B. and Rao, T. I. (1976). Studies on functional and metabolic role of urea cycle intermediates in brain. *J. Neurochem.*, **27**, 785–794

134 Carlton, D. (1977). Disorders of the urea cycle and related diseases. In Vinken, P. J. and Bruyn, G. W. (eds.) *Handbook of Clinical Neurology*. Vol. 29, pp. 87–110. (Amsterdam: North Holland)

135 Shih, V. E. (1978). Urea cycle disorders and other congenital hyperammonemic syndromes. In

Stanbury, J. B., Wyngaarten, J. B. and Fredrickson, D. S. (eds.) *The Metabolic Basis of Inherited Disease*. 4th Edn., pp. 362–386. (New York: McGraw-Hill)

136 Levin, B. (1971). Hereditary metabolic disorders of the urea cycle. In Bodansky, O. and Latner, A. L. (eds.) *Advances in Clinical Chemistry*. Vol. 14, pp. 65–143. (New York: Academic Press)

137 Ebels, E. J. (1972). Neuropathological observations in a patient with carbamylphosphate-synthetase deficiency and in two sibs. *Arch. Dis. Child.*, **47**, 47–51

138 Simmell, O., Perheentupa, J., Rapola, J., Visakorpi, J. K. and Eskelin, L.-E. (1975). Lysinuric protein intolerance. *Am. J. Med.*, **59**, 229–240

139 Columbo, J. P., Richterich, R., Donath, A., Spahr, A. and Rossi, E. (1964). Congenital lysine intolerance with periodic ammonia intoxication. *Lancet*, **1**, 1014–1015

140 Ghadimi, H., Binnington, V. I. and Pecora, P. (1965). Hyperlysinemia associated with retardation. *N. Engl. J. Med.*, **273**, 725–729

141 Dancis, J., Hutzler, J., Cox, R. P. and Woody, N. C. (1969). Familial hyperlysinemia with lysine-ketoglutarate reductase insufficiency. *J. Clin. Invest.*, **48**, 1447–1452

142 Simell, O., Visakorpi, J. K. and Donner, M. (1972). Saccharopinuria. *Arch. Dis. Child.*, **47**, 52–55

143 Gatfield, P. D., Taller, E., Hinton, G. G., Wallace, A. C., Abdelnour, G. M. and Haust, M. D. (1968). Hyperpipecolatemia, a new metabolic disorder associated with neuropathy and hepatomegaly. A case study. *Can. Med. Assoc. J.*, **99**, 1215–1233

144 Stumpf, D. A. and Parks, J. K. (1980). Urea cycle regulation. I. Coupling of ornithine metabolism to mitochondrial oxidative phosphorylation. *Neurology*, **30**, 178–184

145 Gaull, G. E. (1972). Abnormal metabolism of sulfur-containing amino acids associated with brain dysfunction. In Lathja, A. (ed.) *Handbook of Neurochemistry*. Vol. 7, pp. 169–190. (New York: Plenum Press)

146 Gaull, G. E., Bender, A. N., Vulovic, D., Tallan, H. H. and Schaffner, F. (1981). Methioninemia and myopathy: a new disorder. *Ann. Neurol.*, **9**, 423–432

147 Milstein, J. M. (1977). Metabolic derangements of sulfur-containing amino acids. In Vinken, P. J. and Bruyn, G. W. (eds.) *Handbook of Clinical Neurology*. Vol. 29, pp. 111–125. (Amsterdam: North Holland)

148 Mudd, S. H. and Levy, H. L. (1978). Disorders of transsulfuration. In Stanbury, J. B., Wyngaarten, J. B. and Fredrickson, D. S. (eds.) *The Metabolic Basis of Inherited Disease*. 4th Edn., pp. 458–503. (New York: McGraw-Hill)

149 Gibson, J. B., Carson, N. A. and Neill, D. W. (1964). Pathological findings in homocystinuria. *J. Clin. Pathol.*, **17**, 427–437

150 Shih, V. E., Abroms, I. F., Johnson, J. L., Carney, M., Mandell, R., Robb, R. M., Cloherty, J. P. and Rajagopalan, K. V. (1977). Sulfite oxidase deficiency: biochemical and clinical investigations of a hereditary metabolic disorder in sulfur metabolism. *N. Engl. J. Med.*, **297**, 1022–1028

151 Rosenblum, W. I. (1968). Neuropathologic changes in a case of sulfite oxidase deficiency. *Neurology*, **18**, 1187–1196

152 Kivirikko, K. I. and Simila, S. (1977). Aminoacidurias. In Vinken, P. J. and Bruyn, G. W. (eds.) *Handbook of Clinical Neurology*. Vol. 29, pp. 129–148. (Amsterdam: North Holland)

153 Scriver, C. R. (1978). Disorders of proline and hydroxyproline metabolism. In Stanbury, J. B., Wyngaarten, J. B. and Fredrickson, D. S. (eds.) *The Metabolic Basis of Inherited Disease*. 4th Edn., pp. 336–361. (New York: McGraw-Hill)

154 Prusiner, S. B. (1981). Disorders of glutamate metabolism and neurological dysfunction. *Ann. Rev. Med.*, **32**, 521–542

155 Meister, A. (1978). Relation between ataxia and defects of the γ-glutamyl cycle. In Kark, R. A. P., Rosenberg, R. N. and Schut, L. J. (eds.) *Advances in Neurology*. Vol. 21, pp. 289–302. (New York: Raven Press)

156 La Du, B. N., Howell, R. R., Jacoby, G. A., Seegmiller, J. E., Sober, E. K. and Zannoni, V. G. (1963). Clinical and biochemical studies on two cases of histidinemia. *Pediatrics*, **32**, 216–227

157 Neville, B. G., Bentovim, A., Clayton, B. and Shepherd, J. (1972). Histidinaemia. Study of relation between clinical and biological findings in 7 subjects. *Arch. Dis. Child.*, **47**, 190–200

158 Scriver, C. R., Nutzenadel, W. and Perry, T. L. (1978). Disorders of β-alanine and carnosine metabolism. In Stanbury, J. T., Wyngaarten, J. B. and Fredrickson, D. S. (eds.) *The Metabolic Basis of Inherited Disease*. 4th Edn., pp. 528–542. (New York: McGraw-Hill)

159 Terplan, K. L. and Cares, H. L. (1972). Histopathology of the nervous system in carnosinase enzyme deficiency with mental retardation. *Neurology*, **22**, 644–655

160 Sjaastad, O., Berstad, J., Gjesdahl, P. and Gjessing, L. (1976). Homocarnosinosis 2. A familial metabolic disorder associated with spastic paraplegia, mental deficiency and retinal pigmentation. *Acta Neurol. Scand.*, **53**, 275–290

161 Perry, T. L., Kish, S. J., Sjaastad, O., Gjessing, L. R., Nesbakken, R., Schräder, H. and Løken, A. C. (1979). Homocarnosinosis: increased content of homocarnosine and deficiency of homocarnosinase in brain. *J. Neurochem.*, **32**, 1637–1640

162 Leibel, R. L., Shih, V. E., Goodman, S. I., Bauman, M. L., McCabe, E. R. B., Zwerdling, R. G., Bergman, I. and Costello, C. (1980). Glutaric acidemia: a metabolic disorder causing progressive choreoathetosis. *Neurology*, **30**, 1163–1168

163 Wyngaarten, J. B., (1979). The Lesch–Nyhan syndrome (HGPRT deficiency). In Beeson, P. B., McDermott, W. and Wyngaarten, J. B. (eds.) *Cecil Textbook of Medicine*. 15th Edn., pp. 2042–2043. (Philadelphia: Saunders)

164 Nyhan, W. L. (1977). The Lesch–Nyhan syndrome. In Vinken, P. J. and Bruyn, G. W. (eds.) *Handbook of Clinical Neurology*. Vol. 29, pp. 263–278. (Amsterdam: North Holland)

165 Kelley, W. N. and Wyngaarten, J. B. (1978). The Lesch–Nyhan syndrome. In Stanbury, J. B., Wyngaarten, J. B. and Fredrickson, D. S. (eds.) *The Metabolic Basis of Inherited Disease*. 4th Edn., pp. 1011–1036. (New York: McGraw-Hill)

166 Allsop, J. and Watts, R. W. E. (1980). Activities of amidophosphoribosyltransferase (EC 2.4.2.14) and the purine phosphoribosyltransferases (EC 2.4.2.7 and 2.4.2.8), and the phosphoribosylpyrophosphate content of rat central nervous system at different stages of development. *J. Neurol. Sci.*, **46**, 221–232

167 Rockson, S., Stone, R., Van Der Weyden, M. and Kelley, W. N. (1974). Lesch–Nyhan syndrome. Evidence for abnormal adrenergic function. *Science*, **186**, 934–935

168 Nyhan, W. L. (1978). Ataxia and disorders of purine metabolism: defects in hypoxanthine guanine phosphoribosyl transferase and clinical ataxia. In Kark, R. A. P., Rosenberg, R. N. and Schut, L. J. (eds.) *Advances in Neurology*. Vol. 21, pp. 279–287. (New York: Raven Press)

169 Meyer, U. A. and Schmid, R. (1978). The phorphyrias. In Stanbury, J. B., Wyngaarten, J. B. and Fredrickson, D. S. (eds.) *The Metabolic Basis of Inherited Disease*. 4th Edn., pp. 1166–1220. (New York: McGraw-Hill)

170 Bornstein, J. C., Pickett, J. B. and Diamond, I. (1978). Inhibition of the evoked release of acetylcholine by the porphyrin precursor δ-aminolevulinic acid. *Ann. Neurol.*, **5**, 94–96

171 Brennan, M. J. W. and Cantrill, R. C. (1979). The effect of delta-aminolaevulinic acid on the uptake and efflux of amino acid neurotransmitters in rat brain synaptosomes. *J. Neurochem.*, **33**, 721–725

172 Larson, A. W., Wasserstrom, W. R., Felsher, B. F. and Shih, J. C. (1978). Posttraumatic epilepsy and acute intermittent porphyria: effects of phenytoin, carbamazepine and clonazepam. *Neurology*, **28**, 824–828

173 Bosch, E. P., Pierach, C. A., Bossenmaier, I., Cardinal, R. and Thorson, M. (1977). Effect of hematin in porphyria neuropathy. *Neurology*, **27**, 1053–1056

174 McColl, K. E. L., Moore, M. R., Thompson, G. G. and Goldberg, A. (1981). Treatment with haematin in acute hepatic porphyria. *Quart. J. Med.*, **198**, 161–174

175 Weiner, W. J. and Klawans, H. L. (1976). Vitamin B6. In Vinken, P. J. and Bruyn, G. W. (eds.) *Handbook of Clinical Neurology*. Vol. 28, pp. 105–139. (Amsterdam: North Holland)

176 Weiner, W. J. (1976). Vitamin B6 in the pathogenesis and treatment of diseases of the central nervous system. In Klawans, H. L. (ed.) *Clinical Neuropharmacology*. Vol. 1, pp. 107–136. (New York: Raven Press)

177 Lott, I. T., Coulombe, T., Di Paolo, R. V., Richardson, E. P. Jr and Levy, H. L. (1978). Vitamin B6 dependent seizures: pathology and chemical findings in brain. *Neurology*, **28**, 47–54

178 Botez, M. I. and Reynolds, E. H. (eds.) (1979). *Folic Acid in Neurology, Psychiatry and Internal Medicine*. (New York: Raven Press)

179 Kunze, K. and Leitenmaier, K. (1976). Vitamin B12 deficiency and subacute combined

degeneration of the spinal cord. In Vinken, P. J. and Bruyn, G. W. (eds.) *Handbook of Clinical Neurology.* Vol. 28, pp. 141–198. (Amsterdam: North Holland)

180 Fehling, C., Nilsson, B. and Jägerstad, M. (1979). Effect of vitamin B12 deficiency on energy-rich phosphates, glycolytic and citric acid cycle metabolites and associated amino acids in rat cerebral cortex. *J. Neurochem.,* **32,** 1115–1117

181 Mans, A. M., Saunders, S. J., Kirsch, R. E. and Biebuyck, J. F. (1979). Correlation of plasma and brain amino acid and putative neurotransmitter alterations during acute hepatic coma in the rat. *J. Neurochem.,* **32,** 285–292

182 Duffy, T. E., Vergara, F. and Plum, F. (1974). α-Ketoglutaramate in hepatic encephalopathy. *Res. Publ. Assoc. Nerv. Ment. Dis.,* **53,** 39–51

183 Cooper, A. J. L. and Gross, M. (1977). The glutamine transaminase-w-amidase system in rat and human brain. *J. Neurochem.,* **28,** 771–778

184 Ali, F. M., Ansley, J. and Faraj, B. A. (1980). Studies on the influence of portocaval shunt on the metabolism of tyrosine. *J. Pharmacol. Exp. Ther.,* **214,** 546–553

185 Curzon, G., Kantamaneni, B. D., Fernando, J. C., Woods, M. S. and Cavanagh, J. B. (1975). Effects of chronic porto-caval anastomosis on brain tryptophan, tyrosine and 5-hydroxy-tryptamine. *J. Neurochem.,* **24,** 1065–1070

186 Jellinger, K. and Riederer, P. (1977). Brain monoamines in metabolic (endotoxic) coma. A preliminary biochemical study in human postmortem material. *J. Neural Trans.,* **41,** 275–286

187 James, J. H., Hodgman, J. M., Funovics, J. M. and Fischer, J. E. (1976). Alterations in brain octopamine and brain tyrosine following portocaval anastomosis in rats. *J. Neurochem.,* **27,** 223–227

188 Fischer, J. E. (1974). False neurotransmitters and hepatic coma. *Res. Publ. Assoc. Nerv. Ment. Dis.,* **53,** 53–71

189 Fischer, J. E., Funovics, J. M., Falcao, H. A. and Wesdorp, I. C. (1976). L-Dopa in hepatic coma. *Ann. Surg.,* **183,** 386–391

190 Victor, M. (1974). Neurologic changes in liver disease. *Res. Publ. Assoc. Nerv. Dis.,* **53,** 1–12

191 Cavanagh, J. B. (1974). Liver bypass and the glia. *Res Publ. Assoc. Nerv. Ment. Dis.,* **53,** 13–35

192 Schmid, R. and McDonagh, A. F. (1978). Hyperbilirubinemia. In Stanbury, J. B., Wyngaarten, J. B. and Fredrickson, D. S. (eds.) *The Metabolic Basis of Inherited Disease.* 4th Edn., pp. 1221–1257. (New York: McGraw-Hill)

193 Osterberg, K. (1971). Kernicterus (bilirubin encephalopathy). In Minckler, J. (ed.) *Pathology of the Nervous System.* Vol. 2, pp. 1338–1342. (New York: McGraw-Hill)

194 Angelini, C., Philippart, M., Borrone, C., Bresolin, N., Cantini, M. and Lucke, S. (1980). Multisystem triglyceride storage disorder with impaired long-chain fatty acid oxidation. *Ann. Neurol.,* **7,** 5–10

195 Di Donato, S., Cornelio, F., Storchi, G. and Rimoldi, M. (1979). Hepatic ketogenesis and muscle carnitine deficiency. *Neurology,* **29,** 780–785

196 Willner, J. H., Ginsburg, S. and Di Mauro, S. (1978). Active transport of carnitine into skeletal muscle. *Neurology,* **28,** 721–724

197 Carroll, J. E., Brooke, M. H., De Vivo, D. C., Shumate, J. B., Kratz, R., Ringel, S. P. and Hagberg, J. M. (1980). Carnitine 'deficiency': lack of response to carnitine therapy. *Neurology,* **30,** 618–626

198 Hart, Z. H., Chang, C.-H., Di Mauro, S., Farooki, Q. and Ayyar, R. (1978). Muscle carnitine deficiency and fatal cardiomyopathy. *Neurology,* **28,** 147–151

199 Markesbury, W. R., McQuillen, M. P., Procopis, P. G., Harrison, A. R. and Engel, A. G. (1974). Muscle carnitine deficiency. Association with lipid myopathy, vacuolar neuropathy and vacuolated lymphocytes. *Arch. Neurol.,* **31,** 320–324

200 Chapoy, P. R., Angelini, C., Brown, W. J., Stiff, J. E, Shug, A. L. and Cederbaum, S. D. (1980). Systemic carnitine deficiency – a treatable inherited lipid-storage disease presenting as Reye's syndrome. *N. Engl. J. Med.,* **303,** 1389–1394

201 Layzer, R. B., Havel, R. J. and McIlroy, M. B. (1980). Partial deficiency of carnitine palmityl-transferase, physiologic and biochemical consequences. *Neurology,* **30,** 627–633

202 Ionasescu, V., Hug, G. and Hoppel, C. (1980). Combined partial deficiency of muscle carnitine palmitoyl transferase and carnitine with autosomal dominant inheritance. *J. Neurol., Neurosurg. Psychiatry,* **43,** 679–682

126

203 Di Donato, S., Castiglione, A., Rimoldi, M., Cornelio, F., Vendemia, F., Cardace, G. and Bertagnolio, B. (1981). Heterogeneity of carnitine palmitoyltransferase deficiency. *J. Neurol. Sci.*, **50**, 207–215

204 Di Donato, S., Cornelio, F., Pacini, L., Peluchetti, D., Rimoldi, M. and Spreafico, S. (1978). Muscle carnitine palmityltransferase deficiency: a case with enzyme deficiency in cultured fibroblasts. *Ann. Neurol.*, **4**, 465–467

205 Stumpf, D. A. (1979). Mitochondrial multisystem disorders: clinical, biochemical and morphologic features. In Tyler, H. R. and Dawson, D. M. (eds.) *Current Neurology.* Vol. 2, pp. 117–149. (Boston: Houghton Mifflin)

206 Fitzsimmons, R. B. (1981). The mitochondrial myopathies: 9 case reports and a literature review. *Clin. Exp. Neurol.*, **17**, 185–210

207 Pellock, J. M., Behrens, M., Lewis, L., Holub, D., Carter, S. and Rowland, L. P. (1978). Kearns–Sayre syndrome and hypoparathyroidism. *Ann. Neurol.*, **3**, 455–458

208 Frank, J. P., Harrati, Y., Butler, I. J., Nelson, T. E. and Scott, C. I. (1980). Central core disease and malignant hyperthermia syndrome. *Ann. Neurol.*, **7**, 11–17

209 De Vivo, D. C. (1978). Reye syndrome: a metabolic response to an acute mitochondrial insult? *Neurology*, **28**, 105–108

210 Haymond, M. W., Karl, I. E., Keating, J. P. and De Vivo, D. C. (1978). Metabolic response to hypertonic glucose administration in Reye syndrome. *Ann. Neurol.*, **3**, 207–215

211 Ansevin, C. F. (1980). Reye syndrome: serum-induced alterations in brain mitochondrial function are blocked by fatty-acid-free albumin. *Neurology*, **30**, 160–166

212 Trauner, D. A., Brown, F., Ganz, E. and Huttenlocher, P. R. (1978). Treatment of elevated intracranial pressure in Reye syndrome. *Ann. Neurol.*, **4**, 275–278

213 Haas, R., Stumpf, D. A., Bergen, B. J., Parks, J. K. and Eguren, L. (1980). Inhibition of oxidative phosphorylation by sodium valproate. *Neurology*, **30**, 420

214 Trauner, D. A. and Huttenlocher, P. R. (1978). Short chain fatty acid-induced central hyperventilation in rabbits. *Neurology*, **28**, 940–944

215 Huttenlocher, P. R. and Trauner, D. A. (1977). Reye's syndrome. In Vinken, P. J. and Bruyn, G. W. (eds.) *Handbook of Clinical Neurology.* Vol. 29, pp. 331–344. (Amsterdam: North Holland)

216 Vanholder, R., De Reuck, J., Sieben-Praet, M. and De Coster, W. (1979). Reye's syndrome in an adult. *Eur. Neurol.*, **18**, 367–372

217 Trauner, D. A. (1980). Treatment of Reye syndrome. *Ann. Neurol.*, **7**, 2–4

3
Disorders of the metabolism of larger molecules

Neural tissue contains a variety of large molecules which have a number of known biological functions, as well as often providing a basis for cell structure. Proteins function both as enzymes and as receptors for the attachment of other molecules and the subsequent transduction of this attachment into further biochemical events; nucleic acids convey genetic information; carbohydrates (as glycogen) are stores of energy, and lipids appear to have structural, permeability (membrane) and insulating functions (e.g. myelin). In addition, neural tissue contains complex macromolecules formed from admixture of protein, carbohydrate and lipid components, e.g. lipoproteins, glycoproteins, glycosaminoglycans (mucopolysaccharides) and glycolipids. Individual macromolecules may exist in cells bonded into even larger and more complex macromolecule aggregates. For convenience, biological macromolecules are often considered in terms of their individual components which are chemically stable enough to be able to exist in isolation after preliminary separation procedures. Such components are themselves usually comparatively large molecules.

Figures for the chemical composition of adult human brain are set out in Table 3.1 (modified from Norton[1]). To set these figures for brain composition in further perspective, it should be noted that they refer to dry weight of brain. Human white and grey matter contain, respectively, 70% and 83% of water. Therefore, on a wet weight basis, white matter and grey matter contain, respectively, protein 8.5 and 7.5% and lipids 14.8 and 5.8%. In contrast, skeletal muscle[2] contains water 77.5%, protein 19%, lipid 5% and carbohydrate 1%.

While disorders of glycogen metabolism could logically have been considered in the present chapter, these disorders have already been dealt with in relation to their area of functional implication. Similarly, all genetic diseases (due to the occurrence of abnormal DNA and consequently of abnormal enzymes) and immunological disease (related to abnormal antibodies, themselves glyco-

Table 3.1 **Adult human brain percentage composition**

Substance	White matter	Grey matter		
Protein	39	55		} of total
Lipid, including	55	33		
cholesterol	27.5	22		
galactolipids, *including*	26	7		
cerebroside		20	5	
sulphatide		5	2	
phospholipids, *including*	46	70		} of total lipid
ethanolamine phosphatides		15	23	
lecithin		13	27	
sphingomyelin		8	7	
phosphatidyl serine		8	9	
phosphatidyl inositol		1	3	

proteins) could also be regarded as appropriate topics for this chapter. However, in the light of our current ignorance of the detailed structural chemistry of individual enzyme proteins and nucleic acids, it has seemed more useful to consider genetic disorders in relation to their consequent defects in biochemical function. Immunological disease has not been considered further in this book. With the exclusion of the above mentioned disorders of protein and large carbohydrate (glycogen) molecules, the present chapter comes to deal mainly with abnormalities of the following classes of substance:

(1) Lipids, mainly glycosphingolipids,
(2) Glycosaminoglycans (mucopolysaccharides),
(3) Mucolipids,
(4) Sialo-oligosaccharides,
(5) Ceroid-lipofuscin.

On the whole, this listing of the molecular classes tends to correlate with increasing molecular structural complexity. Several of these classes of molecule (glycosaminoglycans, mucolipids and sialo-oligosaccharides) are glycoprotein components. Because disorders of these classes of molecule are nearly always due to enzyme defects, and because the same molecular subcomponents and types of chemical bond may occur in more than one class of large molecule, it is possible for a single enzyme defect to cause altered metabolism of more than one class of large molecule.

In theory, the various disorders considered could involve either the formation of the complex molecules, or their degradation. In practice, molecular degradation is nearly always involved in the diseases. Consequently the disorders here considered usually feature the accumulation of excess quantities of one or more large molecules. Thus the disorders considered are 'storage' diseases.

3.1 DISORDERS OF LIPID METABOLISM

Chemistry of brain lipids – The chemical structures of the classes of lipid mentioned in Table 3.1 are shown below. It is worth indicating that most large molecule disorders involve glycolipids (molecules containing both saccharide and lipid components). More detailed accounts of mammalian lipid chemistry are provided by Stoffel[3] and Rouser and Yamamoto[4].

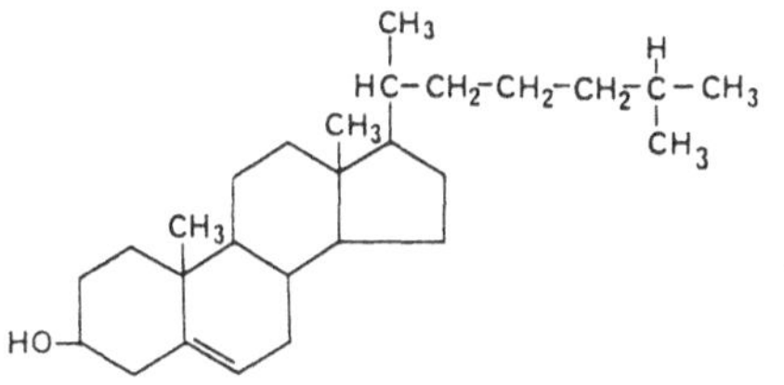

.cholesterol

Galactolipids (galactosphingolipids) – Galactolipids (one variety of glycolipids) are molecules comprising a galactose, a sphingosine and a fatty acid moiety, as a minimum.

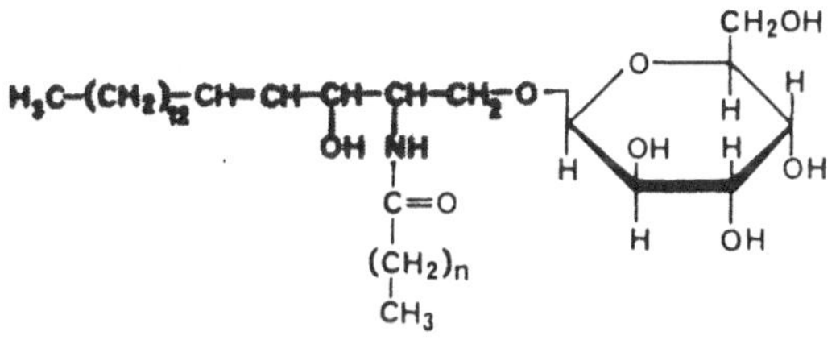

.ceramide

The sphingosine moiety is linked to the fatty acid through an amide bond, and to the sugar through an ester bond. The sphingosine–fatty acid unit is termed ceramide (*N*-acylsphingosine). Ceramide, with the addition of the hexose, becomes a cerebroside (the structure illustrated). The fatty acid is usually a long chain one, often with 16–18 carbon atoms in the case of gangliosides (which also contain *N*-acetyl-neuraminic acid), and usually with 20–24 carbon atoms in the neutral sphingolipids and sphingomyelin[5]. Sulphatides are simply sulphuric acid esters of cerebrosides.

131

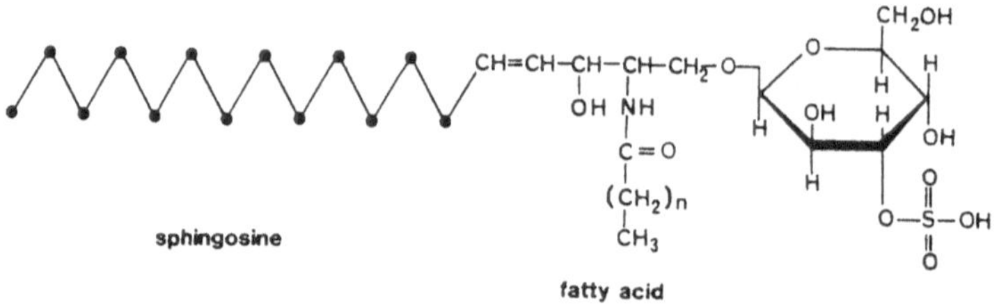

sphingosine

CH=CH-CH-CH-CH$_2$-O-

OH NH

C=O

(CH$_2$)$_n$

CH$_3$

fatty acid

CH$_2$OH

H OH H

OH H OH

O-S-OH

.sulphatide

Phospholipids – Phospholipids are phosphorus-containing lipids. Apart from sphingomyelin (a sphingosine derivative), the phospholipids are glycerol derivatives.

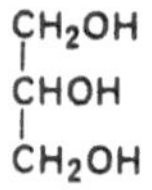

CH$_2$OH

CHOH

CH$_2$OH

.glycerol

The general structure of the glycerophospholipids is as follows:

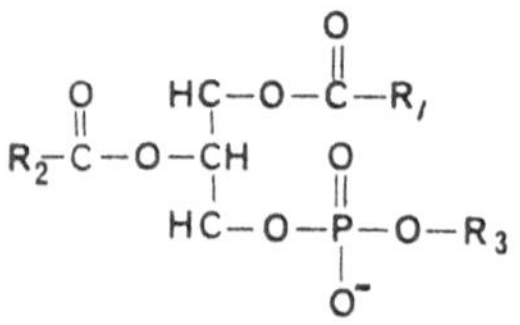

.glycerophospholipid

where R_1 and R_2 represent fatty acid moieties, and R_3 is

ethanolamine (HO—$CH_2.CH_2.NH_3^+$) in the case of phosphatidyl ethanolamines (cephalins),

choline (HO—$CH_2.CH_2.N^+$—$(CH_3)_3$) in the phosphatidyl cholines (lecithins),

serine (HO—$CH_2.CH(NH_2).COOH$) in phosphatidyl serine, and

inositol in phosphatidyl inositols.

132

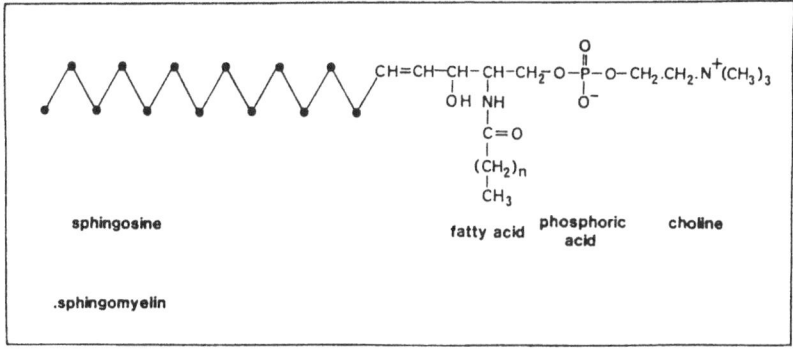

.Inositol

The structure of sphingomyelin is shown below.

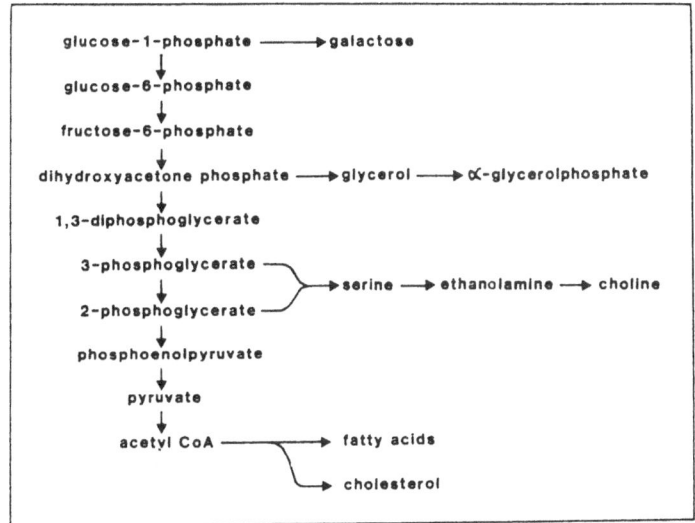

Figure 3.1 Lipid biosynthetic pathways

The synthetic pathways to these various lipids (Figure 3.1) begin in the anaerobic glycolytic pathway (Figure 2.5). The details of the lipid syntheses are not of great importance for the purpose of this book. The only known neurological diseases that are probably due to defective lipid synthesis are adrenoleukodystrophy (in which there may be a defect in the mechanism which controls the length of the fatty acid chains synthesized) and abetalipoproteinaemia. However, there is a report of one instance of disease due to failure of ganglioside synthesis[6]. With these exceptions, the disorders now to be considered are due to defective lipid macromolecule degradation. Apart from the defective catabolism of dietary phytanic acid in Refsum's syndrome, the lipid degradative disorders involve only sphingolipids. There do not appear to be known diseases of the degradation of non-sphingosine containing phospholipids.

3.1.1 Disorder of fatty acid composition

3.1.1.1 Adrenoleukodystrophy, adrenomyeloneuropathy and Schilder's disease

Adrenoleukodystrophy (combined Addison's and Schilder's disease) is an uncommon X-linked recessively inherited degeneration of cerebral myelin and of the adrenals and testes[7]. It is associated with the presence of abnormally long chain fatty acids in cholesterol esters and other complex lipids in the affected tissues. Adrenomyeloneuropathy and classical Schilder's disease appear to be manifestations of the same disease process with different regional dominances of tissue injury within the body. Different manifestations of the biochemical process may occur in the one family[8].

Abnormal biochemistry – In adrenoleukodystrophy, cholesterol esters, gangliosides and galactosylceramide all contain an abnormally high proportion of very long chain (longer than C_{22} and mainly C_{25} and C_{26}) fatty acids[9,10]. There is an accumulation of similar long chain saturated fatty acids in adrenomyeloneuropathy and in atypical adrenoleukodystrophy occurring in females[11]. It is not known whether the biochemical defect resides in the control of chain length during synthesis of fatty acids or in the degradation of the relevant long chain acids, which do occur in normals, though they are uncommon (and have cholesterol esters which are very slowly hydrolysed[12]). The abnormal cholesterol esters and other lipids in the disorder may alter myelin composition so that it is abnormally prone to break down. Abnormal amounts of long chain fatty acids are not found in cholesterol esters from peripheral nerves in adrenoleukodystrophy[13].

Aetiology – The disorder is inherited as an X-linked recessive trait.

134

Structural pathology – The neuropathology is reviewed by Schaumburg *et al.*[14] In Schilder's disease (in which the extra-neural changes are merely less severe than in the other variants of the syndrome), and in adrenoleukodystrophy, there is widespread progressive demyelination of cerebral white matter which tends to spare the subcortical U fibres. The demyelination typically begins in the posterior parts of the hemispheres and extends progressively to involve the entire white matter of cerebrum and cerebellum, and the optic nerves. (The disorder was formerly described as sudanophil diffuse sclerosis, since the products of myelin decomposition stain red with the histochemical reagent Sudan IV.) PAS positive macrophage clusters occur in the brain, testes, liver, lymph nodes and spleen. Characteristic birefringent crystals of cholesterol esters with abnormal long chain fatty acids are seen in the brain, adrenals and testes. In adrenomyeloneuropathy 'dying back' neuropathological changes occur in the long tracts of the spinal cord[14], and there are lamellar cytoplasmic inclusions in the brain, adrenal and testes, similar to the inclusions of adrenoleukodystrophy.

Clinical features – The disorder almost always occurs in males, usually beginning in childhood, though rarely in the neonatal period[15]. There is progressive dementia with behaviour alterations, visual impairment (due to cerebral and optic nerve involvement) and later weakness and spasticity. Clinical and biochemical features of Addison's disease may be superimposed in a minority of cases. Most sufferers die within 2–4 years of the onset of symptoms. In adrenomyeloneuropathy, which may occur in the same family as adrenoleukodystrophy, the presentation is dominantly that of progressive spastic paraplegia, which may begin in the third decade of life[16].

Diagnosis
(1) Clinical: The combination of progressive dementia, neurological signs as described above and evidence of adrenal insufficiency in a young male is virtually diagnostic. In the absence of signs of adrenal insufficiency and of a family history, the diagnosis can only be suspected as one possibility among several.

(2) Laboratory: CT head scanning may reveal the typical diffuse confluent cerebral hemisphere demyelination of Schilder's disease. There are no known diagnostic abnormalities on blood examination, though biochemical changes consistent with Addison's disease may be detected. Biopsy of brain or adrenal shows the characteristic birefringent crystals, and the abnormal fatty acid composition of lipids can be determined chromatographically.

Therapy – Adrenal insufficiency can be corrected by appropriate oral hormonal replacement therapy, but no treatment is known which will prevent the neurological deterioration of the disorder.

135

3.1.1.2 Pelizaeus–Merzbacher disease

In this usually X-linked, recessively inherited, disorder which produces very severe myelin loss, Bourre et al.[17] found a very great decrease in the α-hydroxy fatty acids and long chain ($>C_{18}$) fatty acids in cerebrosides and sulphatides, while Witter et al.[18] noted an absence of long chain fatty acids in cerebrosides. If these findings are confirmed, the Perlizaeus–Merzbacher disease may fall into place beside adrenoleukodystrophy as a progressive demyelinating disorder due to defective metabolism of long chain fatty acids. However, it should be pointed out that Witter et al.[18] considered that the brain lipid profile in their cases was more suggestive of a congenital failure of myelination.

3.1.1.3 Abetalipoproteinaemia

This rare disorder (the Bassen–Kornzweig syndrome) is probably due to an inherited defect in lipoprotein synthesis. It is associated with disordered absorption of dietary lipids, steatorrhoea, altered red blood cell morphology and neurological and retinal abnormalities. Several detailed accounts of the disorder are available[19–21].

Biochemical abnormality – The primary defect seems to be most probably a failure to synthesize β-lipoprotein[22]. Consequently plasma β-lipoprotein level is very low. Lipids from the diet absorb into intestinal mucosal cells, but cannot pass thence into the lymphatics. Infusion of β-lipoproteins into the circulation will not correct this defect of lipid transfer. The explanation for the impaired lipid absorption is uncertain. Red blood cells in the circulation, but not their precursors in the marrow, have an abnormal crenated appearance (acanthocytosis) which can be corrected by detergent infusion. This appearance may be due to an altered membrane lipoprotein content, though other lipids in red cells may also be abnormal. The mechanism responsible for the neurological damage in the disorder is unknown. There has been speculation that steatorrhoea consequent on reduced dietary lipid absorption may interfere with vitamin A absorption, and that the resulting vitamin deficiency may contribute to the retinal changes of the disorder.

Aetiology – The disorder is probably inherited as an autosomal recessive trait, with a predilection for persons of Jewish origin.

Structural pathology – There is demyelination of the posterior columns of the spinal cord, the dorsal and ventral spinocerebellar tracts and the cerebellum, with loss of anterior horn cells and neurons in the deep cerebellar nuclei. Peripheral nerves show focal demyelination. Hepatocytes are vacuolated and lipid laden.

136

Clinical features – The disorder usually presents in adolescence or early adult life. However, a tendency to diarrhoea and steatorrhoea may have been present from birth. Decreased reflexes develop in the legs, and extend to the arms. Cerebellar ataxia appears, with weakness, kyphoscoliosis, pes cavus, Babinski signs and impaired proprioception. Some patients are mentally retarded. There is atypical retinitis pigmentosa, and sometimes nystagmus and opthalmoparesis also occur. The abdomen is distended and diarrhoea and steatorrhoea are usually present. The small bowel is hypotonic and distended. Cardiomyopathic changes have occurred in some cases. Acanthocytosis is present in the peripheral blood. The ESR and serum cholesterol and triglyceride levels are low, and chylomicrons do not appear in serum after a fatty meal.

Diagnosis
(1) Clinical: The neurological features are not specific, and resemble those of Friedreich's ataxia. In the absence of a family history it seems unlikely that the diagnosis could be made clinically, though the association of appropriate neurological features with diarrhoea and steatorrhoea would be suggestive.

(2) Laboratory: The presence of acanthocytosis in a peripheral blood film, the low level of cholesterol in serum, and the absence of β-lipoprotein, are diagnostic. The intestinal mucosa is not atrophic at biopsy, but the mucosal cells show multiple lipid droplets.

Therapy – The steatorrhoea may be relieved by a low fat diet. Medium chain triglyceride and fat soluble vitamin supplements may be necessary. No known treatment definitely benefits the neurological aspects of the disorder, though there is evidence that vitamin E therapy may be of some use. Illingworth *et al.*[23] claimed that a low fat diet with high doses of vitamins A and E prevented neurological deterioration over a 5 year period in two patients with the disorder.

3.1.1.4 *Phytanic acid accumulation – Refsum's syndrome*

Refsum's syndrome (heredopathia atactica polyneuritiformis) is an uncommon inherited neurological and systemic disturbance. The syndrome arises from an inability to metabolize phytanic acid, derived from dietary chlorophyll[24].

Biochemical abnormality – Phytanic acid (3,7,11,15-tetramethylhexadecanoic acid) is derived from chlorophyll through the series of reactions shown in Figure 3.2. The normal degradation of a fatty acid in the human body is by serial β-oxidation, but the presence of a methyl group on the β-carbon atom of phytanic acid prevents the oxidation of this carbon atom. Instead, normally

137

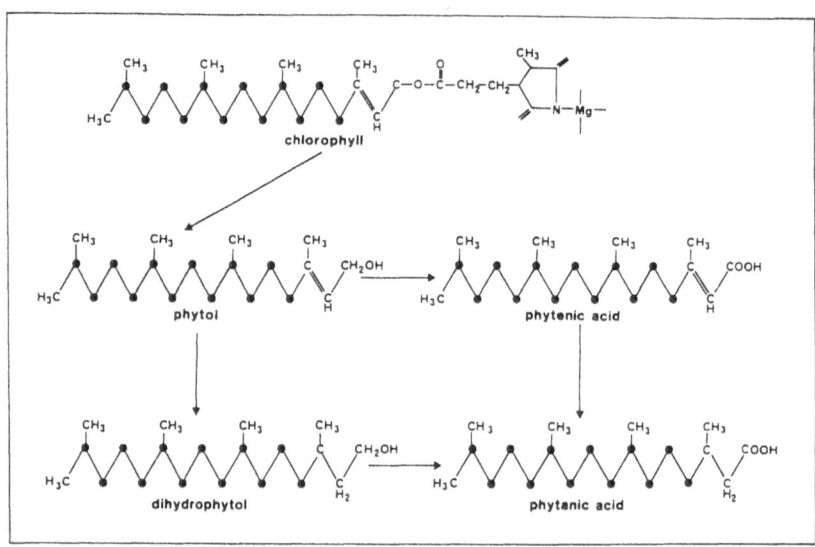

Figure 3.2 Formation of phytanic acid from chlorophyll

the α-carbon atom of phytanic acid is oxidized in a reaction catalysed by the enzyme phytanic acid α-hydroxylase, and a single carbon fragment is thus split off. Following this cleavage, conventional serial β-oxidation of the residue occurs (Figure 3.3). In Refsum's syndrome the enzyme responsible for the α-hydroxylation of phytanic acid is deficient. Consequently phytanic acid of dietary origin cannot be metabolized further, and accumulates in glycolipids and phospholipids[25]. It becomes incorporated into myelin and alters the properties of myelin. Phytanic acid inhibits the metabolism of other fatty acids, and the conversion of palmitic acid to palmityl-CoA. Plasma phytanic acid levels become greatly elevated in Refsum's syndrome, and abnormal lipids occur in the nervous system and other tissues.

Aetiology – Refsum's syndrome is inherited as an autosomal recessive trait.

Structural pathology – There is widespread thickening of peripheral nerves due to a hypertrophic peripheral polyneuritis. Amorphous material separates the individual axons. Schwann cell numbers are increased. The spinal cord tends to be atrophic and there are degenerated neurons in brain stem nuclei with, in some cases, intraneuronal lipid deposition. There is increased lipid in the meninges.

Clinical features – The disorder may appear in childhood or adult life. Its major features are those of a chronic hypertrophic peripheral polyneuritis with cerebellar ataxia, nystagmus, impaired hearing and smell, atypical retinitis

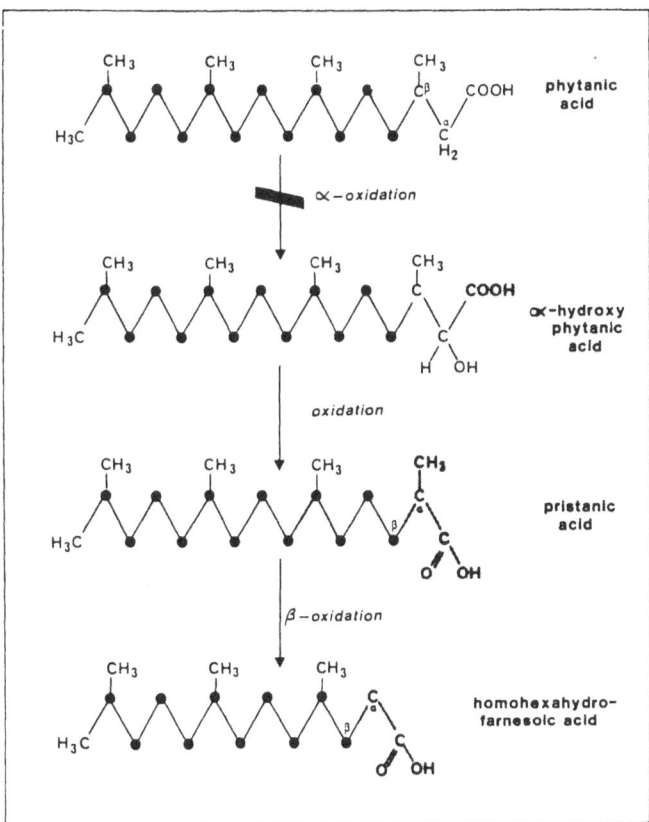

Figure 3.3 Oxidation pathway for phytanic acid

pigmentosa (producing night blindness and concentric constriction of the visual fields), a raised CSF protein content, e.c.g. abnormalities, icthyosis and skeletal malformations. The course of the disorder is generally slowly progressive, though there are sometimes long periods of remission.

Diagnosis
(1) Clinical: The full clinical picture is characteristic. When the disorder is incompletely developed, the possibility of other forms of peripheral poly-neuritis (hereditary or sporadic) may arise and, if cerebellar disturbance is also present, Friedreich's ataxia or one of its variants will enter the different-ial diagnosis.

(2) Laboratory: Considerably raised blood phytanic acid levels, measured by gas chromatography, are diagnostic. Cultured fibroblasts from affected cases lack the capacity to oxidize phytanic acid. Heterozygous carriers'

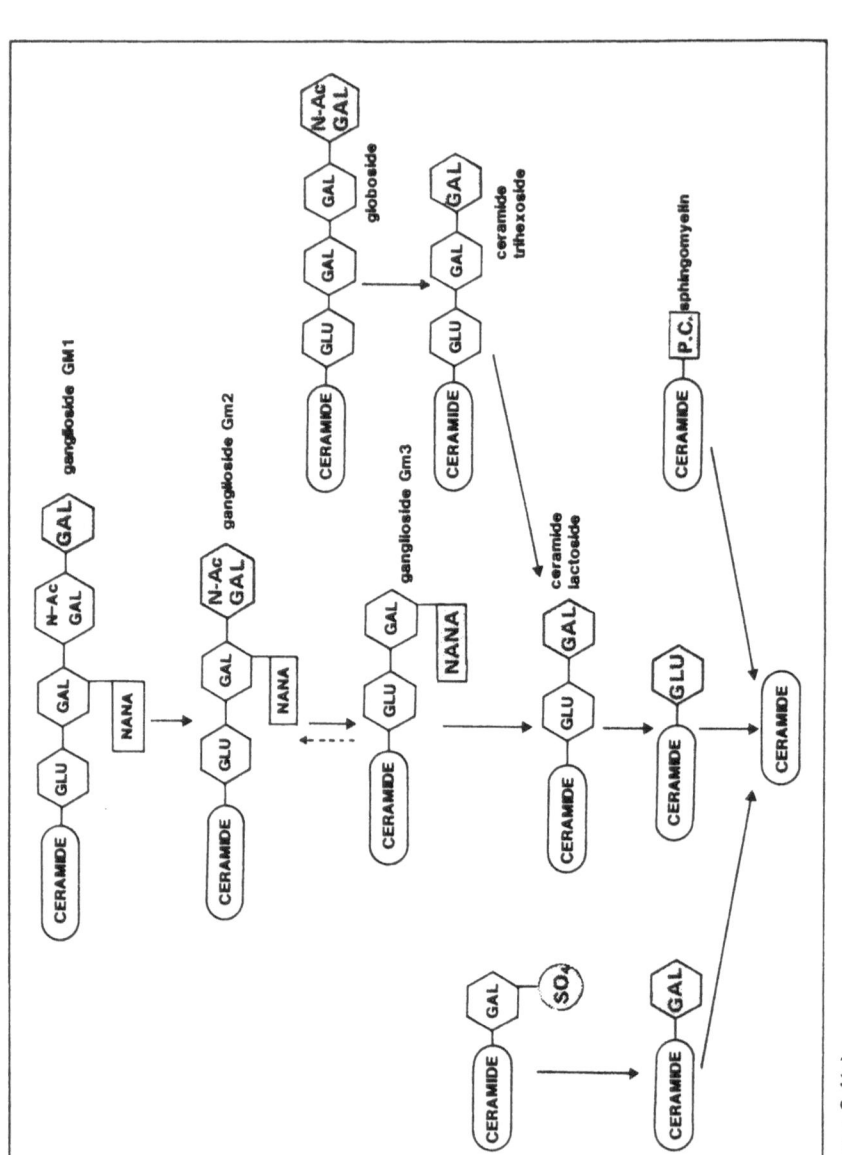

Figure 3.4(a)

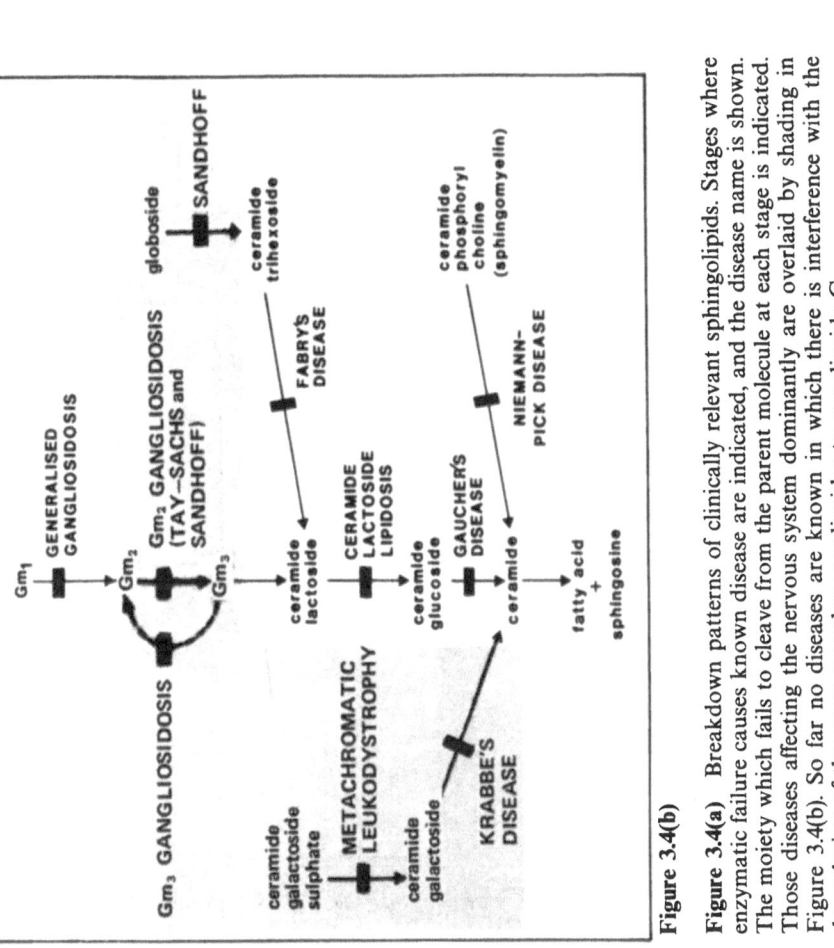

Figure 3.4(b)

Figure 3.4(a) Breakdown patterns of clinically relevant sphingolipids. Stages where enzymatic failure causes known disease are indicated, and the disease name is shown. The moiety which fails to cleave from the parent molecule at each stage is indicated. Those diseases affecting the nervous system dominantly are overlaid by shading in Figure 3.4(b). So far no diseases are known in which there is interference with the degradation of the more complex gangliosides to ganglioside G_{M1}

fibroblasts have about half normal capacity for the oxidation. Nerve conduction velocity measurements provide confirmation of peripheral neuropathy in the disorder, and the histological changes in a sural nerve biopsy may suggest the diagnosis.

Therapy – A diet with a low phytol content may be of benefit for affected persons, particularly if it can be commenced in homozygous siblings (detected by blood phytanate measurement) while their disorder is still subclinical. Carriers of the defect require genetic advice.

3.1.2 Sphingolipidoses

The sphingolipids embrace the sphingosine-containing glycolipids, their sulphate esters, and also sphingomyelin. Various members of these classes of substance may accumulate in cells when the enzymes catalysing their degradations are defective. Ignoring the solitary reported instance of failed ganglioside synthesis, such sphingolipidoses can be seen in a clinical perceptive in relation to the patterns of degradation of several complex sphingolipids (Figure 3.4). The sphingolipids are large molecules, and their structures will nearly always be represented in this book in a condensed form. However, the structural formulae of the individual components of the molecules are shown at various

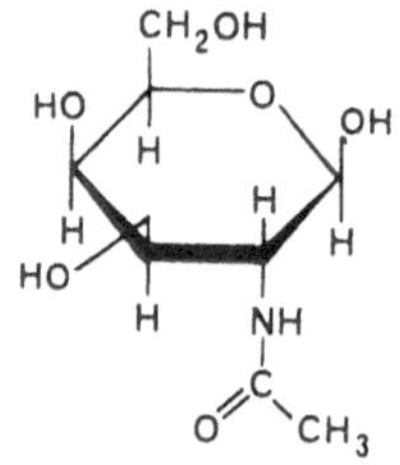

Figure 3.5 Formula of *N*-acetylgalactosamine

places in the book (glucose, p. 9; galactose, p. 32; ceramide, i.e. *N*-acylsphingosine, p. 131; *N*-acetylgalactosamine in Figure 3.5 and *N*-acetylneuraminic acid in Figure 3.6 (lower right hand portion of ganglioside molecule)).

The full structural formula for the main substance stored in the neurologically most important sphingolipidosis, Tay–Sachs' disease, is shown in detail (Figure 3.6), to allow some insight into the complexity of the molecular structures

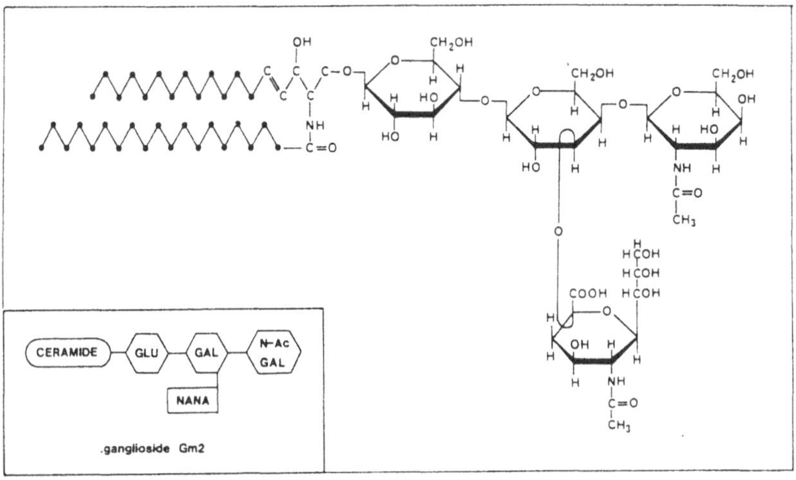

Figure 3.6 Structural formula of ganglioside G_{M2}

involved. This particular stored substance is a ganglioside, i.e. a cerebroside (or ceramide polyhexoside) containing N-acetylneuraminic acid, i.e. sialic acid (or other acid sugar) and is coded as ganglioside G_{M2}. There are, in fact, at least six known sialic acids[26] but N-acetylneuraminic acid is the variety of greatest importance in human medicine, and is often termed sialic acid. It should be noted that the —H and —OH groups attached to the C_1 atom of a hexose may exist in anomeric α- and β-configurations. In the α-configuration the —OH group lies below the plane of the ring as represented in a Haworth projection formula, while in the β-configuration the —OH group lies above the plane of the ring. This point is important, since enzymes cleaving glycosidic bonds may be specific for α- or β-type linkages. In Figure 3.5 N-acetylglucosamine is represented in β-configuration, while in Figure 3.6 the bond between the glucose and the galactose moieties is an α-$1 \rightarrow 4$ one, and the bond between galactose and N-acetylgalactose is a β-$1 \rightarrow 4$ one. Several systems of ganglioside nomenclature exist. That used most commonly, and employed here, is the system of Svennerholm[27], in which the abbreviation G_M refers to monosialogangliosides, the abbreviation G_D to disialoganglio-sides, and the abbreviation G_T to trisialogangliosides. The subscripts 1,2,3 etc. relate to a diminishing complexity of structure within each particular ganglioside subgroup. Although they are synthesized in neuronal cell bodies, gangliosides in the nervous system occur at highest concentration in the membranes of nerve terminals. In the body, gangliosides are degraded by sequential enzymatic cleavage of the terminal moiety from the molecule, except that sialic acid groups (apart from the final one to leave the molecule) can be cleaved at any stage of molecular degradation. The final sialic acid group is cleaved only when it has become the terminal group on the molecule, after removal of other groups[5].

143

As a group of disorders, the sphingolipidoses are reasonably similar clinically and pathologically. In all, complex lipids accumulate within lysosomes in tissue cells and often in cells of the nervous system. The sphingolipidoses are thus lysosomal storage diseases[28,29]. While cellular distension is the most prominent microscopic feature, the reason for failure of cellular function in the disorders is not known with certainty. However, it seems likely that simple mechanical dislocation of normal molecules within cells plays some part. Many of the disorders produce a degree of generalized progressive failure of nervous system function (in particular those disorders in which the stored material is not solely ganglioside or ceramide-galactoside in nature). In some varieties there are obvious extraneural manifestations and abdominal organomegaly. The disorders are due to insufficiency of particular lysosomal enzymes. Since these enzymes often catalyse the hydrolysis of more than one substrate, enzyme failure may result in the accumulation of several substances. Further, secondary alterations in lysosomal enzymes may occur in addition to the primary enzyme defect. In the past these secondary enzymatic alterations sometimes confused the interpretation of the primary enzymatic defect in individual sphingolipidoses. The primary enzyme defect can be identified by studying enzyme activities in (heterozygous) parents' cultured fibroblasts, since nearly all the sphingolipidoses are inherited by autosomal recessive mechanisms. Activity of the primarily affected enzyme, but not enzymes which are altered secondarily, is reduced to about 50% of normal activity in heterozygous parental cells. Measurements of lysosomal enzyme activity in the sphingolipidoses have also sometimes produced misleading results when artificial substrates have been used in enzyme assays. Use of such substrates may simplify laboratory work, but the relevant enzyme may not have the same specificity towards the artificial substrate as towards the natural one.

It should also be realized that secondary alterations in tissue chemical composition may occur in the sphingolipidoses. The component whose metabolism is blocked will accumulate, but the tissue damage in the disorders may itself further alter tissue chemical constitution.

The individual sphingolipidoses will be considered largely in order of decreasing structural complexity of the accumulated lipids. Disturbances of derivatives of ceramide glucoside are dealt with first, then those of ceramide galactoside, and lastly those of ceramide phosphatide.

A number of accounts of the sphingolipidoses are available[30-32].

3.1.2.1 β-Galactosidase deficiency (generalized gangliosidosis)

Inherited deficiency of G_{M1} β-galactosidase leads to accumulation of G_{M1} ganglioside and also keratan-like mucopolysaccharide. Accumulation of the latter adds features of the mucopolysaccharidoses to those of the gangliosidoses. The disorder is considered in the light of a mucopolysaccharidosis in Section 3.3. Detailed descriptions of the disorder are available[33].

Biochemical abnormality – G_{M1} β-galactosidase deficiency causes intralysosomal accumulation of G_{M1} ganglioside and its asialo derivative. The terminal galactose group cannot be split from the G_{M1} ganglioside molecule in the absence of β-galactosidase, and therefore further sequential degradation of the molecule cannot proceed. However, the enzyme N-acetylneuraminidase, which is still present, can split the sialic acid (NANA, i.e. N-acetylneuraminic acid) residue from the molecule, yielding the corresponding asialoganglioside.

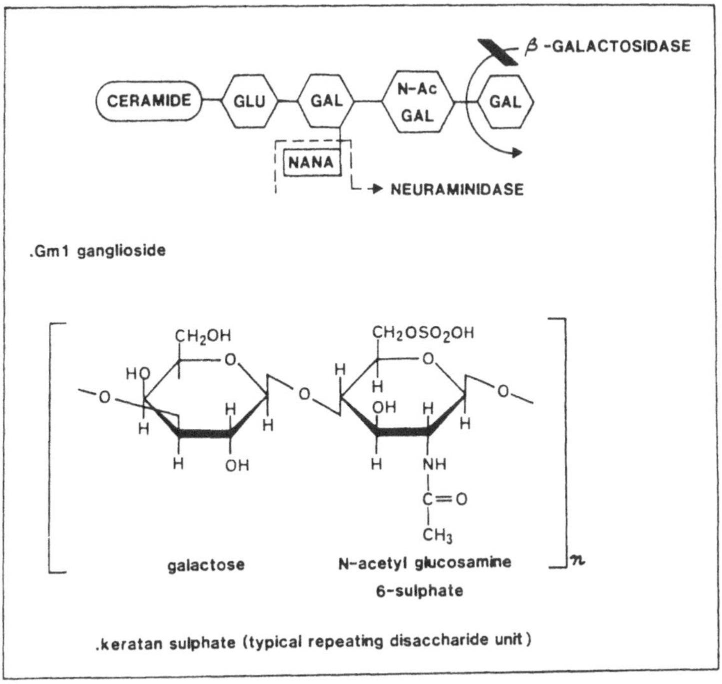

.Gm1 ganglioside

.keratan sulphate (typical repeating disaccharide unit)

Several G_{M1} β-galactosidase isoenzymes are known, and the relative severity of their deficiencies may explain variants of generalized gangliosidosis. O'Brien[34] suggested that the enzyme G_{M1} β-galactosidase had different catalytic sites for its activity towards β-galactosidic bonds in G_{M1} ganglioside, glycosaminoglycans and glycoproteins, and that the clinical expression of deficiency of the enzyme was determined by which of these sites was altered by mutation in G_{M1} gangliosidosis. Farrell and MacMartin[35] demonstrated multiple molecular forms of mutant acid β-galactosidase in affected members of a single family with G_{M1} gangliosidosis. The G_{M1} β-galactosidase isoenzymes deficient in generalized gangliosidosis are not those which cleave β-galactosidic bonds in certain other ceramide-hexoside compounds. Hence ceramide-galactoside and ceramide lactoside do not accumulate in generalized gangliosidosis, though they do accumulate when the different β-galactosidases responsible for their own cleavages are deficient (Section 3.1.2.11).

Keratan sulphate, a mucopolysaccharide (glycosaminoglycan) also accumulates in G_{M1} gangliosidosis. Keratan sulphate is a polymer consisting of repeating galactose and *N*-acetyl glucosamine-6-sulphate units. It is a constituent of certain glycoproteins. Failure of *β*-galactosidase to cleave the terminal galactose unit from the keratan sulphate polymer prevents the further sequential degradation of the macromolecule. Hence keratan sulphate accumulates in the nervous system and viscera in *β*-galactosidase deficiency. Various oligosaccharides also accumulate because of failure to cleave *β*-galactosidic links (see Section 3.3). It appears that the *β*-galactosidase related to generalized gangliosidosis can catalyse the cleavage of both *β*(1 → 3) and *β*(1 → 4) galactosidic bonds. The portion of the keratan sulphate polymer which accumulates in generalized gangliosidosis has the structure shown below.

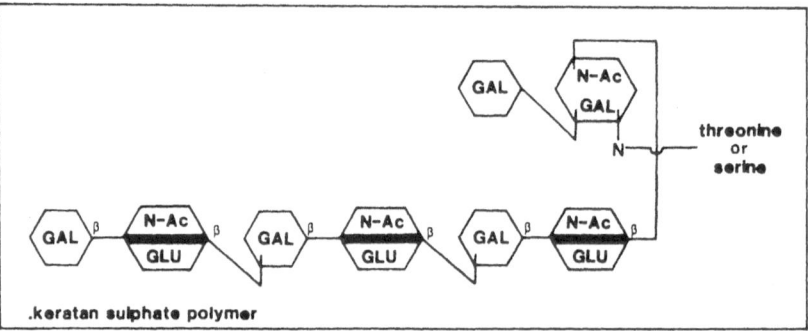

It is, in fact, the region of the molecule by which keratan sulphate is linked to the polypeptide chains of certain glycoproteins. The main oligosaccharide stored in the disorder has the following structure:

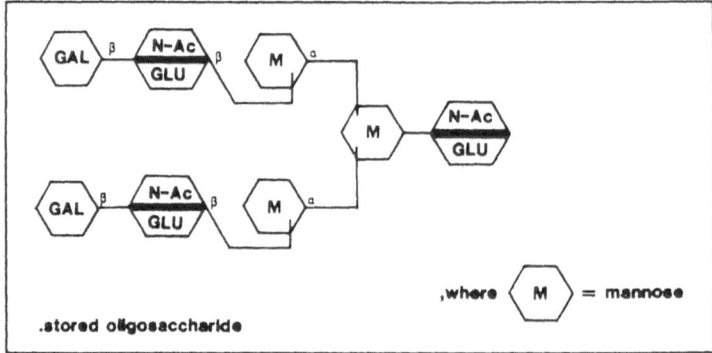

Other galactose-containing glycoproteins also accumulate. How the accumulation of ganglioside and keratan sulphate leads to tissue dysfunction in generalized gangliosidosis is uncertain.

146

Aetiology – The disorder is inherited as an autosomal recessive trait. β-Galactosidase activity is virtually absent in the infantile form, and greatly reduced in the later onset form of the disorder. Both forms have been reported in the one family, an observation supporting the allelic nature of the different clinical forms of the disease[36].

Structural pathology – The heart, liver, spleen and kidneys are enlarged and, particularly in older cases, the brain is atrophic, the cerebral ventricles dilated, and the cerebral white matter often severely demyelinated[30,37]. PAS positive and sudanophil material is stored in neurons, glia, reticuloendothelial, liver and renal cells. The material in neurons is mainly ganglioside, and in glia mucopolysaccharide. At electron microscopy mucopolysaccharide is stored in clear vacuoles in visceral cells and in glia, while the ganglioside accumulates as 'zebra' bodies and membranous cytoplasmic bodies in neurons.

Clinical features – In the infantile variety (Type 1) symptoms may appear soon after birth, with death by the age of 1.5–2 years. There are gargoyle-like features related to mucopolysaccharide accumulation, e.g. bone deformity, facial dysmorphism, macroglossia, abdominal distension. Hepatosplenomegaly develops. Vacuoles occur in lymphocytes and bone marrow cells. Neurologically, there is psychomotor regression which proceeds to spastic tetraparesis, clonus, convulsions, blindness and deafness. About 50% of cases have cherry-red macular changes. A later onset, milder variety (Type II) of the disorder is known, beginning between 6 and 18 months of age. There are no facial, visceral or bony changes and cases may survive to 3–10 years of age. Progressive dementia with seizures and spastic tetraparesis and finally decerebration occur, but vision is not affected. Usually the liver and spleen are not enlarged. Bone changes are less severe than in the infantile variety. Other rare and sometimes mild variants of the disorder are known[38-40].

Diagnosis
(1) Clinical: The gargoyle-like features would suggest a mucopolysaccharoidosis, but if a cherry-red macular appearance were also present, associated gangliosidosis might be suspected, and therefore G_{M1} β-galactosidase deficiency.
Cherry-red changes at the macula seem to occur in disorders in which the stored material is

(a) Sphingomyelin (in Niemann–Pick disease),
(b) Ganglioside (in the gangliosidoses),
(c) Sialo-oligosaccharide (in the sialidoses).

In later onset cases of generalized gangliosidosis lacking mucopolysaccharidosis-type features the various types of gangliosidosis and other forms of early life dementia would enter the differential diagnosis.

(2) Laboratory: Findings such as lymphocyte vacuolation and the presence of

147

membranous cytoplasmic bodies and 'zebra' bodies in ganglion cells in rectal biopsy specimens would suggest the diagnostic category of gangliosidosis. This diagnosis can also be made from the ultrastructural appearance of skin biopsy specimens[41]. The usual methods for demonstrating acid mucopolysaccharides in urine do not detect keratan sulphate, since this contains no hexoseuronic acid. Exact diagnosis of β-galactosidase deficiency depends on demonstrating large amounts of ganglioside G_{M1}, the corresponding asialoganglioside, and also keratan sulphate, in cultured skin fibroblasts or other tissue cells obtained at biopsy. The natures of these materials can be determined by histochemical methods or by chemical techniques, usually thin layer chromatography. Deficient β-galactosidase activity (at pH below 3.6) can also be demonstrated in these tissue specimens, in urine, or in amniotic fluid cells (so that antenatal diagnosis is possible). There is often a secondary increase in the activities of other lysosomal hydrolases.

Therapy – Only symptomatic treatment is available. Progression of the disorder cannot be prevented.

3.1.2.2 *Hexoseaminidase A deficiency (Tay–Sachs disease)*

Tay–Sachs' disease (amaurotic familial idiocy) is the most common of the neurosphingolipidoses. This disorder is due to an inherited deficiency of the enzyme hexoseaminidase A. The main material which accumulates is G_{M2} ganglioside. Reviews of the subject are available[38].

Abnormal biochemistry – Hexoseaminidase A is a hexoseaminidase isoenzyme which catalyses the hydrolytic cleavage of the terminal *N*-acetylgalactosamine moiety from ganglioside G_{M2} (Figure 3.6). There is also a B isoenzyme which cannot catalyse this hydrolysis, and which is functionally intact in Tay–Sachs disease. The substrate specificities of the isoenzymes are discussed further in relation to combined hexoseaminidase A and B deficiency (Section 3.1.2.3). In the absence of the A isoenzyme G_{M2} ganglioside cannot be degraded. It accumulates, as does asialo G_{M2}, since the sialic acid moiety can still be cleaved from the ganglioside by neuraminidase. It is unclear why asialo G_{M2}

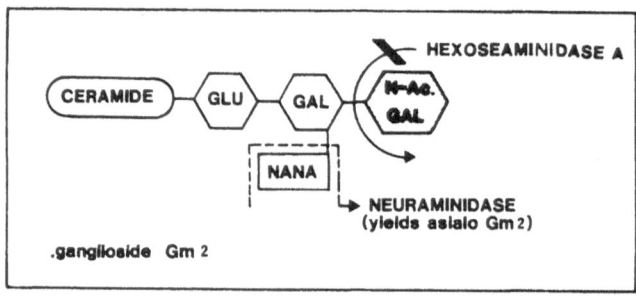

should accumulate, since hexosaminidase B can cleave N-acetylgalactosamine from this molecule. G_{D2} and G_{D1a} gangliosides also accumulate because their N-acetylgalactosamine residues cannot be split off. The exact relation between the ganglioside accumulation and the production of the neurological disturbance is uncertain.

Aetiology – Tay–Sachs' disease is inherited as an autosomal recessive trait. In the infantile form the enzyme defect is almost total; in the juvenile form the deficiency is partial. The biochemical genetics are discussed by Srivistava and Awasthi[39].

Structural pathology – Changes are confined to the nervous system[30,37,42]. In cases who die early, the brain appears atrophic. In those who die later there is megalencephaly (from ganglioside accumulation), with atrophy of the brain stem and cerebellum. The cerebral white matter becomes swollen and multicystic with widespread demyelination and gliosis. Neurons are grossly distended with stored material. Retinal ganglion cells in the macular region are swollen and necrotic and there is optic nerve atrophy. Neurons in the rectal wall are ballooned. At electron microscopy intralysosomal membranous cytoplasmic bodies occur in cells distended with the stored material. These bodies are largely composed of ganglioside G_{M2}.

Clinical features – Manifestations of the typical infantile variety of Tay–Sachs' disease, with almost total absence of hexoseaminidase A activity, often begin between the age of 3 and 7 months, and occur mainly in Jewish children. The infant becomes apathetic, hypotonic, malnourished and retarded. Vision deteriorates and motor abilities decline. Spasticity, contractures and quadriplegia develop. Saliva drools from the mouth, pyramidal signs appear, and there is optic atrophy with a cherry-red macular appearance. At about the age of 2 years head enlargement is apparent. The e.e.g. shows a diffuse excess of high amplitude slow activity, with spike transient at times. There may be foam cells in the CSF. Visceral enlargement does not occur.

A juvenile type of the disorder occurs, due to partial hexoseaminidase A deficiency. The onset is at 5–7 years of age. There is mental deterioration and rapidly progressive visual failure over a period of about 2 years. The optic disc appears yellow grey and retinitis pigmentosa develops. Affected children deteriorate to the level of idiots and have epilepsy, athetosis, dysarthria, signs of pyramidal tract insufficiency and muscle atrophy and contracture. Rarely, more indolent variants of the disorder occur, manifested mainly as a spinocerebellar deficit[43].

Diagnosis
(1) Clinical: Progressive mental deterioration in an infant, with visual failure and a cherry-red macular change is most probably due to G_{M2} gangliosidosis. If the disorder is known to be present in the patient's siblings,

the diagnosis is virtually certain. Before the full picture has evolved, and in juvenile forms, the differential diagnosis may be rather wider. The presence of retinitis pigmentosa in the juvenile form suggests a lipid storage disorder, and may raise the possibility of ceroid lipofuscinosis.

(2) Laboratory: Hexoseaminidase A levels can be measured in serum and also in leukocytes. The enzyme is virtually absent in homozygous patients, and serum levels are reduced in heterozygous carriers, who can thus be detected and given genetic advice. Serum hexoseaminidase assays may yield false positive carrier identification, but the leukocyte assay appears more reliable. Total serum hexoseaminidase activity is little reduced in Tay–Sachs' disease; only the thermolabile A isoenzyme is affected. Prenatal diagnosis is possible by measuring hexoseaminidase A activity in cultured amniotic fluid cells, or in the fluid itself. Levels of G_{M2} ganglioside in CSF are raised in the disease[44]. It is unlikely that the nature of the stored ganglioside in brain would be determined during life, as this would require brain biopsy. Electron microscopy of biopsied rectal wall ganglion cells may show the membranous cytoplasmic bodies suggesting storage of one or other of the gangliosides.

Therapy – A variety of attempts has been made to replace the defective enzyme, thus far without success[31,45]. Genetic counselling for affected families is imperative.

3.1.2.3 Total hexoseaminidase deficiency (Sandhoff–Jackwitz disease; Tay–Sachs variant disease)

This is a less common disorder than Tay–Sachs' disease. It is due to combined deficiency of hexoseaminidases A and B. The disorder presents as a rapidly progressive, Tay–Sachs' like condition. Both G_{M2} ganglioside and globoside accumulate[46].

Abnormal biochemistry – The disorder is due to combined deficiency of hexoseaminidases A and B, though an abnormal hexoseaminidase (hexose-aminidase S) is present[47]. Hexoseaminidase A catalyses the hydrolytic cleavage of N-acetylgalactosamine from G_{M2} ganglioside, from asialo G_{M2}, and from globoside molecules. The B isoenzyme catalyses the cleavage from asialo G_{M2} and from globoside, but not from G_{M2}. Therefore, in combined deficiency of hexoseaminidase A and B, G_{A2} globoside accumulates, as well as G_{M2} ganglio-

CERAMIDE—GLU—GAL—GAL—N-Ac GAL   HEXOSEAMINIDASES A and B

.globoside

side and its asialo-derivative. The storage of the additional material seems to be associated with an accelerated clinical course of the disorder as compared with Tay–Sachs' disease. Accumulated oligosaccharide fragments (from glyco-proteins) may also be isolated from the brain in this disorder[48]. The structure of one of the fragments is shown in Figure 3.8 (p. 174).

Aetiology – The disorder is inherited as an autosomal recessive trait.

Structural pathology – Pathological appearance is generally similar to that of Tay–Sachs' disease[30] though some ganglioside and globoside accumulate in the viscera and some vacuolated histiocytes are present. Membranous cyto-plasmic bodies in neurons may be seen at electron microscopy.

Clinical features – The disorder has less predilection for persons of Jewish origin than Tay–Sachs' disease, but is otherwise similar to it clinically, with typical cherry red spots at the maculae. Rarely, instances of the disorder have begun in later childhood[49], and possibly in adult life[50].

Diagnosis
(1) Clinical: Differentiation from Tay–Sachs' disease is not possible clinically.

(2) Laboratory: Total hexoseaminidase (A plus B) activity is only about 10% of normal in serum, cultured fibroblasts, amniotic fluid and cultured amniotic fluid cells. This finding is diagnostic. At electron microscopy of biopsied rectal wall the presence of membranous cytoplasmic bodies in ganglion cells merely suggests the diagnostic category of gangliosidosis. The stored lipids can be classified by thin layer chromatographic studies on brain biopsy specimens. Raised levels of gangliosides G_{M2} and G_{A2} are found in CSF^{44}.

Therapy – No useful treatment is known.

3.1.2.4 *Hexoseaminidase B deficiency*

Johnson and Chutorian[51] reported a mild, juvenile onset, very slowly progres-sive disorder featuring cerebellar ataxia and cherry red macular changes. Serum, leukocytes and cultured fibroblasts lacked hexoseaminidase B activity, but hexoseaminidase A and S activity was present. Enzyme studies in family members suggested the disorder was inherited as an autosomal recessive trait.

3.1.2.5 N-*acetylgalactoseamine transferase deficiency* (*G_{M3} gangliosidosis*)

This disorder, of which only a solitary instance is known[6,52,53] is the only

151

ganglioside accumulation disorder that appears due to a failure of ganglioside biosynthesis. The relevant synthetic step appears to be the reverse of the degradative one.

Deficiency of the enzyme G_{M3} UDP N-acetylgalactosaminyl transferase caused failure to convert ganglioside G_{M3} to its higher homologues. The latter were deficient in the disorder, and G_{M3} accumulated.

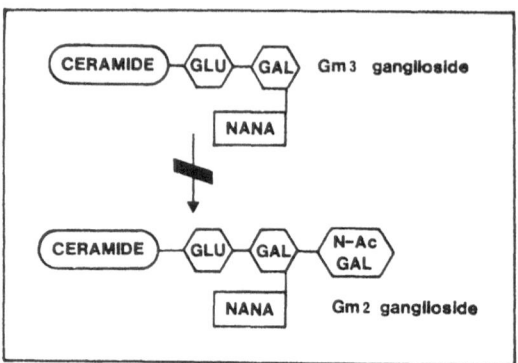

The one reported patient had a progressive neurological deterioration, with epilepsy, and died aged 14 weeks. At autopsy there were no storage vacuoles or inclusion bodies. There was spongy degeneration of the subcortical white matter, brain stem, long tracts and optic nerves. Astrocytic processes were greatly swollen.

Presumably failure of ganglioside synthesis would interfere with the membrane structure of axon terminals and therefore with synaptic function, and thus might disturb the development and function of the nervous system.

3.1.2.6 Ceramide lactoside lipidosis (lactosyl ceramidosis)

Only a single case of this disorder so far appears to be known. Reduced activity of lactosyl ceramidase, i.e. lactosyl ceramide β-galactosidase (neutral β-galactosidase), in liver and cultured skin fibroblasts was associated with accumulation of ceramide lactoside. There was progressive intellectual failure with cerebellar ataxia, epilepsy and spasticity which began in a 2.5 year old child. The liver and spleen were enlarged.

It should be noted that two other β-galactosidases, the variety deficient in Krabbe's disease and that deficient in generalized G_{M1} gangliosidosis, both have the capacity to hydrolyse ceramide lactoside[54]. In the present state of knowledge it is difficult to see why these enzymes did not cleave ceramide lactoside when the specific neutral β-galactosidase was deficient. The question of substrate specificities relative to the effects of deficiencies of the various β-galactosidases needs further clarification. In fact, the existence of lactosyl ceramidosis as a genuine entity is questioned[39].

3.1.2.7 Ceramide trihexoside α-galactosidase deficiency (Fabry's disease)

Fabry's disease results from ceramide trihexoside α-galactosidase deficiency[55,56]. The resulting accumulation of ceramide trihexoside occurs chiefly in extraneural tissues. Neurological disturbance in Fabry's disease is mainly in relation to peripheral structures, or is a consequence of vascular occlusion. Over 250 cases are known[56].

Abnormal biochemistry – The lysosomal enzyme ceramide trihexoside α-galactosidase cleaves the terminal galactose unit (which has an α-glycosidic link to the next galactose unit) from the ceramide trihexoside molecule. This molecule is itself derived from globoside. There are two α-galactosidase isoenzymes, but only the A type is absent in Fabry's disease. The main isoenzyme in brain is the B type. Deficiency of the A isoenzyme therefore causes accumulation of ceramide trihexoside, mainly in extraneural tissues. Substantial amounts of the lipid may be derived from breakdown of red blood cell membranes. Some ceramide digalactoside also accumulates.

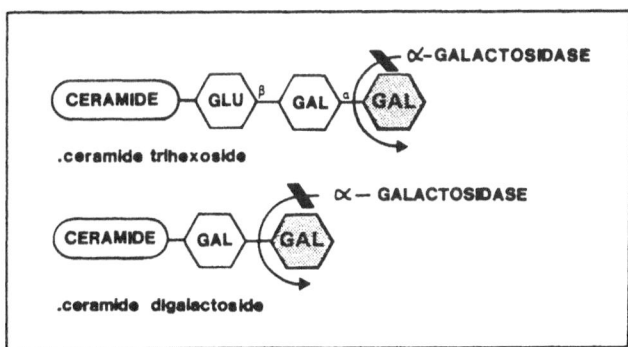

Aetiology – The disorder is inherited as an X-linked recessive trait. X^1X female carriers have mild manifestations of the disorder.

Structural pathology – There is lipid storage in certain groups of neurons, mainly in the autonomic nervous system (including rectal ganglion cells), hypo-thalamus, substantia nigra and certain other brain stem nuclei, but not in the cerebral cortex or cerebellum[30,57]. At electron microscopy the stored material is seen to occur in pleomorphic lamellae in lysosomes or to lie free in the cytoplasm. Foam cells occur in the reticuloendothelial system, while blood vessel walls are thickened due to stored lipid. Cutaneous telangiectasias (angio-keratomas) occur. Glycosphingolipid accumulation in the kidneys is noticeable.

Clinical features – Characteristic skin lesions (angiokeratomas) often occur on the hips, back, thighs and buttocks and appear early in the course of the disorder. There is slight corneal opacity. Painful paraesthesiae (which may be associated with 'crises' of severe pain), vasomotor changes, impaired sweating, fever and anaemia develop. Associated renal and cardiac disease may appear. About one third of cases have acute central nervous system symptoms (e.g. headache, dizziness, cranial nerve palsies, cerebellar disturbance) resulting from cerebral vascular disease. Motor nerve conduction is sometimes slowed[58]. The full clinical picture occurs only in males. Female heterozygotes may exhibit a mild form of the disorder.

Diagnosis
(1) Clinical: In males the full clinical picture, associated with the characteristic skin lesions, is diagnostic. Careful search for the skin lesions may permit the diagnosis in heterozygous females.

(2) Laboratory: Characteristic ultrastructural changes occur in skin bio-psies[41]. There are raised levels of ceramide trihexoside in plasma and in urine. The relevant α-galactosidase levels in plasma, leukocytes, tears, biopsy specimens or cultured skin fibroblasts are greatly reduced in affected males, and less reduced in heterozygous female carriers. Measurement of the stored lipid in biopsy specimens would be feasible. Prenatal diagnosis is possible by measuring α-galactosidase activity in amniotic fluid.

Therapy – The anticonvulsant phenytoin relieves the painful episodes in Fabry's disease. No specific therapy for the metabolic abnormality is available. Genetic counselling is advisable.

3.1.2.8 Glucocerebroside β-glucosidase deficiency (Gaucher's disease)

Gaucher's disease is the most frequently occurring sphingolipidosis, but the nervous system is involved in only the infantile form of the disease. Deficient β-glucosidase (glucosylceramide β-glucosidase) activity causes glucosecerebro-side (ceramide glucoside) accumulation in extraneural (mainly reticuloendo-thelial) tissues. The subject has been reviewed by Schettler and Kahlke[59] and Brady[60].

Abnormal biochemistry – Glucosecerebroside β-glucosidase catalyses the cleavage of ceramide glucoside. This lipid is derived mainly from globoside which is released from the breakdown of effete red and white blood cells. When the β-glucosidase enzyme is deficient the cerebroside accumulates, chiefly in reticuloendothelial cells. Some studies have also shown an increased brain content of ceramide glucoside. This may be derived from breakdown of ganglio-sides. Ganglioside turnover is higher in infancy than in later life, which may explain the tendency for neurological involvement to occur only in the infantile form of Gaucher's disease.

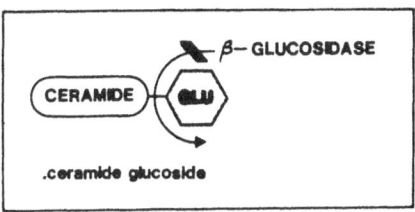

Aetiology – β-Glucosidase deficiency is inherited as an autosomal recessive trait. The adult form of the disorder has a predilection for persons of Jewish origin.

Structural pathology – Characteristic distended foamy Gaucher's cells occur in the reticuloendothelial system where they are associated with liver, spleen and lymph node enlargement[30,42]. In the infantile form of the disease the brain is atrophic, with shrunken neurons. Lipid accumulation in neurons has been reported and some Gaucher's cells may occur in the Virchow–Robin spaces of the brain.

Clinical features – The disorder tends to occur in persons of Jewish origin. One third of cases begin in infancy or childhood. In cases beginning before the age of 6 months there may be feeding difficulty, weight loss, general under-development, squint, opisthotonus, spasticity and severe mental retardation. Hepatosplenomegaly occurs. If the onset is later in childhood, or in adult life, the disease is much milder. Neurological involvement is uncommon in cases beginning after 6 months of age, and does not occur in adult cases. The dominant features of later-onset cases are liver, spleen, and lymph node enlarge-ment, with bone pain in childhood. Radiological changes may be present in the long bones and vertebrae. The skin becomes pigmented and pingueculae occur, but there are no retinal abnormalities. Anaemia and thrombocytopenia may be present. Electroencephalographic features of the disorder are described by Nishimura *et al.*[61]

Diagnosis
(1) Clinical: There are no features specific to the early infantile form which would allow a definitive clinical diagnosis, but the presence of progressive

155

deterioration in cerebral function associated with hepatosplenomegaly raises the diagnostic possibility.

(2) Laboratory: Diagnostic Gaucher's cells may be found in biopsy specimens taken from different sites, e.g. liver, lymph nodes, bone marrow and rectal wall (ganglion cells). Acid phosphatase levels in the plasma are high[62], though this is supportive rather than decisive evidence. The diagnostic decrease in β-glucosidase activity can be measured in peripheral blood, in leukocytes, or in tissue (biopsy) specimens. However, if artificial substrates are used, rather than ceramide glucoside, activity of other β-glucosidases may be measured as well and may confuse the interpretation. The accumulated lipid in Gaucher's disease may be identified and quantitated by thin layer chromatography. The enzyme abnormality and lipid identification are diagnostic.

Therapy – There is no established effective therapy. Enzyme replacement has been attempted, without proven benefit.

3.1.2.9 Arylsulphatase A deficiency: metachromatic leukodystrophy

Hereditary arylsulphatase A isoenzyme deficiency causes the accumulation of metachromatic ceramide galactoside sulphate in various tissues, with disturbance of brain and peripheral nerve function. The disorder is one of the more frequent neurosphingolipidoses, and is discussed extensively in several places in the literature[63,64].

Abnormal biochemistry – Lysosomal arylsulphatase A isoenzyme (cerebroside sulphatase or cerebroside sulphate sulphohydrolase) catalyses the hydrolytic cleavage of certain cerebroside sulphate esters. Farrell *et al.*[65] have shown that there are multiple molecular forms of arylsulphatase A, and that the deficient molecular forms are different in late infantile and juvenile onset cases of metachromatic leukodystrophy. When the sulphatase is absent various sulphatides (notably ceramide-galactoside sulphate, and some ceramide-galactoside-hexosyl sulphates) accumulate in myelin and in other tissue components. Sulphatides formed in the kidneys may spill over into the urine. The altered

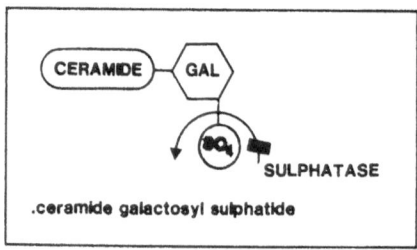

.ceramide galactosyl sulphatide

myelin composition leads to demyelination. The accumulated sulphatides are metachromatic, i.e. they stain red with certain blue dyes, or yellow-brown with certain violet dyes.

Brain cerebroside and sphingomyelin composition is altered in metachromatic leukodystrophy. There is a relative decrease in the content of longer chain (C_{21-26}) fatty acids in these sphingolipids. This may merely be due to demyelination. The fatty acid composition of the accumulated sulphatides is normal.

Aetiology – The disorder is inherited as an autosomal recessive trait. Arylsulphatase A protein is present in affected patients, but lacks enzymatic activity, presumably due to an intramolecular defect.

Structural pathology – Accounts of the pathology are given by Chrome and Stern[30] and Poser[66]. There is diffuse cerebral atrophy, with spongy degeneration and gliosis of the cerebral white matter. Widespread demyelination occurs, sparing the subcortical U-fibres. No sudanophil myelin breakdown products (neutral fats) are present. Neurons in the retina, in spinal grey matter, in the dentate nuclei and parts of the brain stem, and also Schwann cells and macrophages, all show intracytoplasmic storage of metachromatic material, and may become distended with this material. Oligodendroglial loss may be severe. Peripheral nerves show segmental demyelination with onion bulb formation[67,68]. Metachromatic deposits occur in the gall bladder wall, in renal tubule cells and in endocrine tissues.

Clinical features – The disorder may appear at any age from infancy to adult life. It usually runs its course over a few years. There is progressive dementia with a flaccid di- or tetra-plegia, ataxia, tremor and limitation of external eye movements. Later spasticity, myoclonus, epilepsy, optic atrophy, dysarthria and deafness develop, with finally a state of decerebrate idiocy. Peripheral nerve conduction velocity is reduced, the e.e.g. is diffusely abnormal, and CSF protein levels are raised. The CT head scan appearances of this disorder are described by Buonanno *et al.*[69]. Metachromatic material may be found in the urine at microscopy.

Diagnosis
(1) Clinical: There are no diagnostic clinical features and the differential diagnosis is that of progressive dementia in early life.

(2) Laboratory: The association of slowed conduction in peripheral nerves and a non-functioning gall bladder in a child or young adult with progressive dementia is very suggestive of the diagnosis. The presence of metachromatic material in urine, or in a sural nerve or brain biopsy, is very suggestive, though not absolutely diagnostic. The increased sulphatide excretion in urine can be measured. The finding of decreased arylsulphatase A activity in serum or urine is decisive. This enzyme may also be measured in cultured

157

skin fibroblasts, amniotic fluid cells (permitting antenatal diagnosis), or in peripheral leukocytes. Some workers have found decreased levels of the enzyme in peripheral leukocytes from heterozygous carriers[70]. If artificial substrates are used for the assay of arylsulphatase A, care must be taken to exclude the presence of other sulphatases, or to inactivate them before the measurement.

Therapy – There is no known useful treatment for affected persons. Genetic counselling may be offered to the family.

3.1.2.10 Multiple sulphatase deficiency

Inherited deficiency of all three arylsulphatases (A, B and C) may occur. As well as cerebroside sulphates (as in metachromatic leukodystrophy), sulphated glycosaminoglycans (mucopolysaccharides), steroids, and also gangliosides, accumulate. The clinical picture resembles that of metachromatic leuko-dystrophy, but there are also bony abnormalities though not the gargoyle-like features that often occur in the mucopolysaccharide storage disorders. The disorder is considered at greater length in Section 3.3.4, where it is regarded as one of the mucolipidoses.

3.1.2.11 Galactocerebroside-β-galactosidase deficiency: Krabbe's globoid cell leukodystrophy

Galactocerebroside-β-galactosidase deficiency causes retention of ceramide galactoside in neural tissue, with severe demyelination and the development of peculiar 'globoid' cells. The disorder is rare, but at least 150 cases are recorded[71].

Abnormal biochemistry – Galactocerebroside-β-galactosidase is a lysosomal enzyme which specifically cleaves the ester bond between ceramide and galac-tose in galactocerebroside (ceramide galactoside). It differs in substrate specifi-city from the β-galactosidase (Section 3.1.2.1) which is deficient in generalized gangliosidosis[54]. Philippart[72] and Dickerman *et al.*[73] have discussed the pos-sibility that β-galactosyl sphingosine (psychosine), which is probably also degraded by galactocerebroside-β-galactosidase, may accumulate in Krabbe's disease. Psychosine is a known cytotoxin. In deficiency of galactocerebroside-β-galactosidase, galactocerebroside does accumulate in nervous tissue, but almost entirely in the globoid cells and not elsewhere[73]. Ceramide galactosides in humans are largely confined to myelin, so that their altered catabolism affects mainly the nervous system, where myelin becomes prone to premature breakdown.

158

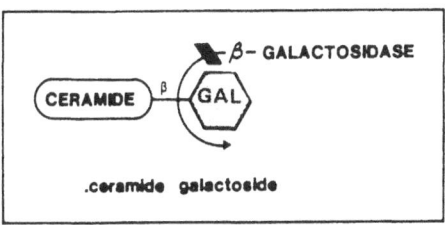

Aetiology – Galactocerebroside-β-galactosidase deficiency is inherited as an autosomal recessive trait.

Structural pathology – The pathological appearances of the disorder have been described by Chrome and Stern[30] and Austin[74]. The cerebrum shows diffuse atrophy with widespread severe demyelination which tends to spare the subcortical U fibres. There is oligodendroglial loss and gliosis. Peripheral nerves are demyelinated. The brain contains globoid cells, macrophages distended with PAS positive material (ceramide-galactoside) and globoid bodies in nearly all, but not in all, cases[75]. No extraneural abnormality occurs.

Clinical features – Symptoms nearly always commence in infancy, rarely later. Affected infants are irritable and stiff, and have episodes of fever. Progressive motor and mental deterioration develops, and leads to spasticity, decerebration and blindness. A peripheral neuropathy is present. The CSF protein content is usually high. There are no extraneural manifestations of the disorder.

Diagnosis
(1) Clinical: In the absence of an established family history, the clinical picture alone is unlikely to permit a definitive diagnosis. Laboratory investigations are required to differentiate the disorder from other causes of dementia in infancy.

(2) Laboratory: Severely reduced galactocerebroside-β-galactosidase levels are found in serum, leukocytes, cultured fibroblasts and amniotic fluid cells (so that prenatal diagnosis is possible). Heterozygous carriers have a lesser degree of reduction in activity of this enzyme, which may be within the normal range in some 50% of obligate carriers[76]. It should be noted that the enzyme must be measured using the natural substrate[39]. Artificial substrates (e.g. β-nitrophenyl-β-galactoside) can produce misleading results. There may be a secondary decrease in cerebroside sulphotransferase activity and other, non-specific, β-galactosidases may exist in the brain[54]. The abnormal stored material in Krabbe's disease can be demonstrated histochemically in brain biopsies, though it gives the same reactions as the ceramide-glucoside which is stored in Gaucher's cells. The nature of the stored cerebroside can be demonstrated by thin layer chromatography.

Therapy – No known therapy is of established benefit.

3.1.2.12 *Sphingomyelinase deficiency (Niemann–Pick disease)*

Sphingomyelinase deficiency causes sphingomyelin accumulation in neural and extraneural tissues, producing a progressive cerebral degeneration with visceral enlargement[77,78]. The disorder is one of the more common sphingolipidoses.

Abnormal biochemistry – Sphingomyelinase catalyses the hydrolytic cleavage of spingomyelin into its components ceramide and phosphatidylcholine. Greatly reduced sphingomyelinase activity in Niemann–Pick disease causes sphingomyelin accumulation in brain, abdominal viscera and other tissues, including the bone marrow and retina. The stored lipid distends neurons and reticuloendothelial cells. Ceramide glucoside and gangliosides G_{M3} and G_{M2} also accumulate to some extent. The biochemical basis for the accumulation of these substances is uncertain. How the sphingomyelin and other lipid accumulation causes the neurological disturbances of the disorder is also unclear.

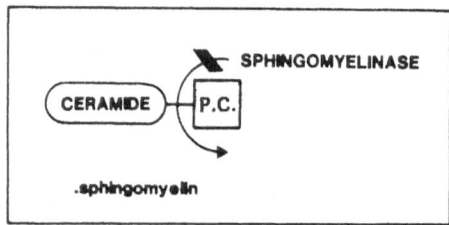

Aetiology – Sphingomyelinase deficiency is inherited as an autosomal recessive trait.

Structural pathology – The changes are described by Crome and Stern[30] and Roizin and Kaufman[42]. The cerebral white matter has a leathery feeling, and there is abnormal softness of the grey matter. There is neuronal loss, with demyelination. Surviving neurons are distended with lipid. Lipid storage also occurs in reticuloendothelial cells. Niemann–Pick cells contain lipofuscin or ceroid material, which is brown coloured and PAS and Sudan Black B positive. Its origin is obscure. Vacuolated lymphocytes and monocytes occur in the peripheral blood.

Clinical features – The disorder usually appears, often in children of Jewish ancestry, in the first year of life. There is hepatosplenomegely and lymphadeno-pathy, with retarded somatic and mental development. Spasticity, incoordina-tion, athetosis, tremor, seizures, cranial nerve palsies and deafness appear. About one case in three has cherry red maculae. The final mental state in affected children is that of idiocy. There may be a yellowish skin pigmentation with xanthomas, and mild anaemia may occur. Occasional cases begin in later childhood, or in adult life. Sometimes there is no neurological disturbance.

Diagnosis

(1) Clinical: Progressive dementia in an infant with hepatosplenomegaly suggests Niemann–Pick or Gaucher's disease. The association of gargoyle-like features may suggest the diagnosis of G_{M1} gangliosidosis.

(2) Laboratory: Characteristic ultrastructural changes occur in skin biopsy specimens[41]. Chemically, Niemann–Pick disease may be diagnosed by finding an increased sphingomyelin content in liver, bone marrow or spleen biopsies. Sphingomyelinase activity is decreased in leukocytes, cultured skin fibroblasts and cultured amniotic fluid cells. Reduced sphingomyelinase activity in leukocytes or cultured skin fibroblasts permits heterozygous carrier detection.

Therapy – No treatment benefits the patient; genetic counselling for the family is advisable.

3.2 GLYCOSAMINOGLYCAN DISORDERS: THE MUCOPOLYSACCHARIDOSES

The mucopolysaccharidoses have undergone some reclassification in recent times, as details of glycosaminoglycan (mucopolysaccharide) chemistry have become clearer. The glycosaminoglycans are long chain polyanionic carbohydrates comprising recurrent polymeric units of pairs of sugars (hexoses). One member of each disaccharide pair is a hexosamine, usually with its amino group sulphated or acetylated, and the other member is either D-glucuronic acid or L-iduronic acid. The chemical structures of the glycosaminoglycans are described in some detail by Kennedy[79] and Lindahl and Hook[80]. The glycosaminoglycans are soluble in water and in formaldehyde. They occur chiefly in connective tissues, where they normally exist bound to proteins, thus forming proteoglycans (glycoproteins). The following glycosaminoglycans are involved in the mucopolysaccharide storage diseases.

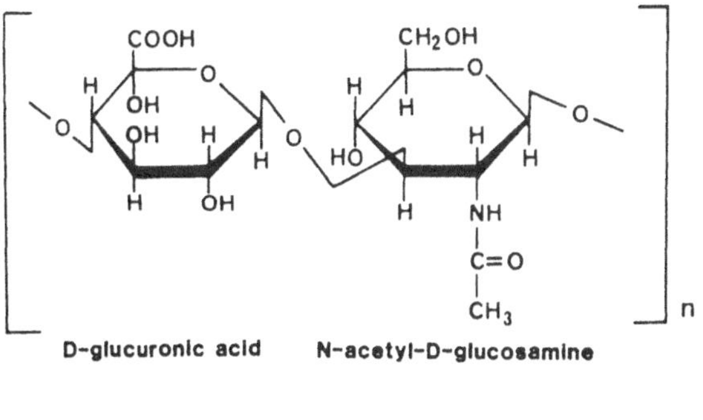

D-glucuronic acid   N-acetyl-D-glucosamine

. hyaluronic acid

161

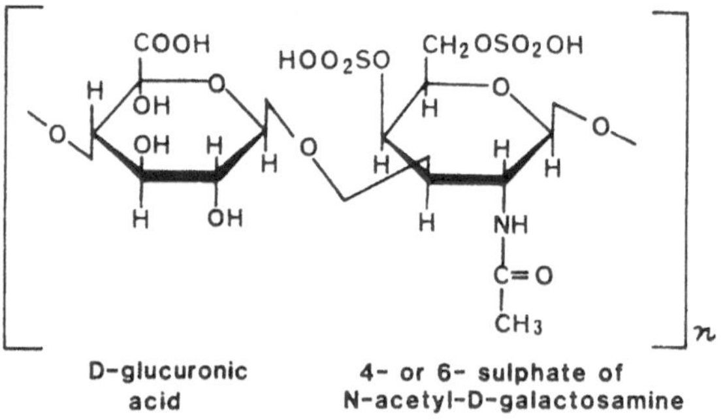
COOH
HOO2SO
CH2OSO2OH
D-glucuronic acid
4- or 6- sulphate of N-acetyl-D-galactosamine

.chondroitin 4- or 6- sulphate

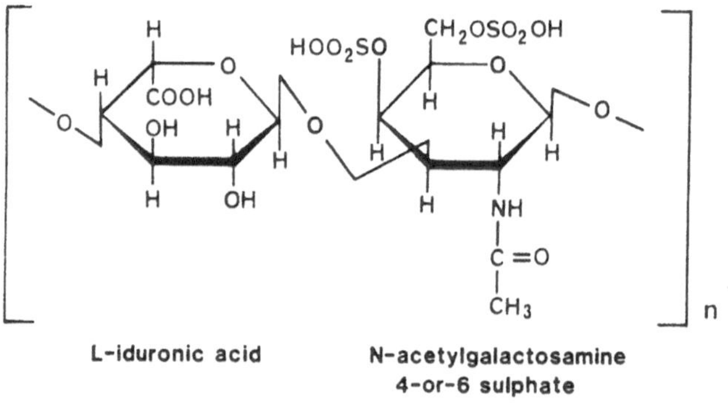
HOO2SO
CH2OSO2OH
COOH
L-iduronic acid
N-acetylgalactosamine 4-or-6 sulphate

.dermatan sulphate

Proteoglycan degradation normally takes place by the molecule being split into a protein residue, glycosaminoglycan polymers and sialo-oligosaccharide portions. The glycosaminoglycan polymers then sequentially break down into their component hexoses. This degradation occurs within lysosomes. Failure of completion of this latter breakdown process causes the various mucopolysaccharidoses. Disturbances in sialo-oligosaccharide degradation causes other patterns of disorder, dealt with later in this chapter (Section 3.3).

Characteristic clinical signs of the mucopolysaccharidoses include coarse facies, dwarfism, hyperteleorism, skeletal abnormalities, cardiac disorders and, in many of the syndromes, severe mental retardation. Incompletely degraded

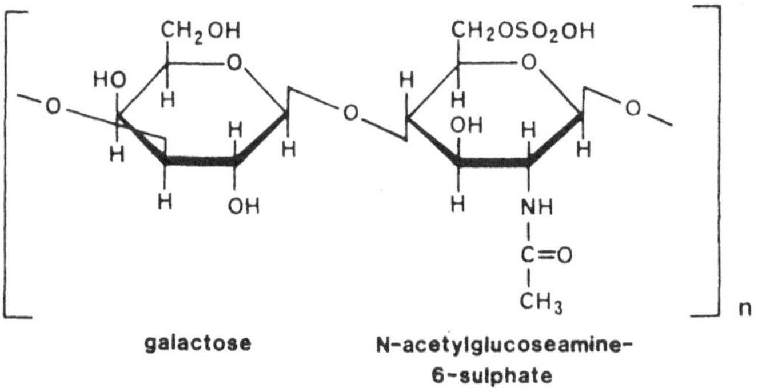

galactose   N-acetylglucoseamine-
6-sulphate

.Keratan sulphate

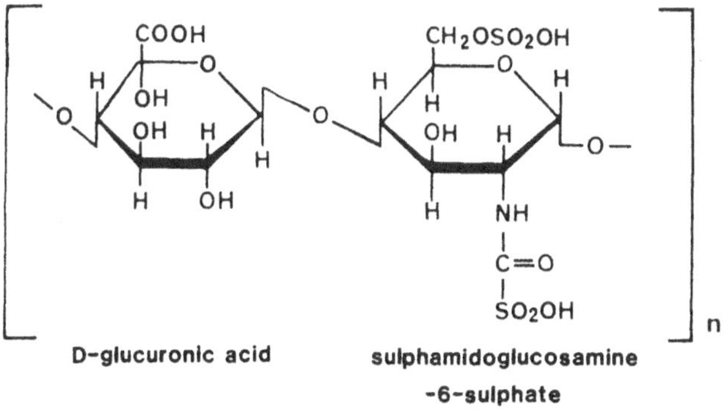

D-glucuronic acid   sulphamidoglucosamine
-6-sulphate

.heparan sulphate

glycosaminoglycan fragments are excreted in urine. Glycosaminoglycan molecules are normally degraded sequentially, hexose after hexose being shed sequentially from the reducing end of the molecule. Once a bond is encountered which should be cleaved by an enzyme which is defective, the polymer fails to break down further. However, in some tissues endoglycosidases are present which may fragment the residual polymers further. The small polymers or oligohexosides resulting from these further degradations also appear in urine. The enzyme deficiency prevents their final breaking down into their component hexoses.

It is uncertain whether the retained glycosaminoglycans themselves disturb neural function. However, it has been suggested that the high negative charge on these molecules may control the entry of water or ions into cells, or cause the molecules to bind (positively charged) biogenic amines. Some evidence suggests that accumulated dermatan sulphate is mainly related to the skeletal abnormalities, and accumulated heparan sulphate to the neurological ones. Normally, very little glycosaminoglycan occurs in the brain (0.02% of brain wet weight in humans). The main glycosaminoglycan polymers normally present in brain are hyaluronic acid, chondroitin sulphate and heparan sulphate.

Following McKusick et al.[81], the mucopolysaccharidoses are classified into six types. However, Type V is currently vacant, the former type V disorder now being regarded as a Type I variant. Since these disorders are due to enzyme defects, and the different glycosaminoglycans have hexose components in common, the one enzyme deficiency may be responsible for the accumulation of more than one glycosaminoglycan. A number of detailed reviews of the mucopolysaccharidoses are available[82-84].

3.2.1 α-L-Iduronidase deficiency: Hurler's disease: type I mucopolysaccharidosis (now termed Type IH)

Hereditary α-L-iduronidase deficiency leads to accumulation of dermatan and heparan sulphates in lysosomes. Mental retardation, facial and skeletal abnormalities develop. The disorder is one of the more common mucopolysaccharidoses.

Abnormal biochemistry – α-L-Iduronidase cleaves the α-glycosidic link between L-iduronic acid and other hexoses, namely N-acetylglucosamine-6-sulphate in dermatan sulphate, and N-sulphated glucosamine-6-sulphate (at sites where L-iduronic acid occurs) in the heparan sulphate polymer. In α-L-iduronidase deficiency degradation of dermatan sulphate and heparan sulphate molecules ceases once a terminal L-iduronic acid residue is encountered, since it cannot be split from the glycosaminoglycan polymer. In certain tissues, including the liver, endoglycosidases can still partly degrade the residual glycosaminoglycan polymer. Incompletely degraded glycosaminoglycan polymers accumulate in tissues and appear in urine. It should be noted that, in contrast to Hurler's disease (Type IH mucopolysaccharidosis), Type Is (formerly V) mucopolysaccharidosis, i.e. Scheie's syndrome, appears to result from a partial deficiency of α-L-iduronidase.

The neurological manifestations of Hurler's disease appear related to the accumulation of various glycosaminoglycans in brain (mainly partly degraded dermatan sulphate). This accumulation appears to lead to a secondary deficiency of a β-galactosidase isoenzyme, which catalyses the hydrolysis of certain terminal β-galactose residues from complex molecules (deficiency of β-galactosidase causes generalized gangliosidosis – G_{M1} gangliosidosis – Section 3.1.2.1). The secondary enzyme deficiency may be due to accumulated

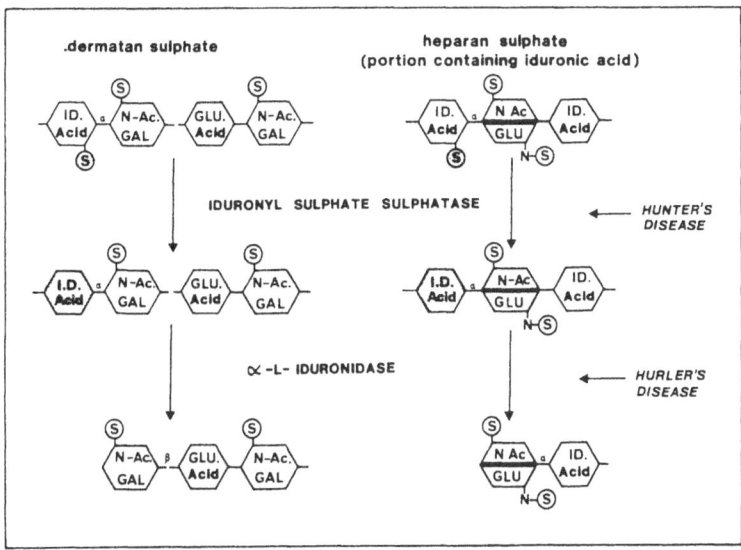

Figure 3.7 The breakdown pathways for L-iduronic acid-containing portions of dermatan sulphate and heparan sulphate polymers

chondroitin sulphate[85]. This β-galactosidase deficiency, together with synthesis of increased amounts of lysosomal membrane, may explain the increased quantities of gangliosides G_{M2} and G_{D3} that are found in the brain in Hurler's syndrome. Whether the mental retardation in the syndrome is related to the altered ganglioside metabolism is conjectural.

Aetiology – α-L-Iduronidase deficiency is inherited as an autosomal recessive trait.

Structural pathology – The pathology of Hurler's disease is described by Crome and Stern[30]. The brain and cerebellum appear atrophic and hydrocephalic, with thickened meninges due to collagen overproduction. Neurons are swollen with a vacuolated cytoplasm yielding positive PAS and Bial reactions, and toludine blue metachromasia. There is little or no demyelination. The liver, heart and spleen are enlarged and contain vacuolated cells with a high content of water-soluble complex carbohydrates. In the heart the endocardium and pericardium are thickened and there are areas of coronary artery narrowing. Stored material occurs in the cells of Bowman's membrane of the cornea. Nodules of swollen fibrous tissue replace and deform bone trabeculae.

At electron microscopy, in the liver there are lysosomal vacuoles which are empty except for a fine granular reticulum. 'Zebra' bodies are seen within the vacuoles in neurons.

Clinical features – Manifestations of the disorder appear in childhood. The neurological features of the disease include mental and motor retardation, and deafness. Dwarfism, skull enlargement, hyperteleorism, depressed nasal bridge and forward opening nostrils with a chronic nasal discharge develop. The lips are thick, the tongue protrudes, the gums are hypertrophic and there are poorly formed and widely spaced teeth. There is a short thorax, and a protruberant abdomen with an enlarged liver and spleen. Umbilical and inguinal herniae are frequent. The long bones are short and thick, the vertebrae and ribs deformed, the hands broad and short-fingered and the joints generally tend to assume a position of flexion. Wedge-shaped vertebrae, deformed ribs and other skeletal abnormalities are present. The skin is coarse, dry and rough, and the hair thick and plentiful. There is often cardiac valvular disease, and corneal clouding is invariable. Lymphocytes in peripheral blood may be vacuolated. Few cases survive beyond the first decade of life.

Diagnosis
(1) Clinical: The facial, connective tissue and skeletal deformities of gargoylism are very suggestive of the diagnostic categories of mucopolysaccharidosis or mucolipidosis. Exact diagnosis of the variety of mucopolysaccharidosis or mucolipodoses depends on laboratory studies.

(2) Laboratory: The urinary excretion of substantial amounts of glycosaminoglycan, detected by the presence of toludine blue metachromasia and increased amounts of hexoseuronic acid, confirms the diagnostic category of mucopolysaccharidosis. Characteristic ultrastructure changes may be seen in skin biopsies[41]. Measurements of α-L-iduronidase activity in liver biopsies, cultured skin fibroblasts or leukocytes provide the basis for exact diagnosis. Enzyme measurement in cultured amniotic cells permits antenatal detection of the affected fetus.

Therapy – There is no useful therapy for affected children, but genetic counselling can be offered to the family.

3.2.2 α-L-Iduronidase partial deficiency: Scheie's syndrome: mucopolysaccharidoses Type Is (formerly Type V)

This disorder is due to a defect in the same enzyme (α-L-iduronidase) that is severely depleted in Hurler's disease. In Scheie's syndrome the enzyme deficiency is partial. Only dermatan sulphate and not heparan sulphate accumulates, and there is little or no mental retardation. However, the skeletal features of a mucopolysaccharidosis occur.

Biochemical abnormality – Details of the relevant biochemical disturbance have been given in relation to Hurler's disease. The main hexuronic acid present in heparan sulphate is glucuronic acid rather than iduronic acid. In Scheie's syndrome there appears to be sufficient residual α-L-iduronidase activity to permit heparan sulphate to be degraded normally. Hence this glycosamino-glycan polymer does not accumulate in the disorder. However, dermatan sulphate, with its much higher iduronic acid content, does accumulate, though not to the same extent as in Hurler's disease, where the enzyme deficiency is more severe. This dermatan sulphate accumulation appears responsible for the connective tissue abnormalities of Scheie's syndrome. Brain glycosamino-glycan content is little increased in Scheie's syndrome.

Aetiology – As in Hurler's disease, inheritance is by autosomal recessive mechanisms, but the deficiency of α-L-iduronidase is incomplete.

Structural pathology – The structural changes are very similar to the non-neurological manifestations of Hurler's disease.

Clinical features – The corneal opacities, skeletal, cardiac and soft tissue abnormalities and facial features of gargoylism are present, but there is no mental retardation.

Diagnosis
(1) Clinical: The facial and skeletal features of the disorder would suggest the diagnostic category of mucopolysaccharidosis or mucolipidosis. The absence of mental retardation might imply that the differential diagnosis could be narrowed down to Scheie's, Morquio's (Section 3.2.8) or the Maroteaux Lamy (Section 3.2.9) syndromes.

(2) Laboratory: The biochemical diagnosis of α-L-iduronidase deficiency is discussed in relation to Hurler's disease, though the severity of the α-L-iduronidase deficiency is less in Scheie's syndrome.

Therapy – No treatment is of proven value.

3.2.3 α-L-Iduronidase deficiency (mixed Hurler–Scheie type): mucopolysaccharidosis I H/S

Cases of α-L-iduronidase deficiency have been reported in which the clinical and pathological features were intermediate between those of Hurler's and those of Scheie's syndromes[86].

3.2.4 α-L-Iduronylsulphate sulphatase deficiency: mucopolysaccharidosis Type II – Hunter's disease

This disorder is very similar clinically and biochemically to Hurler's disease, except that the mode of inheritance and the enzyme defect differ. Corneal

167

clouding does not occur and mental retardation is less severe than in Hurler's syndrome.

Biochemical abnormality – α-L-Iduronylsulphate sulphatase catalyses the intralysosomal hydrolysis of sulphate groups bonded to L-iduronic acid, which itself is a constituent of both keratan sulphate and heparan sulphate. When the sulphatase is defective, polymers of these glycosaminoglycans accumulate. The accumulated polymers have terminal iduronyl sulphate-containing disaccharide moieties, since the glycosaminoglycans cannot break down beyond the first iduronyl sulphate group encountered. The accumulating mucopolysaccharides are slightly different in molecular structure from the glycosaminoglycan polymers which accumulate in Hurler's disease, where the terminal residues in the polymers are desulphated iduronic acid (see Figure 3.7). In some tissues endoglycosidases may be able to degrade in part the glycosaminoglycan residues, thus bypassing the metabolic block to some extent.

Aetiology – α-L-Iduronylsulphate sulphatase deficiency is inherited as an X-linked recessive trait. So far all but one reported case have been male. In terms of the Lyon hypothesis, with random inactivation of one X chromosome in the female, it is possible for a female to be affected if the active X chromosome happens to be the (abnormal) maternal one.

Structural pathology – The changes are very similar to those of Hurler's disease.

Clinical features – The clinical features of Hunter's syndrome are very similar to those of Hurler's disease but the disorder almost always occurs only in males, there is no corneal clouding, and the skeletal and intellectual disturbances are less severe. Survival is longer. There are peculiar nodular skin lesions on the upper back and shoulders. Chronic papilloedema occurs rather frequently.

Diagnosis
(1) Clinical: The clinical appearances suggest mucopolysaccharidosis or mucolipidosis. More exact diagnosis depends on laboratory investigations.

(2) Laboratory: Increased glycosaminoglycan excretion in urine suggests the presence of mucopolysaccharidosis. Until recently there was no convenient assay for α-L-iduronylsulphate sulphatase activity, and the diagnosis of the enzyme deficiency had to be made on the basis that the patient was male, had a Hurler-like syndrome and yet had fibroblasts which could liberate L-iduronic acid from an artificial substrate (e.g. phenyl-α-iduronide) whereas Hurler's disease fibroblasts could not. Now the enzyme activity can be measured directly in serum, amniotic fluid and cells, and in hair roots[87].

Therapy – No useful treatment is available.

168

3.2.5 Sulphamidase deficiency: Sanfillipo A disease: mucopolysaccharidosis Type IIIa

This is one of two variants of Sanfillipo disease, in which there is considerable storage of heparan sulphate, and severe neurological disturbance with relatively little skeletal or visceral abnormality.

Biochemical abnormality – N-Sulphamidase (heparan-N-sulphatase) catalyses the cleavage of the sulphate group from the N atom attached to the 2 position of the glucosamine residues in heparan sulphate. Deficiency of the sulphamidase causes failure of further sequential degradation of heparan sulphate polymers once a sulphamidoglucosamine residue is encountered. N-Sulphated glucosamine does not occur in other glycoaminoglycans, so that heparan sulphate is the only glycosaminoglycan to accumulate in N-sulphamidase deficiency.

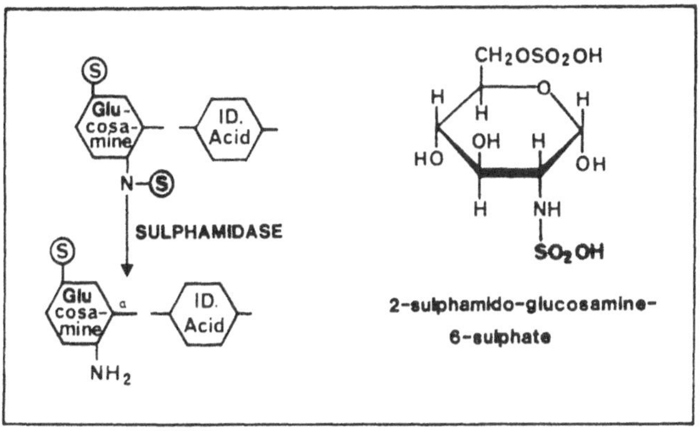

Aetiology – The enzyme deficiency is inherited as an autosomal recessive trait.

Structural pathology – The changes[88,89] are similar to those of Hurler's disease, though the extraneural abnormalities are less severe.

Clinical features – The clinical picture is Hurler-like, with severe progressive mental retardation but with less severe skeletal abnormalities and no corneal changes.

Diagnosis
(1) Clinical: Clinical diagnosis is unlikely to be taken beyond the category of mucopolysaccharidosis or mucolipidosis.

(2) Laboratory: Skin biopsies show diagnostic ultrastructural changes[41]. Increased glycosaminoglycan excretion in urine (detected as uronic acid

excretion, and metachromasia) confirms the diagnosis of mucopolysaccharidosis. Sanfillipo A skin fibroblasts will not release radiolabelled —HSO$_3$ from heparan sulphate specifically tagged with ^{35}S attached to the N atom at the 2 position of the glucosamine residues. No simple convenient assay for the enzyme is yet available.

Therapy – No useful therapy is yet possible.

3.2.6 α-*N*-Acetylglucosaminidase deficiency: Sanfillipo B disease: mucopolysaccharidosis Type IIIb

This is the second variant of Sanfillipo disease. It differs from Sanfillipo A disease only in the nature of the causative enzyme deficiency, and in the composition of the heparan sulphate polymer that accumulates. It is less frequent than the A type.

Biochemical abnormality – α-*N*-Acetylglucosaminidase cleaves the *N*-acetylglucosamine group from uronic acid residues in heparan sulphate. Deficiency of the enzyme therefore prevents the complete breakdown of heparan sulphate polymer. Other glycosaminoglycans contain *N*-acetylglucosamine but their cleavages are unaffected because of the linkage of *N*-acetylglucosamine to other disaccharide units is of β type. As in Sanfillipo A disease, heparan sulphate accumulates, though it has different terminal residues to those of the heparan sulphate which accumulates in Sanfillipo A disease, where the terminal residues are *N*-sulphated. There is also some ganglioside accumulation in grey matter in Sanfillipo B disease[90].

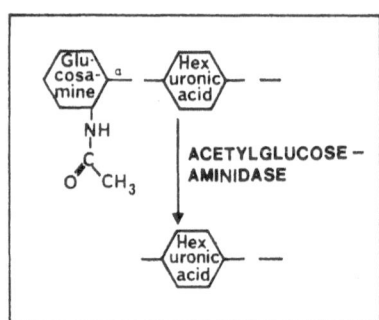

Aetiology – The α-*N*-acetylglucosaminidase deficiency is inherited as an autosomal recessive trait.

Structural pathology – The neuropathological changes[91] include cerebral atrophy with lipid storage in neurons and the presence of membranous cytoplasmic bodies (gangliosides). There are no established neuropathological distinctions among the mucopolysaccharidosis subtypes.

Clinical features – The clinical picture appears identical with that of Sanfillipo A disease, namely severe mental retardation, with the presence of relatively mild connective tissue and skeletal features of gargoylism. Corneal clouding is absent.

Diagnosis
(1) Clinical: The patient's appearance may suggest the category of mucopoly-saccharidosis or mucolipidosis, but more exact diagnosis depends on biochemical studies.
(2) Laboratory: A high level of uronic acid derivative in urine confirms the diagnostic category of mucopolysaccharidosis. Measurement of α-N-acetylglucosaminidase activity in cultured skin fibroblasts permits the definitive diagnosis. Characteristic ultrastructural appearances may be seen in skin biopsies[41].

Therapy – No known therapy is of use. Intravenous administration of the enzyme α-N-acetylglucosaminidase has not produced benefit.

3.2.7 Acetyl CoA (α-glucosaminidase N-acetyl transferase) deficiency (Sanfillipo C disease)

This rare mucopolysaccharidosis variant has been described[91]. Its clinical and neuropathological features are similar to those of the other Sanfillipo variants, though the causal enzyme deficiency is different.

3.2.8 N-Acetylhexosamine-6-sulphate sulphatase deficiency: Morquio's disease: mucopolysaccharidosis Type IV

This disorder has very little effect on the nervous system clinically, and hence will be dealt with briefly. However, it should be noted that, at autopsy in Morquio's disease, neurons in the cortex, Ammon's horn, basal ganglia and thalamus may appear swollen and contain PAS positive granular inclusions[92]. The disease is due to the autosomal recessive inheritance of a deficiency of a very specific sulphatase which hydrolyses sulphate groups linked to the C_6 atom of various hexoses contained in glycosaminoglycan polymers. Such 6-sulphates occur in keratan sulphate, which accumulates in Morquio's disease. The main features of the disorder, including the dwarfism, relate to the skeletal abnormalities.

3.2.9 Arylsulphatase B deficiency: Maroteaux–Lamy syndrome: mucopolysaccharidosis Type VI

The Maroteaux–Lamy syndrome is associated with great accumulation of dermatan sulphate, and skeletal and corneal abnormalities, but no mental retardation. The deficient enzyme is arylsulphatase B. It should be noted that arylsulphatase A deficiency causes accumulation of ceramide galactoside sulphate, and produces metachromatic leukodystrophy (Section 3.1.2.9); multiple arylsulphatase, i.e. A, B and C deficiency, is also known, and causes a syndrome with combined features of a mucopolysaccharidosis and metachromatic leukodystrophy (Section 3.3.4).

Biochemical abnormality – The natural substrate of arylsulphatase B is uncertain, but is thought to be probably *N*-acetylgalactosamine-4-sulphate, which is present in chondroitin sulphate and dermatan sulphate.

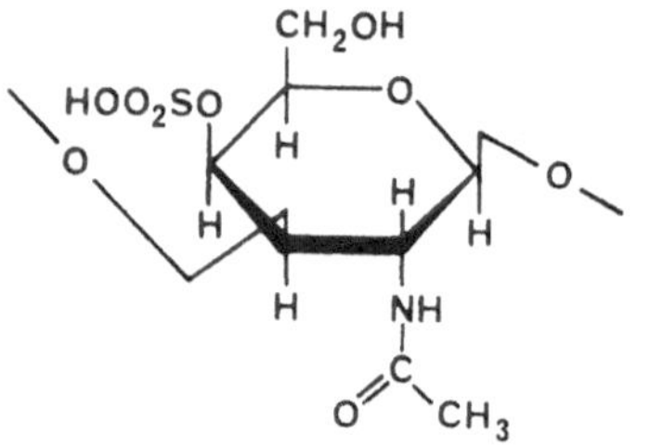

.N–acetylgalactosamine–4–sulphate

In arylsulphatase B deficiency, dermatan sulphate polymer accumulates, mainly in extraneural tissues (where its presence appears related to the tissue alterations) and is excreted in excess in urine.

Aetiology – Arylsulphatase B deficiency is inherited as an autosomal recessive trait.

Structural pathology – The extraneural manifestations of mucopolysaccharidosis occur, but detailed autopsy data are not available.

Clinical features – The skeletal, soft tissue and corneal abnormalities of mucopolysaccharidosis are well developed in the disorder, but there is no intellectual retardation. Occasionally atlanto-occipital subluxation can cause neurological problems. Rarely, the disorder occurs in adults[93].

Diagnosis
(1) Clinical: The patient's appearance should suggest the diagnostic category of mucopolysaccharidosis or mucolipidosis, in which case the absence of mental retardation narrows the diagnostic possibilities to the Maroteaux–Lamy, Scheie or Morquio syndromes.

172

(2) Laboratory: Increased urine excretion of hexuronic acid derivative confirms the diagnostic category. Measurement of greatly decreased arylsulphatase B activity in cultured skin fibroblasts permits the definitive diagnosis. Prenatal diagnosis is possible.

Therapy – No known treatment is of benefit.

3.2.10 β-Glucuronidase deficiency: mucopolysaccharidosis Type VII

A single case of almost total absence of β-glucuronidase activity was reported by Shy *et al.*[94] A few similar cases have since been described. The presentation in the original case was similar to that of the Hunter–Hurler syndrome and included mental retardation, though subsequent cases have shown rather different features, and not all have been mentally backward. The glycosaminoglycan excreted in excess in urine in these cases has been heparan sulphate or dermatan sulphate, but not both as would be expected from the fact that both glycosaminoglycans contain half their carbohydrate as β-linked glucuronic acid residues. The defect in enzyme activity can be measured in fibroblasts, leukocytes or serum.

3.3 THE MUCOLIPIDOSES

The term 'mucolipidosis' has been applied by Spranger and Wiedemann[95] to a group of disorders which have mixed features of the glycosphingolipidoses and the mucopolysaccharidoses. Sufferers show Hurler-like clinical features, without greatly increased glycosaminoglycan excretion in urine, yet with visceral storage of glycosaminoglycans and glycosphingolipids. Since the original descriptions of these disorders, the condition previously designated mucolipidosis I has come to be recognized as a sialidosis[96]. It is dealt with in Section 3.4, in relation to the sialidoses.

It has already been pointed out that, in the body, glycosaminoglycans occur linked to protein molecules (proteoglycans, i.e. glycoproteins). Most body proteins are glycoproteins. In glycoproteins the saccharide portion of the molecule is often joined to the polypeptide portion through an amide bond between acetylglucosamine in the sugar and asparagine in the protein, forming an aspartylacetylglucosamide linkage (Figure 3.8). However, in mucins the link is between *N*-acetylgalactosylamine in the saccharide portion and serine or threonine in the polypeptide portion. The various stages in the degradation of one of the oligosaccharide units linked to the protein moiety provide a setting against which the biochemical abnormalities of the mucolipidoses (and to some extent the glycosphingolipidoses) can be appreciated (Figure 3.8).

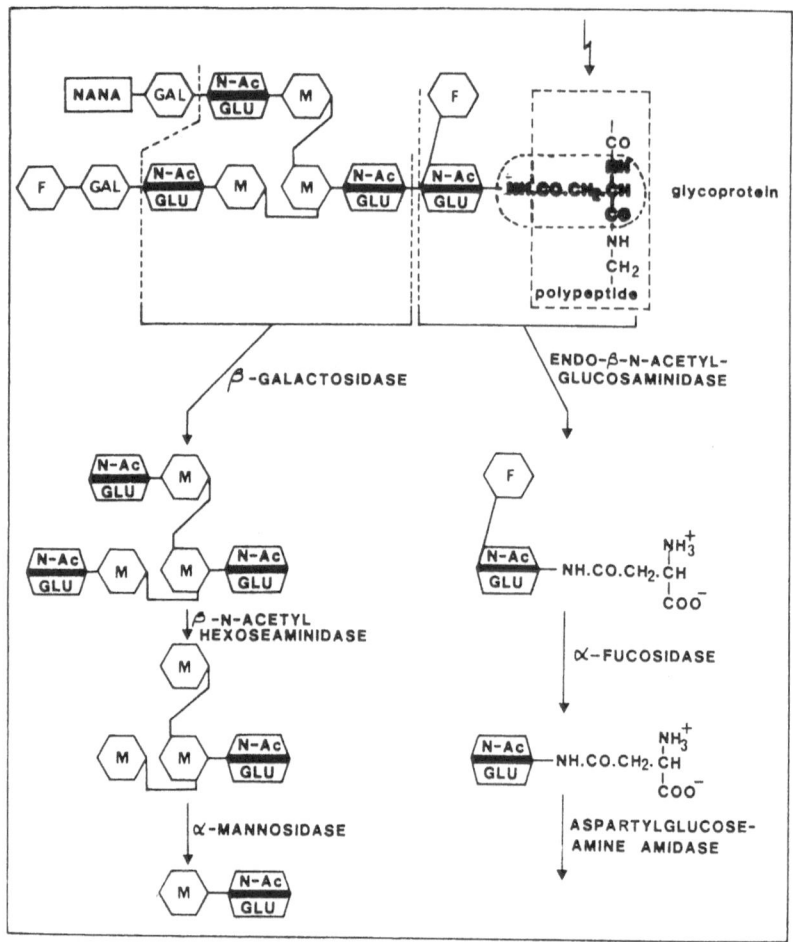

Figure 3.8 Pathways for the metabolic degradation of a commonly occurring oligosaccharide side chain linked to the polypeptide portion of a glycoprotein molecule through an N-acetylglucosamine-aspartate linkage. Failure to cleave the terminal sialic acid (N-acetylneuraminic acid: NANA) residue from the oligosaccharide causes sialidosis (Section 3.4) rather than mucolipidosis

The structures of the various hexose molecules shown in Figure 3.8 are:

N-acetylneuraminic acid, i.e. sialic acid (p. 143),
galactose (p. 32),
N-acetylglucosamine (pp. 145 and 161),
D-mannose (p. 175),
L-fucose (p. 175).

. α -D-mannose   .L-fucose

The combined actions of β-galactosidase and an endo-N-acetylglucosamini-dase cleave the oligosaccharide portion from the glycoprotein, removing with it the aspartate moiety linked to a disaccharide unit. This leaves behind the protein residue. Then the oligosaccharide fragment is further cleaved at β-galactoside links. Deficiencies in the various enzymes of the degradative path-ways produce several diseases (in some of which several different materials may accumulate), as follows:

(1) β-Galactosidase deficiency, which causes generalized gangliosidosis (see Section 3.1.2.1). In this disorder there is accumulation of G_{M1} ganglioside, asialo-G_{M1} ganglioside, the galactose-containing glycosaminoglycan kera-tan sulphate, and various oligosaccharides such as that illustrated in Figure 3.8.

(2) β-N-Acetylhexoseaminidase (A and B isoenzymes) deficiency, causing Sandhoff's disease (Section 3.1.2.3). In this disorder there is accumulation of G_{M2} ganglioside, asialo-G_{M2} ganglioside and globoside (all glyco-sphingolipids), but also oligosaccharides such as the hexasaccharide illus-trated in Figure 3.8.

(3) α-Mannosidase deficiency, causing mannosidosis. In this disorder mannose-rich oligosaccharides accumulate (Section 3.3.1).

(4) α-Fucosidase deficiency, causing fucosidosis, in which fucose-rich oligosac-charides and fucose-containing glycosphingolipids accumulate (Section 3.3.2).

The above enzyme deficiency disorders, together with multiple sulphatase deficiency (Section 3.3.4), and the ill-understood conditions termed muco-lipidosis II (I-cell disease) and mucolipidosis III, comprise the mucolipidoses. β-Galactosidase deficiency and β-N-acetylhexoseaminidase B deficiency will not be described again at this stage. It is convenient to deal with aspartyl-glucosamineamidase deficiency, which causes aspartylglucosaminuria (Section 3.3.3) in relation to this group of disorders. While aspartylglycosaminuria could be regarded as a disorder of amino-acid metabolism, the enzyme defect is a lysosomal one and follows in chemical sequence the defects causing mucolipidoses (Figure 3.8).

3.3.1 α-Mannosidase deficiency: mannosidosis

α-Mannosidase deficiency in man is rare. The disorder causes accumulation of mannose-containing oligosaccharides, with neurological, skeletal and visceral abnormalities[97].

Biochemical abnormality – The lysosomal enzyme α-mannosidase catalyses the cleavage of α-glycosidic bonds between mannose and other hexoses in the oligosaccharide core of many proteoglycans (Figure 3.8). In α-mannosidase deficiency tri-, tetra- and penta-saccharides accumulate in tissues and appear in urine. These oligosaccharides contain respectively 2, 3 and 4 molecules of mannose, and one molecule of *N*-acetylglucosamine. These oligosaccharide fragments occur linked to peptides. The causal relation between the stored material and the production of neural and skeletal defects is unclear.

Aetiology – α-Mannosidase deficiency is probably inherited as an autosomal recessive trait but the genetics of the disorder are as yet not established beyond doubt.

Structural pathology – The brain shows widespread myelin loss with gliosis. Neurons have ballooned cell bodies and contain PAS positive material. There are vacuoles in glial and endothelial cells[98].

Clinical features – In early life affected children are tall, and have large extremities, enlarged livers and spleens and show slow psychomotor development. Later progressive dementia, declining motor performance, coarse facies and lumbar kyphosis appear, while the hepatosplenomegaly disappears. Signs of pyramidal tract insufficiency are usually present by the fourth year of life, and affected children may later become unable to sit unsupported. The skull bones are thick, there are Hurler-like changes in the vertebral bodies and radiographically the long bones are porotic with poor trabeculation. Vacuolated lymphocytes occur. There is no increase in urine glycosaminoglycan excretion. Attacks of severe ketoacidosis occur by the age of 4–6 years, and lead to death.

Diagnosis
(1) Clinical: The clinical features of the disorder would suggest a mucopolysaccharidosis or mucolipidosis, but diagnosis could hardly be taken beyond this level without laboratory investigations.

(2) Laboratory: The absence of increased urinary excretion of hexoseuronic acid derivatives would tend to exclude mucopolysaccharidosis. Determining the chemical nature of the material excreted in urine, and the material

retained in tissue biopsies and cultured skin fibroblasts, and measurement of α-mannosidase activity in cellular material, provide the definitive diagnosis. However, these are matters for specialized laboratories.

Therapy – No known treatment is of value.

3.3.2 α-Fucosidase deficiency; fucosidosis

α-Fucosidase deficiency is a rare disorder in which accumulation of fucose-containing glycolipid is associated with neurological and skeletal abnormalities[99].

Biochemical abnormality – The lysosomal enzyme α-fucosidase catalyses the cleavage of α-glycosidic bonds between fucose and other hexoses in glycoproteins. The full biochemical ramifications of α-fucosidase deficiency are not yet worked out but in several patients the major stored material appears to have been the H antigen, a glycolipid, which has the structure shown below. Another possible chemical consequence of α-fucosidase deficiency can be inferred from Figure 3.8. Dawson and Lenn[83] claim to have identified fucose α-(1 → 6)-N-acetylglucosamine-aspartate in fibroblasts, brain and

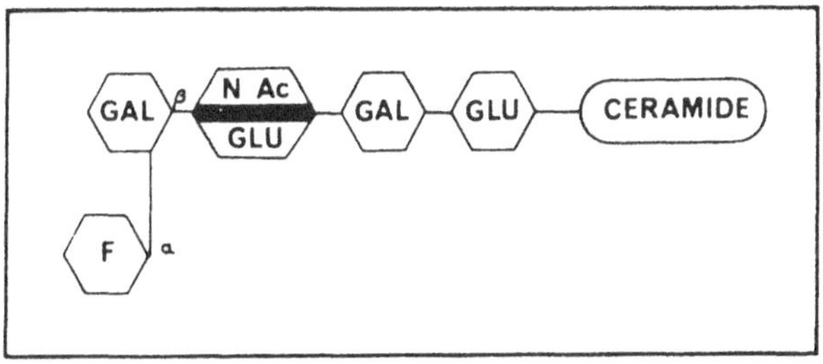

urine from cases of fucosidosis. An oligosaccharide (fucose-galactose-N-acetylglucosamine-mannose[fucose-galactose-N-acetylglucosamine-mannose]-mannose-N-acetylglucosamine) has also been identified[48].

Aetiology – α-L-Fucosidase deficiency is probably inherited as an autosomal recessive trait.

Structural pathology – Neurons and glial cells in the disorder are vacuolated and there is loss of cerebral neurons with severe demyelination. Hepatocytes contain storage vacuoles. At electron microscopy there are small clear vacuoles in the cytoplasm of liver and brain cells.

Clinical features – Affected infants show a progressive and rapid decline in intellectual and motor function, with the development of decerebrate or decorticate rigidity and gross dementia by the age of 2 years. Hurler-like skeletal and soft tissue features are present, and there may be hepatosplenomegaly.

Diagnosis
(1) Clinical: The full clinical picture would suggest mucopolysaccharidosis or mucolipidosis. The further stages of diagnosis are biochemical.

(2) Laboratory: Definitive diagnosis depends on the measurement of decreased α-L-fucosidase activity in cultured skin fibroblasts.

Therapy – No useful therapy is available.

3.3.3 Aspartylglucosaminuria

Aspartylglucosaminuria is a lysosomal storage disorder due to deficiency of the enzyme aspartylglucosylamine amidase[100]. The disorder causes mental retardation. Nearly 60 cases were reported up to 1977[101]. The great majority of cases have been of Finnish origin[102].

Biochemical abnormality – Aspartylglucosylamine amidase catalyses the hydrolysis of aspartylglucosamine (β-aspartyl-N-acetylglucosamine) within lysosomes. Aspartylglucosamine peptide accumulates inside lysosomes in cells of various tissues, including neurons. The distribution of the retained materials is fairly uniform throughout the brain[102]. It is not known whether the mechanical effects of the retained material, or some as yet ill-understood biochemical disturbance, explains the nervous system dysfunction. There is a great increase in aspartylglucosamine concentration in urine and to some extent in CSF, though there is little increase in blood concentration of aspartylglucosamine.

Aetiology – The disorder is inherited as an autosomal recessive trait.

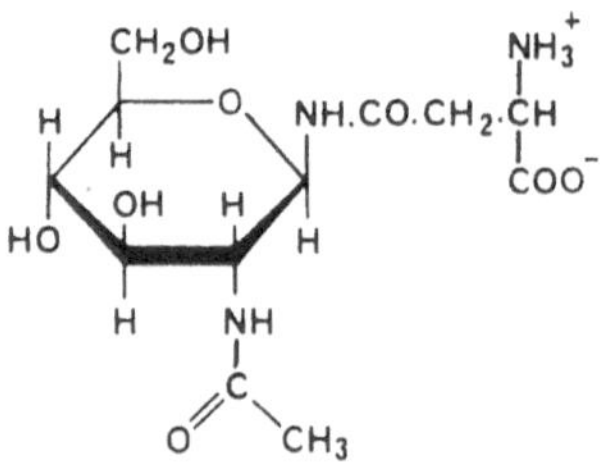

·β–aspartyl-N-acetylglucosamine

Structural pathology – The brain is atrophic with distended neurons in the cortex and basal ganglia. Little gliosis occurs. The liver is enlarged and there is considerable vacuolation of the hepatocytes. Abnormally large lysosomes occur in the liver, brain and kidney, but not in the skin.

Clinical features – The disorder leads to progressive psychomotor retardation, convulsions, e.e.g. disturbance and connective tissue and skeletal abnormalities (coarse facies, sagging cheeks, broad nose, short neck, brachycephaly, kyphoscoliosis, hypermobile joints, thin calvarium and thin cortices of long bones). Patients suffer recurrent skin and respiratory infections. Vacuolated lymphocytes occur in peripheral blood.

Diagnosis
(1) Clinical: the clinical features would suggest the diagnosis of one of the storage diseases, most probably a mucopolysaccharidosis or mucolipidosis.
(2) Laboratory: Raised urine levels of aspartylglucosamine, and reduced levels of aspartylglucosylamine amidase in plasma, seminal fluid and tissue, with characteristic lysosomal appearance at electron microscopy of biopsy specimens, permit the diagnosis.

Therapy – There is no known useful therapy.

3.3.4 Multiple sulphatase deficiency: juvenile sulphatidosis

Combined deficiency of arylsulphatases A, B and C is rare. It causes a disorder with features of metachromatic leukodystrophy together with some features of gargoylism[103].

Biochemical abnormality – Arylsulphatase A catalyses the cleavage of the sulphate ester of galactose ceramide. Deficiency of this enzyme causes metachromatic leukodystrophy, with accumulation of galactosylceramide sulphate. With the additional absences of arylsulphatases B and C, there is also failure to cleave the sulphate esters of various glycosaminoglycans, and these also accumulate, as do gangliosides G_{M3}, G_{M2}, G_{D3} and ceramide lactoside and trihexoside[104]. Presumably the glycosaminoglycan accumulation is related to the connective tissue abnormality in the disorder.

In multiple sulphatase deficiency cholesterol sulphatase, dehydroepiandrosterone sulphatase, cerebroside sulphatase and psychosine sulphatase may all also be deficient[104]. The related sulphates may accumulate.

Aetiology – Multiple sulphatase deficiency is probably inherited as an autosomal recessive trait.

Structural pathology – Reported cases have shown atrophy of the cerebrum and the cerebellar cortex, with thickened, cloudy meninges. Some cases have had nodular thickenings on the mitral valve. There was mild hepato-splenomegaly in some cases but, unlike in metachromatic leukodystrophy, the gallbladder has not been affected[103].

Clinical features – The affected infant is retarded from birth, and there is progressive psychomotor deterioration in the second and third years of life with increasing spasticity, myoclonic seizures, coarse tremor and nystagmus. The facial features of gargoylism are not conspicuous and there are no corneal opacities, but the thorax is deformed and the stature short. The liver and spleen are not enlarged. There may be optic disc pallor. Radiological studies show the changes of mild dysostosis multiplex. Lymphocytes are vacuolated with abnormal granulation. CSF protein levels are raised and peripheral nerve conduction is slowed. The disorder progresses to a final vegetative state in later childhood.

Diagnosis
(1) Clinical: The clinical features may suggest gargoylism, in which case muco-polysaccharidosis or mucolipidosis may be suspected. It is unlikely that diagnosis can be taken further without biochemical studies.
(2) Laboratory: The diagnosis depends on demonstrating the deficiency of all three arylsulphatases (A, B and C) in cultured skin fibroblasts or tissue biopsy specimens.

Therapy – No treatment is available.

3.3.5 Mucolipidosis II: I-cell disease

This rare disorder has an inadequately understood biochemical pathogene-sis but causes a progressive neurological degeneration with gargoyle-like features[81].

Biochemical abnormality – The basic biochemical defect in the disorder is uncertain. No mucopolysacchariduria occurs. Levels of most glycosylhydrolase enzymes in serum are greatly elevated. Brain and liver contents of certain glycosphingolipids, mainly ganglioside G_{M3} (ceramide dihexoside) are raised, though the levels of the brain enzymes responsible for their degradation are normal. Liver and brain galactosidase levels are 20–30% (or less) of normal though it is uncertain whether this reduction is primary or secondary[105]. Miller et al.[106] obtained tentative evidence that the enzyme molecule in this disorder contains an altered carbohydrate moiety. In cultured skin fibroblasts gangliosides G_{M3} and G_{D3} accumulate, together with dermatan sulphate, and there is a general reduction in activity of all lysosomal hydrolases except acid phosphatase and β-glucosidase, due to leakage of the enzymes from lysosomes. This leakage also occurs *in vivo* and may cause the high levels of the enzymes found in serum.

Aetiology – The disorder, whatever its fundamental biochemical nature, appears to be inherited as an autosomal recessive trait.

Structural pathology – Cultured fibroblasts and other cells, including neurons in some cases, contain numerous inclusions which at electron microscopy prove to be enlarged lysosomes[107].

Clinical features – The disorder produces early onset severe and progressive psychomotor retardation with a very short stature, striking dysostosis multiplex, a slowly developing gargoyle-like facies and sometimes corneal opacities. Cardiac murmurs occur and cardiac failure may develop. Mucopolysaccharides are not found in urine but there are vacuolated lymphocytes and monocytes in peripheral blood.

Diagnosis
(1) Clinical: The combination of gargoyle-like facies, bony dysostosis and progressive dementia in a young child raises the question of mucopolysaccharidosis or mucolipidosis. More exact diagnosis depends on laboratory investigations.
(2) Laboratory: The absence of increased glycosaminoglycan excretion in urine narrows the diagnostic possibilities to the category of mucolipidosis. The presence of the characteristic large inclusions in skin fibroblasts strongly suggests the diagnosis, as does the high levels of multiple glycohydrolase enzymes in serum (except β-glucosidase).

Therapy – No useful treatment is available.

3.3.6   Mucolipidosis III

The biochemical basis of this disorder and the nature of the storage substance in the disorder are unknown[81]. The condition is rare. There is mild mental retardation, early limitation of joint mobility, unusual bony abnormalities (short stature, short neck, scoliosis and hip dysplasia) and no increase in urinary excretion of glycosaminoglycan. Vacuolated bone marrow cells occur, though peripheral lymphocytes are normal. Inheritance is probably by autosomal recessive mechanisms.

3.4 α-N-ACETYL NEURAMINIDASE DEFICIENCY: SIALIDOSIS

This condition, formerly called mucolipidosis I, has since been shown due to α-N-acetyl neuraminidase deficiency[96]. Retained material consists of sialyl oligosaccharides. The disorder produces cherry-red maculae with visual failure, myoclonus and epileptic phenomena (the cherry-red spot myoclonus syndrome). In addition, in one group of cases there have been Hurler-like features. The nosological category of sialidosis is a comparatively new one, and has been reviewed in some detail by Lowden and O'Brien[108].

Biochemical abnormality – The enzyme α-N-acetylneuraminidase catalyses the cleavage of $\alpha2 \rightarrow 6$ glycosidic bonds between N-acetylneuraminic acid (sialic acid) and the adjacent hexose in sialyl oligosaccharides. It also catalyses the cleavage of $\alpha2 \rightarrow 3$ bonds between sialic acid and the adjacent hexose in this class of compound. The enzyme does not catalyse the cleavage of sialic acid from gangliosides (in which the linkage to hexoses is often of a glycosidic $\alpha2 \rightarrow 3$ type). In α-N-acetylneuraminidase deficiency there remains virtually no capacity to cleave $\alpha2 \rightarrow 6$ bonds, but some capacity to cleave $\alpha2 \rightarrow 3$ bonds in sialo-oligosaccharides, and full capacity to cleave sialo-hexose glycosidic bonds in gangliosides. Hence, in neuraminidase deficiency there is no accumulation of gangliosides, but sialyloligosaccharide accumulates and particularly fragments with sialo-hexose $\alpha2 \rightarrow 6$ glycosidic bonds.

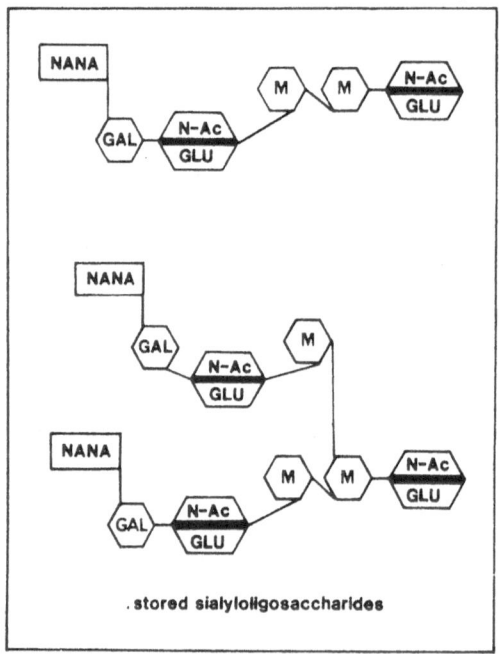

. stored sialyloligosaccharides

In human neuraminidase deficiency, at least 12 sialyloligosaccharides[109] normally present in urine in quantities of less than 1 mg/l are excreted in great excess. The structure of the two main sialyloligosaccharides identified in urine from cases of neuraminidase deficiency are shown opposite.

The chemical structures of all the stored substances in the disorder are not yet known. They probably derive from the sialyloligosaccharide side chains of proteoglycans, and are cleaved from the protein by an endo-N-acetylglucosaminidase (see Figure 3.8). In some cases lipofuscin accumulation also has occurred.

Sialidosis can be considered to resemble the mucolipidoses in its chemical pathogenesis. The stored material has a similar origin, but the stored molecules tend to be large (sialided glycopeptides and glycolipids) and have different chemical properties, because terminal N-acetylneuraminic acid residues cannot be removed from the oligosaccharide chain (compare Figure 3.8).

β-Galactosidase activity is also reduced in some patients with sialidosis (in general those with mental retardation and features resembling gargoylism). However, the available evidence suggests that this enzyme deficiency does not cause accumulation of G_{M1} ganglioside or galactose-containing oligosaccharide (unlike the situation in the β-galactosidase deficiency that causes G_{M1} gangliosidosis). The basis of the β-galactosidase deficiency in sialidosis is unknown.

Aetiology – The neuraminidase deficiency is transmitted as an autosomal recessive trait.

Structural pathology – Neurons at brain biopsy have contained vacuoles which either appeared empty or contained a little amorphous granular material; lipofuscin bodies have been prominent. The cerebral cortex has been thin. At biopsy hepatocytes also have contained vacuoles, and there have sometimes been foamy histiocytes in bone marrow.

Clinical features – α-N-Acetyl neuraminidase deficiency sufferers appear to fall into two clinical groups[108]. In both groups there are visceral failure, cherry-red maculae, myoclonus and epileptic manifestations, and grossly excessive excretion of sialyloligosaccharide in urine. In the first group the onset is in the teenage or early adult years[110], there is no skeletal, facial or growth disturbance, and intelligence remains normal. The second group comprises juvenile and infantile cases. Here, superimposed on the clinical picture of the first type there are mental retardation, long tract signs, ataxia, corneal clouding, lens opacity, deafness and skeletal and facial alterations resembling those of Hurler's syndrome, with enlarged abdominal viscera in the infantile cases. Foam cells (histiocytes) in the bone marrow, and e.e.g. changes, are frequent in both varieties of the disorder, and peripheral motor nerve conduction velocities may be mildly slowed[111].

Serum and tissue β-galactosidase activity tends to be normal in the first type of sialidosis, but reduced in the second. Serum levels of other lysosomal lycohydrolases are not raised (unlike the situation in mucolipidosis II).

Diagnosis
(1) Clinical: The presence of the so-called 'cherry-red myoclonus' syndrome suggests the diagnosis of sialidosis, particularly if there are Hurler-like features also. The other main causes of cherry-red maculae, the gangliosidoses (again disorders in which sialic acid-containing material accumulates) lack Hurler-type features (except in β-galactosidase deficiency, i.e. generalized G_{M1} gangliosidosis) and begin much earlier in life than sialidosis (except for the infantile cases). Myoclonic and other epileptic manifestations are not so prominent a feature in the gangliosidoses, and the adolescent-onset cases of sialidosis, with no mental retardation, do not closely resemble typical gangliosidosis clinically.

(2) Laboratory: Changes such as the presence of vacuolated lymphocytes and foamy histiocytes in bone marrow merely suggest a storage disorder. High urine levels of sialyloligosaccharide, demonstrated by thin layer chromatography, are very suggestive of the diagnosis, though concentrations of these substances are raised (but less elevated) in mucolipidosis II. Measurement of the specific neuraminidase which is deficient in leukocytes or cultured skin fibroblasts depends on the availability of substrates with $\alpha 2 \rightarrow 6$ glycosidic linkages, but is diagnostic. The presence of considerably diminished β-galactosidase activity supports the diagnosis of the type II variety of sialidosis.

Therapy – No curative treatment is available. Use of anticonvulsants might help control any epileptic phenomena which may be present.

3.5   THE CEROID LIPOFUSCINOSES

Since the exact chemical natures of ceroid and lipofuscin are incompletely understood, and the biochemical basis of the ceroid lipofuscinoses is obscure, these disorders must be discussed largely from the clinical viewpoint.

The ceroid-lipofuscinoses are a group of inherited disorders in which there is progressive degeneration of the nervous system with accumulation of ceroid and lipofuscin in neurons. These disorders were formerly designated the infantile (Haltia–Santavudri), late infantile (Bielschowsky–Jansky), juvenile (Speilmeyer–Vogt: Batten–Mayou) and adult (Kufs) varieties of amaurotic familial idiocy, the conventional infantile variety of amaurotic familial idiocy now being recognized as a chemically different disorder, G_{M2} gangliosidosis (Tay–Sachs' disease). A detailed account of the ceroid lipofuscinoses is given by Zeman and Siakotos[112].

Biochemical abnormality – Ceroid is a yellowish autofluorescent pigment which accumulates in neuronal perikarya. It is insoluble in water and also in non-polar solvents (e.g. xylol). It has been suggested that ceroid is formed by the development of cross linkages between storage lipids, these linkages arising from abnormalities of vitamin E metabolism or action. Some hypotheses relevant to this question are mentioned by Swaimann, Garg and Lockman[113].

Aetiology – The disorder appears to be inherited as an autosomal recessive trait.

Structural pathology – Chrome and Stern[30], Roizin and Kaufman[42] and Ryan et al.[114] provide accounts of the pathological changes found in ceroid lipofuscinosis. There is intraneuronal granular lipid storage, but often little loss of cortical neurons. There is usually fibrillary gliosis of the white matter of the cerebrum and spinal cord. In one rare variant there is extracellular non-melanin pigmentation of the basal ganglia[115]. The cerebellum is severely degenerated and gliosed. The retina shows degeneration of the outer nuclear layer with loss of rods and cones. The viscera may contain collections of lipid-storing macrophages.

Clinical features – There are three main types of the disorder. The Jansky–Bielschowsky variety begins at 2–4 years of age with convulsions. It rapidly progresses to dementia and blindness with optic atropy, and sometimes retinal pigmentation and macular degeneration. The Speilmeyer–Sjogren type begins later (4–8 years) and runs a longer course (average 11 years). Blindness is the earliest manifestation, with macular degeneration and retinitis pigmentosa. Later a moderate degree of dementia develops. In the Kufs variety the onset is over the age of 20 years, there is usually no retinal disorder or blindness, and mental deterioration is mild. This variant runs its course over 5–35 years. Atypical cases occur[115]. There is also a group of early onset cases in which the dominant feature is an extrapyramidal disorder, without blindness (familial juvenile dystonic lipidosis[116]). Some cases of ceroidosis present with myoclonus. Swaiman et al.[113] described a type of familial ceroid lipofuscinosis with 'sea-blue' histiocytes in the bone marrow and progressive posterior column and sometimes pyramidal tract degeneration.

Diagnosis
(1) Clinical: In the presence of typical retinal pigmentary changes and optic atrophy, with the presence of other consistent clinical features, a presumptive diagnosis of ceroidosis is possible. The clinical picture of ceroid lipofuscinosis does not closely resemble those of the other neurological disorders in which retinitis pigmentosa occurs, e.g. Refsum's syndrome (Section 3.1.1.4) and abetalipoproteinaemia (Section 3.1.1.3).

(2) Laboratory: Vacuolated circulating lymphocytes[117] and vacuolated neurons (from rectal or brain biopsies) containing autofluorescent PAS positive pigment permit the definitive diagnosis of ceroidosis. Characteristic ultrastructural changes may be found in skin biopsies[41]. 'Sea-blue' histiocytes are sometimes found in bone marrow, but are not specific[118].

Reduced levels of leukocyte peroxidase have been reported, but these changes are not consistently present[119,120].

Therapy – No known therapy is of proven benefit.

3.6 DISORDERS OF PROTEIN COMPOSITION

3.6.1 Amyloidosis

The term amyloidosis refers to a group of disorders, hereditary or acquired, in which abnormalities of protein conformation lead to a progressive deposition of abnormal fibrillary protein which comes to interfere with the functioning of various tissues. Neural elements are involved only in certain familial amyloid syndromes, and in some acquired cases associated with plasma cell dyscrasias.

Biochemical abnormality – The biochemical abnormalities of amyloidosis are described in some detail by Glenner, Ignaczak and Page[121] and amyloid protein composition is discussed by Shoji and Okano[122]. Amyloid deposits comprise β-plated sheets of fibrillary proteins. Some of these proteins appear to originate from abnormal immunoglobulin protein chains, possibly from other serum proteins and from polypeptides of endocrine tissue origin. When polypeptide fragments from these proteins take up a β-plated conformation they become less soluble and more resistant to proteolysis. If these fragments are deposited in tissues, they tend to remain there. The deposits grow as further β-plated polypeptide sheets accumulate. Whether the tissue damage results from direct mechanical effects of the deposits, from the deposits interfering with the local blood supply, or because toxic molecules are formed in relation to the deposits, are unsettled questions.

Aetiology – There are many known causes of amyloidosis, but the varieties in which nervous system involvement occurs are the four types of hereditary (autosomal dominant) neuropathic amyloidosis, and some sporadic cases in which there is evidence of plasma cell dysfunction.

Structural pathology – Amyloid deposits occur in blood vessel walls in various viscera, particularly the kidneys and heart. Nervous system involvement is largely confined to the peripheral nervous system, though the meninges may also be infiltrated by amyloid.

Clinical features – Neural amyloidosis presents with the features of slowly progressive sensori-motor peripheral neuropathy. In Types I and II, the dominant changes are those of a peripheral polyneuropathy, more severe in the lower limbs than the upper, and in Type I with associated severe autonomic involvement. In Type II, median nerve compression in the carpal tunnels tends to dominate the clinical picture, while in Type IV the neuropathy is predomin-

186

antly cranial (facial). Associated abnormalities include vitreous opacities and e.c.g. changes (Type II), renal insufficiency, duodenal ulcer and e.c.g. changes (Type III), and lattice dystrophy of the cornea and a pendulous skin (Type IV).

Diagnosis
(1)   Clinical: The presence of chronic progressive peripheral neuropathy can be established clinically, and confirmed by nerve conduction studies. While the associated features, or the presence of thickened peripheral nerves, might suggest the possibility of amyloidosis, in the absence of a family history of proven amyloidosis it seems unlikely that a definitive diagnosis could be achieved without laboratory investigation.

(2)   Laboratory: Diagnosis depends on the histological demonstration of amyloid deposits, preferably in a peripheral nerve (e.g. sural nerve) biopsy.

Treatment – No treatment is known which will halt, or reverse, the progression of amyloid peripheral neuropathy. Management is symptomatic only, though carpal tunnel decompression may relieve the manifestations of any median nerve compression.

REFERENCES

1   Norton, W. T. (1976). Formation, structure and biochemistry of myelin. In Siegel, G. J., Albers, R. W., Kratzmen, R. and Agranoff, B. W. (eds.) *Basic Neurochemistry*, 2nd Edn., pp. 74–99. (Boston: Little, Brown & Co.)
2   McIlwain, H. and Bachelard, H. S. (1971). *Biochemistry and the Central Nervous System*. 4th Edn. (Edinburgh: Churchill-Livingstone)
3.   Stoffel, W. (1967). The chemistry of mammalian lipids. In Schettler, G. (ed.) *Lipids and Lipidoses*. pp. 1–39. (Berlin: Springer)
4   Rouser, G. and Yamamoto, A. (1969). Lipids. In Lathja, A. (ed.) *Handbook of Neurochemistry*. Vol. 1, pp. 121–169. (New York: Plenum Press)
5   Awasthi, Y. C. and Srivastava, S. K. (1980). Structure, function and metabolism of glycosphingolipids. In Kumar, S. (ed.) *Biochemistry of Brain*. pp. 1–20. (Oxford: Pergamon Press)
6   Max, S. R., MacLaren, N. K., Brady, R. O., Bradley, R. M., Rennels, M. B., Tanaka, J., Garcia, J. H. and Cornblath, M. (1974). G_{M3} (hematoside) sphingolipodystrophy. *N. Engl. J. Med.*, **291**, 929–931
7   Schaumburg, H. H., Powers, J. M., Raine, C. S., Suzuki, K. and Richardson, E. P. Jr. (1975). Adrenoleucodystrophy. A clinical and pathological study of 17 cases. *Arch. Neurol.*, **32**, 577–591.
8   O'Neill, B. P., Marmion, L. C. and Feringa, E. R. (1981). The adrenoleukomyeloneuropathy complex: expressions in four generations. *Neurology*, **31**, 151–156
9   Igarashi, M., Schaumburg, H. H., Powers, J., Kishimoto, Y., Kolodney, E. and Suzuki, K. (1976). Fatty acid abnormality in adrenoleukodystrophy. *J. Neurochem.*, **26**, 851–860
10   Menkes, J. H. and Corbo, L. M. (1977). Adrenoleucodystrophy. Accumulation of cholesterol esters with very long chain fatty acids. *Neurology*, **27**, 926–932
11   Molzer, B., Bernheimer, H., Budka, H., Pilz, P. and Toifl, K. (1981). Accumulation of very long chain fatty acids is common to 3 variants of adrenoleukodystrophy (ALD). 'Classical' ALD, atypical ALD (female patient) and adrenomyeloneuropathy. *J. Neurol. Sci.*, **51**, 301–310
12   Ogino, T. and Suzuki, K. (1981). Specificities of human and rat brain enzymes of cholesterol ester metabolism towards very long chain fatty acids: implication for biochemical pathogenesis of adrenoleucodystrophy. *J. Neurochem.*, **36**, 776–779

13 Yahara, S., Moser, H. W., Kolodny, E. H. and Kishimoto, Y. (1980). Reverse phase high-performance liquid chromatography of cerebrosides, sulfatides and ceramides: microanalysis of homolog composition without hydrolysis and application to cerebroside analysis in peripheral nerves of adrenoleucodystrophy patients. *J. Neurochem.*, **34**, 694–699

14 Schaumburg, H. H., Powers, J. M., Raine, C. S., Spencer, P. S., Griffin, J. W., Prineas, J. W. and Boehme, D. M. (1977). Adrenomyeloneuropathy: a probable variant of adrenoleucodystrophy. II. General pathologic, neuropathologic, and biochemical aspects. *Neurology*, **27**, 1114–1119

15 Manz, H. J., Schuelein, M., McCullogh, D. C., Kishimoto, Y. and Eiben, R. M. (1980). New phenotypic variant of adrenoleucodystrophy. Pathologic, ultrastructural and biochemical study in two brothers. *J. Neurol. Sci.*, **45**, 245–260

16 Griffin, J. W., Goren, E., Schaumburg, H., Engel, W. R. and Loriaux, L. (1977). Adrenomyeloneuropathy: a probable variant of adrenoleukodystrophy. 1. Clinical and endocrinologic aspects. *Neurology*, **27**, 1107–1113

17 Bourre, J. M., Bornhofen, J. H., Araoz, C. A., Daudu, O. and Baumann, N. A. (1978). Pelizaeus–Merzbacher disease: brain lipid and fatty composition. *J. Neurochem.*, **30**, 719–727

18 Witter, B., Debuch, H. and Klein, H. (1980). Lipid investigation of central and peripheral nervous system in connatal Pelizaeus–Merzbacher's disease. *J. Neurochem.*, **34**, 957–962

19 Kahlke, W. (1967). A- β-lipoproteinemia (Bassen–Kornzweig syndrome). In Schettler, G. (ed.) *Lipids and Lipidoses.* pp. 382–400. (Berlin: Springer)

20 Bruyn, G. W. (1977). Bassen–Kornzweig disease. In Vinken, P. J. and Bruyn, G. W. (eds.) *Handbook of Clinical Neurology.* Vol. 29, pp. 401–429. (Amsterdam: North Holland)

21 Herbert, P. N., Gotto, A. M. and Fredrickson, D. S. (1978). Familial lipoprotein deficiency (abetalipoproteinemia, hypobetalipoproteinemia, and Tangier disease). In Stanbury, J. B., Wyngaarten, J. B. and Fredrickson, D. S. (eds.) *The Metabolic Basis of Inherited Disease.* 4th Edn., pp. 544–588. (New York: McGraw-Hill)

22 Scanu, A. M. (1978). Abetalipoproteinemia and hypobetalipoproteinemia: what is the primary defect? In Kark, R. A. P., Rosenberg, R. N. and Schut, L. J. (eds.) *Advances in Neurology*, Vol. 21, pp. 125–130. (New York: Raven Press)

23 Illingworth, D. R., Connor, W. E. and Miller, R. G. (1980). Abetalipoproteinemia. Report of two cases and review of therapy. *Arch. Neurol.*, **37**, 659–662

24 Steinberg, D. (1978). Phytanic acid storage disease: Refsum's syndrome. In Stanbury, J. B., Wyngaarten, J. B. and Fredrickson, D. S. (eds.) *The Metabolic Basis of Inherited Disease.* 4th Edn., pp. 688–706. (New York: McGraw-Hill)

25 Steinberg, D. (1978). Evaluation of the metabolic error in Refsum's disease: strategy and tactics. In Kark, R. A. P., Rosenberg, R. N. and Schut, L. J. (eds.) *Advances in Neurology.* Vol. 21, pp. 113–124. (New York: Raven Press)

26 Warren, L. (1967). The metabolism of sialic acids. In Aronson, S. M. and Volk, B. W. (eds.) *Inborn Errors of Sphingolipid Metabolism.* pp. 251–259. (Oxford: Pergamon)

27 Svennerholm, L. (1972). Gangliosides, isolation. In Whistler, R. L. and Bemiller, J. N. (eds.) *Methods in Carbohydrate Chemistry.* Vol. 4, pp. 464–474. (New York: Academic Press)

28 Hers, H. G. and Van Hoof, F. (1973). *Lysosomes and Storage Diseases.* (New York: Academic Press)

29 Neufeld, E. F., Lim, T. W. and Shapiro, L. J. (1975). Inherited disorders of lysosomal metabolism. *Annu. Rev. Biochem.*, **44**, 357–376

30 Crome, L. and Stern, J. (1976). Inborn lysosomal enzyme deficiencies. In Blackwood, W. and Corsellis, J. A. N. (eds.) *Greenfield's Neuropathology.* 3rd Edn., pp. 500–580. (London: Arnold)

31 Brady, R. O. (1978). Sphingolipidoses. *Annu. Rev. Biochem.*, **47**, 687–713

32 O'Brien, J. S. (1978). The gangliosidoses. In Stanbury, J. B., Wyngaarten, J. B. and Fredrickson, D. S. (eds.) *The Metabolic Basis of Inherited Disease*, 4th Edn., pp. 841–865. (New York: McGraw-Hill)

33 Van Hoof, F. (1973). G_{M1} gangliosidosis. In Hers, H. G. and Van Hoof, F. (eds.) *Lysosomes and Storage Diseases.* pp. 305–321. (New York: Academic Press)

34 O'Brien, J. S. (1975). Molecular genetics of G_{M1}-galactosidase. *Clin. Genet.*, **8**, 303–313

35 Farrell, D. F. and MacMartin, M. P. (1981). G_{M1} gangliosidosis: enzymatic variation in a single family. *Ann. Neurol.*, **9**, 232–236

36 Farrell, D. F. and Ochs, U. (1981). G_{M1} gangliosidosis: phenotypic variation in a single family. *Ann. Neurol.*, **9**, 225–231

37 Volk, B. W., Adachi, M. and Schneck, L. (1971). The gangliosidoses. In Zimmerman, H. (ed.) *Progress in Neuropathology*. Vol. 1, pp. 232–254. (New York: Grune and Stratton)

38 O'Brien, J. S. (1973). Tay–Sachs' disease and juvenile G_{M2} gangliosidoses. In Hers, H. G. and Van Hoof, F. (eds.) *Lysosomes and Storage Diseases*. pp. 323–344. (New York: Academic Press)

39 Srivastava, S. K. and Awasthi, Y. C. (1980). Metabolic disorders in sphingolipidoses. In Kumar, S. (ed.) *Biochemistry of Brain*. pp. 21–47. (Oxford: Pergamon Press)

40 Goldman, J. E., Katz, D., Rapin, I., Purpura, D. P. and Suzuki, K. (1981). Chronic G_{M1} gangliosidosis presenting as dystonia. 1. Clinical and pathological features. *Ann. Neurol.*, **9**, 465–475

41 O'Brien, J. S., Bernett, J., Veath, M. L. and Paa, D. (1975). Lysosomal storage disorders. Diagnosis by ultrastructural examination of skin biopsy specimens. *Arch. Neurol.*, **32**, 592–599

42 Roizin, L. and Kaufman, M. A. (1971). Dyslipidoses. In Minckler, J. (ed.) *Pathology of the Nervous system*. Vol. 2, pp. 1284–1314. (New York: McGraw-Hill)

43 Rapin, J., Suzuki, K., Suzuki, K. and Valsamis, M. P. (1976). Adult (chronic) G_{M2} gangliosidosis. Atypical spinocerebellar degeneration in a Jewish sibship. *Arch. Neurol.*, **33**, 120–130

44 Bernheimer, H., Molzer, B. and Deisenhammer, E. (1977). Sandhoff disease: ganglioside G_{M2} and asialo-G_{M2} accumulation in the cerebrospinal fluid. *J. Neurochem.*, **29**, 351–352

45 Von Specht, B. U., Geiger, B., Arnon, R., Passwell, J., Keren, G., Goldman, B. and Padeh, B. (1979). Enzyme replacement in Tay–Sachs' disease. *Neurology*, **29**, 848–854

46 Sandhoff, K. and Harzer, K. (1973). Total hexosaminidase deficiency in Tay–Sachs' disease (Variant O). In Hers, H. G. and Van Hoof, F. (eds.) *Lysosomes and Storage Diseases*. pp. 346–356. (New York: Academic Press)

47 Johnson, W. G., Chutorian, A. and Miranda, A. (1977). A new juvenile hexoseaminidase deficiency disease presenting as cerebellar ataxia. *Neurology*, **27**, 1012–1018.

48 Tsay, G. C. and Dawson, G. (1976). Oligosaccharide storage in brains from patients with fucosidosis, G_{M1}-gangliosidosis and G_{M2}-gangliosidosis (Sandhoff's disease). *J. Neurochem.*, **27**, 733–740

49 MacLeod, P. M., Wood, S., Jan, J. E., Applegarth, D. A. and Dolman, C. L. (1977). Progressive cerebellar ataxia, spasticity, psychomotor retardation and hexoseaminidase deficiency in a 10-year-old child: juvenile Sandhoff disease. *Neurology*, **27**, 571–573

50 O'Neill, B., Butler, A. B., Young, E., Falk, P. M. and Bass, N. H. (1978). Adult-onset G_{M2} gangliosidosis. *Neurology*, **28**, 1117–1123

51 Johnson, W. G. and Chutorian, A. M. (1978). Inheritance of the enzyme defect in a new hexoseaminidase deficiency disease. *Ann. Neurol.*, **4**, 399–403

52 MacLaren, N. K., Max, S. R., Cornblath, M., Brady, R. O., Ozand, P. T., Campbell, J., Rennels, M., Mergner, W. J. and Garcia, J. H. (1976). G_{M3} gangliosidosis: a novel human sphingolipody-strophy. *Pediatrics*, **57**, 106–110

53 Tanaka, J., Garcia, J. H., Max, S. R., Viloria, J. E., Kamijo, Y., MacLaren, N. K., Cornblath, M. and Brady, R. O. (1975). Cerebral sponginess and G_{M3} gangliosidosis. Ultrastructure and probable pathogenesis. *J. Neuropathol. Exp. Neurol.*, **34**, 249–262

54 Carter, T. P., Beblowski, D. W., Savage, M. H. and Kanfer, J. N. (1980). Human brain cerebroside β-galactosidase: deficiency of transgalactosidic activity in Krabbe's disease. *J. Neurochem.*, **34**, 189–196

55 Kint, J. A. and Carton, D. (1973). Fabry's disease. In Hers, H. G. and Van Hoof, F. (eds.). *Lysosomes and Storage Diseases*. pp. 357–380. (New York: Academic Press)

56 Desnick, R. J., Klionsky, B. and Sweeley, C. C. (1978). Fabry's disease (α-galactosidase A deficiency). In Stanbury, J. B., Wyngaarten, J. B. and Fredrickson, D. S. (eds.) *The Metabolic Basis of Inherited Disease*. 4th Edn., pp. 810–840. (New York: McGraw-Hill)

57 Grunnet, M. L. and Spilsbury, P. R. (1973). The central nervous system in Fabry's disease. *Arch. Neurol.*, **28**, 231–234

58 Sheth, K. J. and Swick, H. M. (1980). Peripheral nerve conduction in Fabry's disease. *Ann. Neurol.*, **7**, 319–323

59 Schettler, G. and Kahlke, W. (1967). Gaucher's disease. In Schettler, G. (ed.) *Lipids and Lipidosis*. pp. 260–287. (Berlin: Springer)

60 Brady, R. O. (1978). Glucosyl ceramide lipidosis, Gaucher's disease. In Stanbury, J. B., Wyngaarten, J. B. and Frederickson, D. S. (eds.) *The Metabolic Basis of Inherited Disease*. 4th Edn., pp. 731–746. (New York: McGraw Hill)

61 Nishimura, R., Omos-Lau, N., Ajmone-Marsan, C. and Barranger, J. A. (1980). Electroencephalographic findings in Gaucher's disease. *Neurology*, **30**, 152–159

62 Robinson, D. B. and Glew, R. H. (1980). Acid phosphatase in Gaucher's disease. *Clin. Chem.*, **26**, 371–382

63 Austin, J. (1973). Metachromatic leucodystrophy (sulfatide lipidosis). In Hers, H. G. and Van Hoof, F. (eds.) *Lysosomes and Storage Diseases*. pp. 412–437. (New York: Academic Press)

64 Dulaney, J. T. and Moser, H. W. (1978). Sulfatide lipidosis: metachromatic leukodystrophy. In Stanbury, J. B., Wyngaarten, J. B. and Fredrickson, D. S. (eds.) *The Metabolic Basis of Inherited Disease*. 4th Edn., pp. 770–809. (New York: McGraw-Hill)

65 Farrell, D. F., MacMartin, M. P. and Clark, A. F. (1979). Multiple molecular forms of arylsulphatase A in different forms of metachromatic leucodystrophy (MLD). *Neurology*, **29**, 16–20

66 Poser, C. M. (1968). Diseases of the myelin sheath. In Minckler, J. (ed.) *Pathology of the Nervous System*. Vol. 1, pp. 767–821. (New York: McGraw-Hill)

67 Percy, A. K., Kaback, M. M. and Herndon, R. M. (1977). Metachromatic leukodystrophy: comparison of early and late onset forms. *Neurology*, **27**, 933–941

68 Haltia, T., Palo, J., Haltia, M. and Icen, A. (1980). Juvenile metachromatic leukodystrophy. Clinical, biochemical and neuropathologic studies in nine new cases. *Arch. Neurol.*, **37**, 42–46

69 Buonanno, F. S., Ball, M. R., Laster, D. W., Moody, D. M. and McLean, W. T. (1978). Computed tomography in late-infantile metachromatic leukodystrophy. *Ann. Neurol.*, **4**, 43–46

70 Raghavan, S. S., Gajewski, A. and Kolodny, E. H. (1981). Leukocyte sulfatidase for the reliable diagnosis of metachromatic leukodystrophy. *J. Neurochem.*, **36**, 724–731

71 Suzuki, K. and Suzuki, Y. (1978). Galactosylceramide lipidoses: globoid cell leukodystrophy (Krabbe's disease). In Stanbury, J. B., Wyngaarten, J. B. and Fredrickson, D. S. (eds.) *The Metabolic Basis of Inherited Disease*. 4th Edn., pp. 747–769. (New York: McGraw-Hill)

72 Philippart, M. (1978). Clinical and biochemical pathophysiology of ataxia in the sphingolipidoses. In Kark, R. A. P., Rosenberg, R. N. and Schut, L. J. (eds.) *Advances in Neurology*. Vol. 21, pp. 131–149. (New York: Raven Press)

73 Dickerman, L. H., Kurczynski, T. W. and MacBride, R. G. (1981). The effect of psychosine upon growth of human skin fibroblasts from patients with globoid cell leukodystrophy. *J. Neurol. Sci.*, **50**, 181–190

74 Austin, J. H. (1968). Globoid (Krabbe) leukodystrophy. In Minckler, J. (ed.) *Pathology of the Nervous System*. Vol. 1, pp. 843–858. (New York: McGraw-Hill)

75 Dunn, H. G., Dolman, C. L., Farrell, D. F., Tischler, B., Hasinoff, C. and Woolf, L. I. (1976). Krabbe's leukodystrophy without globoid cells. *Neurology*, **26**, 1035–1041

76 Svennerholm, L., Vanier, M-T., Hakansson, G. and Mansson, J-E. (1981). Use of leukocytes in diagnosis of Krabbe disease and detection of carriers. *Clin. Chim. Acta*, **112**, 333–342

77 Brady, R. O. and King, F. M. (1973). Niemann–Pick's disease. In Hers, H. G. and Van Hoof, F. (eds.) *Lysosomes and Storage Diseases*. pp. 439–452. (New York: Academic Press)

78 Brady, R. O. (1978). Sphingomyelin lipidosis: Niemann–Pick disease. In Stanbury, J. B., Wyngaarten, J. B. and Fredrickson, D. S. (eds.) *The Metabolic Basis of Inherited Disease*. 4th Edn., pp. 718–730. (New York: McGraw-Hill)

79 Kennedy, J. F. (1976). Chemical and biochemical aspects of the glycosaminoglycans and proteoglycans in health and disease. *Adv. Clin. Chem.*, **18**, 1–101

80 Lindahl, U. and Höök, M. (1978). Glycosaminoglycans and their binding to biological macromolecules. *Annu. Rev. Biochem.*, **47**, 385–417

81 McKusick, V. A., Neufeld, E. F. and Kelly, T. E. (1978). The mucopolysaccharide storage diseases. In Stanbury, J. B., Wyngaarten, J. G. and Fredrickson, D. S. (eds.) *The Metabolic Basis of Inherited Disease*. 4th Edn., pp. 1282–1307. (New York: McGraw-Hill)

82 Van Hoof, F. (1973). Mucopolysaccharidoses. In Hers, H. G. and Van Hoof, F. (eds.) *Lysosomes and Storage Diseases.* pp. 218–259. (New York: Academic Press)

83 Dawson, G. and Lenn, N. J. (1976). Polysaccharide metabolism disorders. In Vinken, P. J. and Bruyn, G. W. (eds.) *Handbook of Clinical Neurology.* Vol. 27, pp. 143–168. (Amsterdam: North Holland)

84 Dorfman, A. and Matalon, R. (1976). The mucopolysaccharidoses (a review). *Proc. Natl. Acad. Sci., USA*, **73**, 630–637

85 Kint, J. A., Dacremont, G., Carton, D., Orye, E. and Hooft, C. (1973). Mucopolysaccharidosis: secondarily induced abnormal distribution of lysosomal isoenzymes. *Science*, **181**, 352–354

86 Winters, P. R., Harrod, M. J., Molenich-Heetred, S. A., Kirkpatrick, J. and Rosenberg, R. N. (1976). α-L-Iduronidase deficiency and possible Hurler–Scheie genetic compound. Clinical, pathologic, and biochemical findings. *Neurology*, **26**, 1003–1007

87 Nwokoro, N. and Neufeld, E. F. (1979). Detection of Hunter heterozygotes by enzymatic analysis of hair roots. *Am. J. Hum. Genet.*, **31**, 42–49

88 Taori, G. M., Iyer, G. V., Mokashi, S., Balasubramanian, K. A., Cherian, R., Chandi, S., Job, C. K. and Bachhawat, B. K. (1972). Sanfilipo syndrome (mucopolysaccharidosis – III). *J. Neurol. Sci.*, **17**, 323–345

89 Dekaban, A. S. and Constantopoulos, G. (1977). Mucopolysaccharidosis Types I, II, IIIA and V. Pathological and biochemical abnormalities in the neural and mesenchymal elements of the brain. *Acta Neuropathol.*, **39**, 1–7

90 Constantopoulos, G., Eiben, R. M. and Schafer, I. A. (1978). Neurochemistry of the mucopolysaccharidoses: brain glycoaminoglycans, lipids and lysosomal enzymes in mucopolysaccharoidosis type IIIB (α-*N*-acetylglucosamidase deficiency). *J. Neurochem.*, **31**, 1215–1222

91 Hadfield, M. G., Ghatak, N. R., Nakonezna, I., Lippman, H. R., Myer, E. C., Constantopoulos, G. and Bradley, R. M. (1980). Pathologic findings in mucopolysaccharidosis type IIIB (Sanfillipo's syndrome B). *Arch. Neurol.*, **37**, 645–650

92 Koto, A., Horwitz, A. L., Suzuki, K., Tiffany, C. W. and Suzuki, K. (1978). The Morquio syndrome: neuropathology and biochemistry. *Ann. Neurol.*, **4**, 26–36

93 Pilz, H., Von Figura, K. and Goebel, H. H. (1979). Deficiency of arylsulfatase B in 2 brothers aged 40 and 38 years (Maroteaux–Lamy syndrome, type B). *Ann. Neurol.*, **6**, 315–325

94 Sly, W. S., Quinton, B. A., McAlister, W. H. and Rimoin, D. L. (1973). Beta-glucuronidase deficiency. Report of clinical, radiologic and biochemical features of a new mucopolysaccharidosis. *J. Pediatr.*, **82**, 249–257

95 Spranger, J. and Wiedermann, H. (1970). The genetic mucolipidoses. *Hum. Genet.*, **9**, 113–139

96 Spranger, J., Gehler, J. and Cantz, M. (1977). Mucolipidosis 1 – a sialidosis. *Am. J. Med. Genet.*, **1**, 21–29

97 Ockerman, P. A. (1973). Mannosidosis. In Hers, H. G. and Van Hoof, F. (eds.) *Lysosomes and Storage Diseases.* pp. 291–304. (New York: Academic Press)

98 Sung, J. H., Hayano, M. and Desnick, R. J. (1977). Mannosidosis: pathology of the nervous system. *J. Neuropathol. Exp. Neurol.*, **36**, 807–820

99 Van Hoof, F. (1973). Fucosidosis. In Hers, H. G. and Van Hoof, F. (eds.) *Lysosomes and Storage Diseases.* pp. 277–290. (New York: Academic Press)

100 Van Hoof, F. and Hers, H. G. (1973). Other lysosomal storage disorders. In Hers, H. G. and Van Hoof, F. (eds.) *Lysosomes and Storage Diseases.* pp. 554–573. (New York: Academic Press)

101 Shih, V. (1977). Miscellaneous metabolic disorders involving amino acids and organic acids. In Vinken, P. J. and Bruyn, G. W. (eds.) *Handbook of Clinical Neurology.* Vol. 29, pp. 195–243. (Amsterdam: North Holland)

102 Maury, C. P. J., Haltia, M. and Palo, J. (1981). Regional distribution of glycoasparagine storage material in the brain in aspartylglycosaminuria. *J. Neurol. Sci.*, **50**, 291–298

103 Austin, J. H. (1973). Studies in metachromatic leukodystrophy. XII. Multiple sulfatase deficiency. *Arch. Neurol.*, **28**, 258–264

104 Eto, Y., Meier, C. and Herschkowitz, N. N. (1976). Chemical composition of brain and myelin in two patients with multiple sulphatase deficiency (a variant from metachromatic leukodystrophy). *J. Neurochem.*, **27**, 1071–1076

105 Eto, Y., Owada, M., Kitagawa, T., Kokubun, Y. and Rennert, O. M. (1979). Neurochemical

abnormality in I-cell disease: chemical analysis and a possible importance of beta-galactosidase deficiency. *J. Neurochem.*, **32**, 397–405

106 Miller, A. L., Levitt, P., Ingraham, H., Converse, J. and Lewis, L. (1979). Properties of acid-β-D-galactosidase isolated from I-cell disease brain and spleen. *J. Neurochem.*, **32**, 1479–1485

107 Martin, J. J., Leroy, J. G., Farriaux, J. P., Fontaine, G., Desnick, R. J. and Cabello, H. (1975). I-cell disease (mucolipidosis II). A report on its pathology. *Acta Neuropathol.*, **33**, 285–305

108 Lowden, J. A. and O'Brien, J. S. (1979). Sialidosis: a review of human neuramidase deficiency. *Am. J. Hum. Genet.*, **31**, 1–18

109 Federico, A., Cecio, A., Apponi Battim, G., Michalski, J. C., Strecker, G. and Guazzi, G. C. (1980). Macular cherry-red spot and myoclonus syndrome. Juvenile form of sialidosis. *J. Neurol. Sci.*, **48**, 157–169

110 Franceschetti, S., Uziel, G., Di Donato, S., Caimi, L. and Avanzini, G. (1980). Cherry-red spot myoclonus syndrome and α-neuraminidase deficiency: neurophysiological, pharmacological and biochemical study in an adult. *J. Neurol. Neurosurg. Psychiatry.*, **43**, 934–940

111 Steinman, L., Tharp, B. R., Dorfman, L. J., Forno, L. S., Sogg, R. L., Kelts, K. A. and O'Brien, J. S. (1980). Peripheral neuropathy in the cherry-red spot-myoclonus syndrome (sialidosis type 1). *Ann. Neurol.*, **7**, 450–456

112 Zeman, W. and Siakotos, A. N. (1973). The neuronal ceroid-lipofuscinoses. In Hers, H. G. and Van Hoof, F. (eds.) *Lysosomes and Storage Diseases.* pp. 519–551. (New York: Academic Press)

113 Swaiman, K. F., Garg, B. P. and Lockman, L. A. (1975). Sea-blue histiocyte and posterior column dysfunction: a familial disorder. *Neurology*, **25**, 1064–1067

114 Ryan, G. B., Anderson, R. McD., Menkes, J. H. and Dennett, X. (1970). Lipofuscin (ceroid) storage disease of the brain. Neuropathological and neurochemical studies. *Brain*, **93**, 617–628

115 Jervis, G. A. and Pullarkat, R. K. (1978). Pigment variant of lipofuscinoses. *Neurology*, **28**, 500–503

116 Greenwood, R. S. and Nelson, J. S. (1978). Atypical neuronal ceroid-lipofuscinosis. *Neurology*, **28**, 710–717

117 Noonan, S. M., Desousa, J. and Riddle, J. M. (1978). Lymphocyte ultrastructure in two cases of neuronal ceroid-lipofuscinosis. *Neurology*, **28**, 472–477

118 Miley, C. E., Gilbert, E. F., France, T. D., O'Brien, J. F. and Chun, R. W. M. (1978). Clinical and extraneural histologic diagnosis of neuronal ceroid-lipofuscinosis. *Neurology*, **28**, 1008–1012

119 Pilz, H., Schwendemann, G. and Goebel, H. H. (1978). Diagnostic significance of myeloperoxidase assay in neuronal ceroid-lipofuscinoses (Batten–Vogt syndrome). *Neurology*, **28**, 924–927

120 Schwerer, B. and Bernheimer, H. (1978). Leukocyte PPD-peroxidase activity with polyunsaturated fatty acid hydroperoxides: normal values in Batten's disease. *J. Neurochem.*, **31**, 457–460

121 Glenner, G. G., Ignaczak, T. F. and Page, D. L. (1978). The inherited systemic amyloidoses and localized amyloid deposits. In Stanbury, J. B., Wyngaarten, J. B. and Fredrickson, D. S. (eds.) *The Metabolic Basis of Inherited Disease.* 4th Edn., pp. 1308–1339. (New York: McGraw-Hill)

122 Shoji, S. and Okano, A. (1981). Amyloid fibril protein in familial amyloid polyneuropathy. *Neurology*, **31**, 186–190

192

4
Disorders involving inorganic ions

Inorganic ionic (Na^+, Cl^-, K^+) concentration gradients across the outer limiting membranes of neurons and muscle cells determine the excitability of these cells, while calcium ion (Ca^{2+}) concentrations influence neurotransmitter release and also excitation–contraction coupling in muscle. Disturbances in the concentration of these small inorganic anions and cations which are ubiquitously present in body fluids therefore alter neural and muscular functions. The disorders produced by alterations in individual ions will be considered below.

4.1 Na^+

4.1.1 Hyponatraemia and hypo-osmolality

Since Na^+ is the main anion in extracellular fluid, clinically significant hyponatraemia is almost inevitably associated with extracellular hypo-osmolality unless some osmotically active substance (e.g. a hexose) is present simultaneously at abnormally high concentrations in extracellular fluid. Hypo-osmolality may have implications for ions other than Na^+ in extracellular fluid. Both hyponatraemia and hypo-osmolality affect nervous system function. Further, certain types of neurological dysfunction may themselves cause extracellular hypo-osmolality throughout the body.

The subject of hyponatraemia is discussed in detail by Fishman[1].

Abnormal biochemistry – Na^+ ionic concentration gradients across the neuronal cell membrane (142 mEq/l outside, 5 mEq/l inside) are a major factor in maintaining the electrical potential difference of some 90 mV that exists across the resting membrane. This potential difference makes the neuron an excitable cell, and makes possible the dendrito-somal-axonic conduction of nerve impulses.

The magnitude of the potential difference across a cell membrane is determined by

193

(1)   The ionic concentration gradients (mainly Na^+, K^+, Cl^-) across the membrane, and

(2)   The differential permeabilities (P) of the membrane to these ions (in the resting state $K^+:Na^+:Cl^-$ permeabilities are as 1.0:0.04:0.45).

The resting membrane potential (E_R) is described by the Hodgkin–Huxley modification of the Nernst equation, namely

$$E_R = \frac{RT}{F} \log \frac{P_{K^+}.[K^+]_{in} + P_{Na^+}.[Na^+]_{in} + P_{Cl^-}.[Cl^-]_{out}}{P_{K^+}.[K^+]_{out} + P_{Na^+}.[Na^+]_{out} + P_{Cl^-}.[Cl^-]_{in}}$$

The numerical values for this at 20° C, are

$$E_R = 58 \log_{10} \frac{(1 \times 140) + (0.04 \times 4) + (0.45 \times 120)}{(1 \times 4) + (0.04 \times 140) + (0.45 \times 6)}$$

The calculated E_R works out at approximately 70 mV. During activation of the membrane, Na^+ relative permeability rises from 0.04 to about 0.2, associated with an inflow of Na^+ along its ionic concentration gradient. This inflow is associated with a fall in cell membrane potential, since the denominator in the Nernst equation has been increased more than the numerator by the change in P_{Na^+}. The persistence of ionic concentration gradients across the neuronal cell membrane, in the presence of ionic permeability barriers in the membrane which are not complete, depends on active work being done by the cell to maintain the concentration gradients. It appears that there is an enzymatic transport mechanism which is orientated across the cell membrane so that it can actively transport Na^+ from the inside of the cell to the outside. This transport is linked to the transport of K^+ from extracellular fluid to the inside of the cell. This 'sodium pump' system appears to be a Na^+, K^+-linked adenosine triphosphatase, which simultaneously catalyses the hydrolysis of adenosine triphosphate, formed intracellularly during mitochondrial electron transport (Figure 2.16). Thus the maintenance of ionic concentration gradients, and of neuronal excitability, is linked to the production of ATP in the energy-yielding pathway.

Reduction in extracellular Na^+ concentration will reduce the denominator term in the equation above, and the potential difference across the neuronal cell membrane will be increased. Neural tissue will then become less excitable, since a greater fall in membrane potential must be induced before the potential reaches the voltage threshold at which a spontaneous action potential will be propagated. A fall in extracellular Na^+ concentration, without change in the concentrations of other ions, may lead to an osmotic shift of water to the intracellular compartment, with a consequent fall in intracellular K^+ concentration. This would further tend to increase the potential difference across the resting membrane. It is possible that intracellular overhydration has other biochemical consequences, but these are not accurately known. It is also possible that similar electrolyte disturbances occurring in muscle may contribute to some of the neurological manifestations of hyponatraemia, e.g. the weakness and lethargy.

The extent to which neural function is altered by hyponatraemia depends not only on the extent of the decrease in extracellular Na^+ concentration but also on the rate of development of the decrease. Rate of decrease in plasma osmolality (normally 285–295 mOsm/l) is more important than extent of decrease in determining altered neural function. If the brain has sufficient time, compensatory ionic changes may develop to minimize the ion and volume disturbances which might otherwise have occurred.

Aetiology – There are numerous causes of hyponatraemia and hypo-osmolality. The more important aetiological categories include the following:

(1)   Disturbances which cause insufficient water excretion relative to Na^+ excretion:
   (a)   Following severe Na^+ and H_2O depletion, with H_2O replacement only, as may occur in treating diarrhoea or vomiting,
   (b)   Inappropriate (excessive) antidiuretic hormone secretion, associated with
      – carcinoma in various sites
      – lung disease
      – neurological disease, e.g. encephalitis, head injury, glioma, cerebral infarction, polyneuritis, and
   (c)   Hypoadrenalcorticism (Addison's disease).

(2)   Drug effects, e.g. carbamazepine, cyclophosphamide, vincristine, chlorpropamide, diuretics.

(3)   Late stages of congestive heart failure and hepatic cirrhosis (mechanisms uncertain).

(4)   Acute oliguric renal failure or following water overloading in chronic renal failure.

(5)   Psychogenic polydypsia.

(6)   Following solute accumulation (e.g. glucose, mannitol) in extracellular fluid – here plasma osmolality may not be reduced though Na^+ concentration is.

Structural pathology – The pathological changes resulting from hyponatraemia and hypo-osmolality occur at a molecular rather than a microscopically or macroscopically visible level.

Clinical features – Plasma Na^+ concentrations below 120 mEq/l are likely to be associated with neurological disturbance[2] which begins with anorexia and lethargy and develops into delirium, confusion, stupor and coma. Weakness, nausea, vomiting, headache, muscle cramping, myoclonus, asterixis and, in some instances, epilepsy may occur. If the fall in Na^+ concentration has been

195

rapid enough, brain oedema and raised intracranial pressure may be present.

Diagnosis
(1) Clinical: The clinical picture of hyponatraemia is itself non-specific. If the typical symptoms develop in circumstances in which hyponatraemia is likely, the nature of the disturbance may be inferred. Otherwise hyponatraemia can only be one clinical diagnostic possibility among others. The differential diagnosis then includes drug overdosage, various forms of toxic, metabolic and infective encephalopathy, and the various causes of raised intracranial pressure.

(2) Laboratory: Measurement of plasma Na^+ concentration and plasma osmolality will establish the diagnosis.

Therapy – The underlying cause of the disorder should be treated, where possible. Intravenous infusion of hypertonic Na^+-containing fluids in a quantity sufficient to restore plasma Na^+ levels to about 130 mEq/l, will usually correct the neurological disturbance. It may be unwise to achieve full biochemical correction too rapidly. Inappropriate antidiuretic hormone secretion is sometimes treated with demeclocycline.

4.1.2 Hypernatraemia

In practice hypernatraemia is commonly associated with plasma hyperosmolality, and is a consequence of disease which causes loss of water in excess of loss of Na^+. The neurological aspects of the subject are dealt with at some length by Swanson[3].

Biochemical abnormality – At a cellular level, the effects of extracellular hypernatraemia would be expected to be the converse of the effects of hyponatraemia, namely intracellular dehydration with a fall in the resting membrane potential. Hypernatraemia sometimes occurs because depressed brain function disturbs central osmotic autoregulation. Thus, a primary cerebral illness which causes electrolyte alteration may itself secondarily change brain biochemical function. It may sometimes be difficult to distinguish the cerebral molecular effects of hypernatraemia alone from the effects of the causative brain disease.

Aetiology – The main causes of severe hypernatraemia are:

(1) Failure of adequate water intake, due to confusion or coma.

(2) Diabetes insipidus, with inadequate water replacement.

(3) Osmotic diuresis, e.g. after intravenous mannitol therapy given to reduce intracranial pressure.

(4) Diabetes mellitus.

Structural pathology – There may be no morphological changes. In pure hypernatraemia produced by the deliberate administration of hypertonic saline or the accidental entry into the circulation of hypertonic saline solution administered into the uterus to produce abortion, the picture may be compounded by the occurrence of multiple intracranial haemorrhages. These haemorrhages are probably due to sudden massive water translocations causing rupture of smaller blood vessels[4].

Clinical features – A degree of depression of consciousness, or abnormality of mental functioning, often determines the failure of water replacement which causes hypernatraemia. Therefore, it is often difficult to distinguish the clinical consequences of hypernatraemia from the clinical manifestations of its cause. In the hypertonic saline intrauterine infusions mentioned above, neurological manifestations were probably largely determined by the sites and size of the intracranial haemorrhages. Hypernatraemia (Na^+ above 150 mE/l) causes thirst, nausea, vomiting, tremors, seizures and coma, and rarely chorea and myoclonus[5].

Diagnosis
(1) Clinical: The possibility of hypernatraemia should be suspected whenever a patient does not maintain his expected required water intake.

(2) Laboratory: Plasma Na^+ concentration and osmolality measurements settle the diagnostic issue.

Therapy – If possible, the cause of hypernatraemia should be corrected. Infusion of fluids with a low or negligible Na^+ content, e.g. 5% glucose, or increased oral water intake, will correct the disturbance. Too rapid correction of the hypertonicity may produce the hazard of water intoxication.

4.2 K^+

4.2.1 Hypokalaemia

Disturbances of body potassium metabolism affect muscle tissue very much more than neural tissue. The effects of potassium disturbance in neurology are considered by Reynolds[6].

Biochemical abnormality – Brain K^+ concentration shows little change relative to change in blood K^+ concentration. Glial cells have high intracellular K^+ concentrations, and are capable of depolarizing in response to raised K^+ levels in extracellular fluid. However, the glial depolarization does not set up a propagating action potential. Glial cells have high activities of Na^+, K^+-linked adenosine triphosphatase. Consequently glia may be able to act as a sink to take up K^+ released by neuronal activity. Thus they may be able to keep extracellular K^+ levels in the brain near physiological values despite the repeated K^+ fluxes occasioned by neuronal activity. The same K^+ uptake

197

action of glia may explain why changes in general extracellular K^+ levels have so little effect on brain K^+ levels.

In muscle the situation is different. Here there appears to be no mechanism for buffering the effects of altered extracellular K^+ concentrations. When extracellular K^+ concentration falls, one would expect hyperpolarization of the resting muscle cell membrane (see the modified Nernst equation, p. 194). Such hyperpolarization would decrease muscle excitability. This would particularly be the case if there was also K^+ transfer from extracellular to muscle intracellular fluid, as is thought to be the case in one variety of hypokalaemia, namely hypokalaemic periodic paralysis. During attacks of hypokalaemic periodic paralysis muscle is inexcitable, apparently due to cell membrane dysfunction[7]. However, contrary to the above prediction, the potential difference across the muscle cell membrane is not increased. It is not known whether altered muscle cell membrane properties explain the absence of occurrence of the expected hyperpolarization of the inexcitable cell membrane, and also the weakness, that occur in hyperkalaemic periodic paralysis. The reason for the periodicity of the attacks of weakness in the disorder is also obscure. It is difficult to obtain data on muscle cell membrane potential measurements in other forms of hypokalaemia.

Aetiology – There are several major causes of hypokalaemia, including the following:

(1) Excess renal K^+ loss, due to
   (a) Excess glucocorticoid activity, from
       – primary hyperaldosteronism
       – secondary hyperaldosteronism
       – Cushing's syndrome
       – iatrogenic factors, e.g. in treating diabetic ketoacidosis,
   (b) Diuretic intake, without adequate K^+ replacement, and
   (c) Renal tubular disease.

(2) Excessive alimentary K^+ loss, due to
   (a) Vomiting,
   (b) Gastric or intestinal drainage,
   (c) Diarrhoea.

(3) Hypokalaemic periodic paralysis, in which the hypokalaemia is due to intracellular transfer of K^+, there being no increased renal K^+ loss. The disorder is inherited as an autosomal dominant trait, often not expressed in the female, so that its inheritance may be misinterpreted as an autosomal recessive one.

Structural pathology – There is no major microscopic or macroscopic pathological change resulting directly from hypokalaemia. During (and sometimes between) attacks of hypokalaemic periodic paralysis, there may be clear or PAS positive vacuoles in the sarcoplasm[8,9].

Clinical features – Hypokalaemia severe enough even to produce cardiac asystole does not of itself alter nervous system function. However, other biochemical disturbances (e.g. acidosis) which are consequences of the cause of the hypokalaemia may be present simultaneously and may alter neural functioning.

Severe hypokalaemia (plasma K^+ concentration below 2–2.5 mE/l) is associated with muscle weakness which may progress to an areflexic paralysis and respiratory insufficiency should the K^+ concentration become low enough. E.c.g. changes occur. The S–T segment sags, the T wave flattens and the U wave becomes prominent. Hypokalaemia may also cause paralytic ileus.

Hypokalaemic periodic paralysis is a hereditary disorder in which recurrent attacks of severe widespread areflexic skeletal muscular weakness occur. The onset is usually in the second decade of life. In each episode weakness increases over an initial period of 1 hour or so and then remains static. The weakness spares facial muscles, external eye muscles and respiratory muscles. After 12–24 hours the weakness usually resolves rapidly. Attacks occur at variable intervals in different individuals, and may be provoked by heavy exercise followed by sleep or rest, by emotional stress, physical trauma, alcohol intake or high carbohydrate meals. Serum K^+ levels are reduced only during attacks. The weakness may appear at K^+ levels of 3 mE/l, a lesser degree of hypokalaemia than that associated with weakness in other varieties of hypokalaemia.

Diagnosis
(1) Clinical: The development of weakness in circumstances where there may have been excessive K^+ loss raises the possibility of hypokalaemia. When there have been several attacks of fully reversible areflexic widespread paresis, one of the periodic paralyses (hypo-, normo- or hyperkalaemic) is likely. Thyrotoxicosis is a rare cause of a hypokalaemic periodic paralysis.

(2) Laboratory: Typical e.c.g. changes suggest the diagnosis of hypokalaemia, and measurement of serum K^+ level confirms it. Attacks of hypokalaemic periodic paralysis can be induced in sufferers by the intake of glucose and insulin, and the e.c.g. and serum K^+ concentration may be monitored serially to confirm the diagnosis.

Therapy – Where possible, the underlying disorder should be treated. Oral potassium supplements, or intravenous K^+ administration, will correct hypokalaemia, but there is the well-known danger of overcorrection, with hyperkalaemia. Oral K^+ therapy can be used to shorten attacks of hypo-kalaemic periodic paralysis. There are reports that continuous acetazolamide therapy is of benefit in preventing attacks in some patients. How the drug acts in this disorder is obscure.

4.2.2 Hyperkalaemia

Hyperkalaemia itself produces no disturbance in neural function and little alteration in skeletal muscle function, though one variety of periodic paralysis in man is associated with hyperkalaemia.

Biochemical abnormality – The expected effect of raised extracellular K^+ concentration, as judged from Hodgkin–Huxley modification of the Nernst equation (p. 194), would be to decrease potential differences across the cell membranes of excitable tissues. In nervous tissue this effect does not occur, because of the K^+ buffering effects of the glial K^+ sink which minimizes changes in pericellular K^+ levels. However, in hyperkalaemia, muscle cells become partly depolarized, and hyperexcitable, with heightened irritability to percussion. Skeletal muscle also becomes weakened, but only at very high K^+ levels which usually gravely compromise cardiac function. In hyperkalaemic periodic paralysis, much lower, though still raised, serum potassium levels are associated with attacks of weakness. The patient's muscle seems abnormally sensitive to the effects of K^+[10].

Aetiology – Hyperkalaemia may have several causes, as follows:

(1) Renal disease, mainly acute renal failure.

(2) Mineralocorticoid deficiency, in Addison's disease and in hypoaldo-steronism.

(3) Use of potassium-sparing diuretics, plus the intake of dietary K^+ supplements.

(4) Familial hyperkalaemic periodic paralysis (adynamia episodica hereditaria), inherited as an autosomal dominant trait with fairly complete penetrance.

Structural pathology – Hyperkalaemia *per se* produces no structural changes. Muscle usually appears histologically normal in hyperkalaemic periodic paralysis.

Clinical features – Hyperkalaemia does not affect nervous system function. At very high serum K^+ levels muscle weakness develops, with some evidences of muscle hyperexcitability, e.g. Chvostek's sign. By this K^+ level there is likely to be severe cardiac disturbance, with characteristic e.c.g. changes (tall T waves, low voltage P waves, atrial asystole, intraventricular conduction block and finally ventricular standstill). Hyperkalaemia may cause death from cardiac asystole before there is detectable skeletal muscle weakness.

In hyperkalaemic periodic paralysis episodes of weakness appear in early life. Attacks are provoked by exercise, and usually last about 1 hour, but sometimes longer. Some cases develop a degree of permanent weakness. Myotonia is present in most cases, often involving muscles in the face, eyes, tongue and hands.

Diagnosis

(1) Clinical: Hyperkalaemia is unlikely to present neurologically except as a periodic paralysis syndrome in which the associated presence of myotonia and the shorter duration of the attacks will help in the differentiation from hypokalaemic periodic paralysis. Whether the very rare normokalaemic periodic paralysis (with relief of attacks by giving Na^+) is a genuine entity, remains uncertain[9].

(2) Laboratory: Hyperkalaemia during attacks of paralysis confirms the diagnosis.

Therapy – Continued acetazolamide or chlorothiazide therapy appears to decrease the frequency of attacks of hyperkalaemic periodic paralysis. Severe hyperkalaemia may be treated by the intravenous infusion of K^+-free fluids, so long as there is no contraindication, e.g. anuria.

4.3 Ca^{2+}

The effects of disordered calcium metabolism on the nervous system are discussed by Davis and Schauf[11].

4.3.1 Hypocalcaemia

Hypocalcaemia, arising from various causes, leads to increased excitability of nervous tissue and muscle.

Biochemical abnormality – In nervous tissue and muscle, calcium ion (Ca^{2+}) is concerned with the institution of further biochemical events following membrane depolarization. In the neuronal cell membrane, depolarization seems to alter the binding of Ca^{2+} at its binding sites. Following depolarization, increased inward Na^+ flux along concentration gradients begins, and if the flux attains a sufficient magnitude a propagating action potential develops. At nerve terminals the depolarization from the action potential causes Ca^{2+} entry into the cell and it there facilitates neurotransmitter release. In muscle, Ca^{2+} is actively concentrated in sarcoplasmic reticulum by virtue of Ca^{2+}-linked adenosine triphosphatase activity. This 'Ca^{2+} pump' requires Mg^{2+} for its functioning. When muscle is depolarized by the arrival of a sufficient number of acetylcholine molecules at motor end plate receptors, the depolarization causes Ca^{2+} release from sarcoplasmic reticulum into the muscle cytosol. Here the Ca^{2+} activates trophonin, which in turn activates trophomyosin. This then activates actin which subsequently interacts with myosin, the interaction producing muscle shortening, i.e. contraction, which persists until the released Ca^{2+} is pumped back into the sarcoplasmic reticulum.

Hypocalcaemia, with decreased pericellular Ca^{2+} concentrations, might be expected to alter the propagation of action potentials, and to decrease neuro-transmitter release at axon terminals. Because intracellular Ca^{2+} is at a lower concentration than extracellular Ca^{2+}, reduced Ca^{2+} concentration in extra-cellular fluid will decrease membrane potentials, and hence increase membrane excitability. In the central nervous system the overall consequence of all the above processes appears to be heightened neural excitability, possibly because much neurotransmitter activity in the brain seems to mediate inhibitory processes. With deficiency of Ca^{2+} in the fluid around muscle cells, excitation–contraction coupling would be disturbed. The effect of these several effects of hypocalcaemia is to produce increased excitability of the central and peripheral nervous systems, and of skeletal muscle.

Aetiology – Reduced extracellular fluid levels of Ca^{2+} occur in

(1) Hypoparathyroidism.

(2) Calcium malabsorption due to alimentary disease or 25-hydroxy-cholecalciferol deficiency.

(3) Vitamin D deficiency.

(4) Uraemia.

(5) Acute pancreatitis.

(6) Osteoblastic bone metastases.

(7) Hypoalbuminaemia, as in cirrhosis or nephrosis.

(8) Neonatal hypocalcaemia
    (a) In the first two days of life,
    (b) In the 4–10th day of life, when it is due to a relative phosphate overload from cow's milk intake.

(9) Hyperventilation – a dubious cause.

Structural pathology – Hypocalcaemia *per se* does not cause structural changes in nervous tissue or muscle, unless it produces raised intracranial pressure which may itself lead to morphological alterations.

Clinical features – Hypocalcaemia causes tetany, with peripheral and cir-cumoral paraesthesiae, cramps, carpopedal spasm, laryngeal stridor, hyper-reflexia and convulsions. Sometimes these symptoms may persist in an attenuated form for many years. The e.e.g. shows a poorly developed or absent alpha rhythm, paroxysmal slow rhythms and later bursts of spike-wave activity. The ST interval is lengthened in the e.c.g. Occasionally raised intracranial pressure develops. Its pathogenesis is obscure.

Diagnosis
(1) Clinical: In overt tetany, particularly when an appropriate cause is present,

the existence of hypocalaemia may be inferred with reasonable confidence, though hypomagnesaemia can produce similar manifestations. When the manifestations of tetany are mild and chronic, there is a considerable risk that a misdiagnosis of psychoneurosis will be made.

(2) Laboratory: Measurement of serum Ca^{2+} level confirms the diagnosis.

Therapy – Whenever possible, the causative mechanism should be corrected. Oral calcium therapy will help restore serum Ca^{2+} levels.

4.3.2 Hypercalcaemia

Hypercalcaemia, irrespective of its aetiology, may disturb neural and muscular function. It is not a common disorder, and the diagnosis may be missed clinically.

Biochemical abnormality – The biochemical and electrophysiological consequences of raised extracellular concentrations of Ca^{2+} are the opposite of those produced by hypocalcaemia. Their overall effect is to produce decreased excitability of nerve tissue and muscle.

Aetiology – The causes of hypercalcaemia include the following:

(1) Hyperparathyroidism,

(2) Multiple osteolytic bone lesions (commonly from breast carcinoma metastases or myeloma),

(3) Vitamin D overdosage,

(4) Vitamin A overdosage,

(5) Sarcoidosis,

(6) Hyperthyroidism,

(7) Adrenal insufficiency,

(8) Prolonged immobilization,

(9) Thiazide diuretic administration,

(10) Acute renal failure – diuretic phase,

(11) Following renal transplantation,

(12) Milk–alkali syndrome.

Structural pathology – Hypercalcaemia of itself causes no microscopic or macroscopic structural abnormality.

Clinical features – Hypercalcaemia causes various patterns of mental disturbance, e.g. confusion, lethargy, bizarre mental states, stupor or coma, together with a generalized myopathic disorder with weakness, hypotonia and fatigueability. The e.e.g. shows generalized slowing of normal cerebral rhythms with bilateral frontal slow transients.

Diagnosis
(1) Clinical: In circumstances in which hypercalcaemia might be expected, the presence of appropriate symptoms would suggest the diagnosis. However, hypercalcaemia often seems to be detected fortuitously on routine biochemical screening investigations.

(2) Laboratory: Serum calcium concentration measurement will establish the diagnosis of hypercalcaemia.

Therapy – As far as possible, the underlying cause of the hypercalcaemia should be treated. A variety of pharmacological measures is available to deal with hypercalcaemia due to specific disorders, e.g. the use of glucocorticoids in hypercalcaemia due to sarcoidosis or breast carcinoma or alkylating agents in hypercalcaemia due to myeloma. In life-threatening hypercalcaemia, rehydration with isotonic saline and then parenteral sodium sulphate or intravenous phosphate administration may be used to deal with the emergency situation.

4.4 Mg^{2+}

The subject of magnesium metabolism has been reviewed by Paymaster[12], and the effects of altered magnesium metabolism on neurological function discussed by Durlach[13].

4.4.1 Hypomagnesaemia

When disturbed magnesium metabolism occurs, there is also often altered metabolism of another divalent cation, calcium. Altered Mg^{2+} levels have similar clinical effects to altered Ca^{2+} levels; the biochemical mechanisms involved are not identical.

Biochemical abnormality – Mg^{2+} activates Na^+, K^+-adenosine triphosphatase. Therefore Mg^{2+} depletion causes decreased Na^+ pumping from within cells to extracellular fluid. This causes decreased Na^+ concentration gradients across neuronal cell membranes, which become partly depolarized and more excitable. Mg^{2+} is also required for normal activity of the Ca^{2+} pump (Ca^{2+}-

adenosine triphosphatase) in muscle. Decreased extracellular Mg^{2+} concentrations will therefore lead to some depolarization of neurons, and have a similar effect to that of hypocalcaemia in increasing muscle excitability.

Aetiology – The major causes of hypomagnesaemia are:

(1) Increased faecal Mg^{2+} loss in early renal failure (in more severe renal failure Mg^{2+} accumulates).

(2) Malabsorption syndromes.

(3) Chronic excess ethanol intake, which impairs renal tubular reabsorption of Mg^{2+}.

Structural pathology – There is no evidence that hypomagnesaemia causes any structural changes by its direct effects.

Clinical features – Hypomagnesaemia causes anxiety, a sense of a lump in the throat, chest tightness and pain, headache, tremor, dizziness, insomnia, faintness, abdominal cramps, and paraesthesiae. The e.e.g. may show paroxysmal disturbances.

Diagnosis
(1) Clinical: The differentiation between hypocalcaemia and hypomagnesaemia cannot be made with any certainty on purely clinical grounds.

(2) Laboratory: Measurement of plasma Mg^{2+} level (normally 1.6–2.1 mE/l) permits the diagnosis of hypomagnesaemia. If hypocalcaemia is also present, it may be necessary to correct one or other electrolyte disturbance before the relative roles of the deficiencies of the two divalent cations in producing the patient's symptoms can be determined.

Therapy – The underlying disorder should be corrected. Oral magnesium sulphate can be used, if necessary, to correct hypomagnesaemia.

4.4.2 Hypermagnesaemia

Raised extracellular Mg^{2+} concentrations have similar effects to raised extracellular Ca^{2+} levels on nervous system and muscular function.

Biochemical abnormality – The biochemical effects of hypermagnesaemia are the converse of those described for hypomagnesaemia. Their consequence is decreased neural and muscular excitability, and impaired neuromuscular transmission.

Aetiology – The causes of hypermagnesaemia include the following:

(1)   Chronic renal failure.

(2)   Excess absorption of Mg^{2+} salts, e.g. from magnesium sulphate enemas.

(3)   Dialysis procedures where a deionizer failure and the use of hard water allows too much Mg^{2+} to diffuse into the patient's blood.

Clinical features – Hypermagnesaemia causes drowsiness, hyporeflexia, hypotension, nausea, bradycardia and oliguria.

Diagnosis
(1)   Clinical: In circumstances where Mg^{2+} accumulation is likely, the occurrence of appropriate symptoms should raise suspicion of the disturbance.

(2)   Laboratory: The diagnosis of hypermagnesaemia can be confirmed by measuring the plasma magnesium level.

Therapy – Correction of the causative disorder should relieve the clinical disturbance.

4.5 H^+

The effects of altered H^+ concentration on the nervous system are considered at some length by Lockman[14].

4.5.1   Acidosis

Although acidosis is a reasonably frequent metabolic abnormality, the specific effects of acidosis itself on neurological and muscle function are unclear, usually because of the simultaneous presence of other metabolic consequences of the disorder which has produced the acidosis.

Biochemical abnormality – Acidosis, an increase in H^+ concentration in extracellular fluid (i.e. a fall in extracellular fluid pH), may arise because of changes in the parameters of the Henderson–Hasselbalch equation.

$$pH = pK_a + \log \frac{[HCO_3^-]}{[H_2CO_3]}$$

Thus either a rise in pCO_2 ($CO_2 + H_2O \rightleftharpoons H_2CO_3$) or a fall in $[HCO_3^-]$ may produce acidosis. The body has available a number of mechanisms which buffer pH changes, e.g. extracellular HCO_3^-, plasma proteins, respiratory compensation (which alters H_2CO_3 concentration), and the more slowly developing mechanism of enhanced renal excretion of acid and NH_4^+ (the latter replacing HCO_3^- in urine). When, despite the buffering mechanisms, H^+ concentration increases in extracellular fluid, H^+ enters cells and displaces intracellular Na^+

and K^+ to extracellular fluid. Little information is available regarding the extent to which change in intracellular H^+ concentration alters the rates of metabolic reactions or whether such reactions are affected differentially. In acute, experimentally-induced, CO_2 retention in animals, brain glucose-6-phosphate and fructose-6-phosphate levels rise, pyruvate and lactate levels fall, and concentrations of Krebs cycle intermediates, amino acids and adenosine phosphates remain unaltered[15]. With more prolonged hypercapnia, concentrations of Krebs cycle intermediates, and later amino-acid concentrations, fall. The acidosis may inhibit phosphofructokinase and thus impair glycolysis, so that energy sources other than glucose may become increasingly involved in maintaining cellular function. In the diseased patient, interpretation of the biochemical situation is often confounded by other chemical abnormalities which arise as consequences of the cause of the acidosis.

The pH of CSF is 0.083 units lower than the pH of plasma. In acidosis, CSF pH falls further, causing an increase in respiration. This increase produces a fall in pCO_2 and consequently decreased amounts of H_2CO_3 in blood and CSF. These changes tend to restore pH values. HCO_3 does not diffuse into or out of CSF as rapidly as H_2CO_3 does. Consequently CSF H^+ concentration changes more rapidly in response to alterations in general extracellular H_2CO_3 concentration than it does in response to HCO_3^- concentration changes. These differential rates of entry into CSF of H_2CO_3 and HCO_3^- may cause problems, as when extracellular acidosis is corrected rapidly by infusion of HCO_3^-. The consequent rise in extracellular H_2CO_3 level (from altered respiration) may be followed rapidly by a rise in CSF H_2CO_3 concentration. However, HCO_3^- level in CSF rises more slowly. Consequently CSF H^+ concentration rises for a short time, even though general extracellular H^+ concentration falls. CSF acidosis appears to be associated with impairment of consciousness and coma, and these symptoms may worsen temporarily in the circumstances described immediately above. The biochemical mechanism whereby CSF acidosis is associated with impaired consciousness is not understood.

Aetiology – The conventional primary division of the causes of acidosis is as follows:

(1) Respiratory – a failure of adequate excretion of CO_2 due to a variety of pulmonary diseases or respiratory control disorders.

(2) Metabolic – due to:

    (a) Excessive loss of base from the body, as in diarrhoea, pancreatic fistulae, uretero-sigmoid anastomosis with stasis.

    (b) Renal inability to excrete NH_4^+, resulting in excess loss of HCO_3^-, as may occur in renal failure.

    (c) Excess ingestion of, or metabolic production of, acid. Metabolic overproduction of H^+ may be due to:
    – Diabetes mellitus.
    – Excess ethanol intake, without food intake.

– Lactic acidosis, due to tissue anoxia, a variety of severe illnesses, and certain drugs (e.g. phenformin, isoniazid) in overdosage. Here anaerobic glycolysis outstrips metabolic capacity to handle pyruvate, which is then diverted to lactate.
– Leigh's disease (Section 2.2.5.5).
– Branched chain ketoacidosis (Section 2.5.2.1).
– Isovaleric acidaemia (Section 2.5.2.1).
– β-Methyl crotonic aciduria (Section 2.5.2.1).
– Ketotic hyperglycinaemic syndromes (Section 2.5.2.2.2), e.g. propionic acidemia, methylmalonic aciduria.

Structural pathology – It is not known for certain whether acidosis of itself produces any structural pathological changes in man.

Clinical features – The only clinical manifestation unequivocally due to acidosis is hyperventilation, produced by the effects of increased H^+ concentrations on the respiratory centre in the brain stem. As mentioned above, it seems likely that an increased CSF H^+ concentration is associated with depression of consciousness.

Diagnosis
(1) Clinical: Clinically, the presence of hypoventilation raises the possibility of respiratory acidosis, and hyperventilation the possibility of metabolic acidosis.

(2) Laboratory: Measurement of reduced arterial blood pH is diagnostic for acidosis. In respiratory acidosis pCO_2 is raised and in metabolic acidosis it is lowered, but with a more than corresponding lowering in blood HCO_3^- concentration.

Therapy – As far as possible the underlying disorder should be corrected. Parenteral basic fluids (Na HCO_3 or Na lactate), or oral base intake, may be necessary to restore extracellular HCO_3^- concentration. Management of acidosis involves frequent monitoring of extracellular electrolyte concentrations in case there are secondary alterations in ions, e.g. in K^+ concentration, during pH correction.

4.5.2 Alkalosis

Alkalosis is a not infrequent metabolic abnormality though it appears to produce no specific signs or symptoms, as distinct from the signs and symptoms of the disorder which causes the alkalosis.

Biochemical abnormality – Biochemical consequences of an abnormally low extracellular H^+ concentration will tend to be the converse of those due to an excess H^+ concentration (i.e. acidosis). The same mechanisms for buffering any pH changes apply, though in the opposite direction. There is little clear information available about the effect of extracellular pH changes on the velocities of intracellular chemical reactions.

Aetiology – Like acidosis, alkalosis may be of respiratory or metabolic origin. Respiratory alkalosis is due to overbreathing. Metabolic alkalosis occurs in:

(1) Loss of gastric acid.

(2) Use of thiazides and other diuretics.

(3) Cushing's syndrome.

(4) Conn's syndrome.

(5) Continued glucocorticoid intake.

(6) Following the relief of chronic hypercapnia, when there is a temporary persistence of the renal mechanisms which compensated for the previous respiratory acidosis.

(7) Excess intake of base, e.g. $NaHCO_3$.

Structural pathology – Alkalosis *per se* produces no structural abnormality.

Clinical features – Alkalosis produces no specific symptoms or signs. The manifestations of overbreathing (e.g. parasthesia in the extremities and around the mouth, chest tightness, clouding of consciousness and emotional disturbance) have sometimes been attributed to decreased extracellular concentrations of ionized calcium, but it is likely that hypocapnia-induced cerebral vasoconstriction makes the major contribution to the production of these symptoms.

Diagnosis
(1) Clinical: Alkalosis may be suspected when disorders occur in which alkalosis is likely to develop.

(2) Laboratory: Measurement of extracellular fluid pH will permit the diagnosis of alkalosis; the arterial pCO_2 will be lowered in respiratory, but raised in metabolic, alkalosis.

Therapy – If possible, the underlying disorder should be corrected. The patient should be rehydrated with sodium chloride-containing solutions; occasionally it may be necessary to add more acid substances (e.g. NH_4Cl, HCl itself) to an intravenous infusion to correct the pH of extracellular fluid.

4.6 COPPER

4.6.1 Copper toxicity: Wilson's disease

Wilson's disease is an uncommon inherited disorder of copper metabolism affecting the brain, liver, kidney, cornea and other tissues[16].

Biochemical abnormality – The primary defect in Wilson's disease appears to be a failure of the liver to synthesize the specific Cu binding plasma globulin caeruloplasmin, together with decreased biliary excretion of the metal. Because of the caeruloplasmin deficiency, relatively more copper than normal is carried in blood loosely bound to albumin though the total serum copper content is decreased. Since the portion of blood copper bound to albumin is dissociable with abnormal ease, excess copper enters tissues and comes to be excreted in urine. Copper accumulates in the liver first, and later in other organs, e.g. the brain, renal tubules, cornea. The copper appears to cause tissue damage through chemical mechanisms which are not yet understood. Copper released from damaged hepatocytes may be translocated into other tissues. Renal tubular damage from copper deposition causes aminoaciduria.

Aetiology – The disorder is inherited as an autosomal recessive trait.

Structural pathology – Wilson's disease (hepatolenticular degeneration) affects many tissues[17,18]. The brain shows cerebral cortical atrophy and in particular shrinkage of the lenticular and caudate nuclei which in some cases may undergo cavitation. Alzheimer type astroglia occur in the basal ganglia. The liver is cirrhotic, renal tubular damage occurs and in the eyes there is copper deposition at the corneo-scleral limbus region (causing the appearance of Kayser–Fleischer rings).

Clinical features – When Wilson's disease develops in childhood there may be no neurological disturbance. The presentation is then as an atypical or pro-longed hepatitis. More often the disease appears in adolescence or early adult life. There is then progressive tremor, ataxia of gait and arm movement, dysarthria, choreo-athetosis, dystonia and hypertonus. Later inanition, mental deterioration and severe rigidity develop. Cirrhosis is present, but clinically silent. The cornea-scleral limbus region shows brownish copper deposition, visible to the naked eye or to the slit lamp microscope (Kayser–Fleischer rings). Aminoaciduria is commonly present.

Diagnosis
(1) Clinical: Kayser–Fleischer rings are pathognomonic. When they are not seen on clinical examination, but signs of dyskinesia, rigidity or cerebellar ataxia have appeared in a young person, slit lamp microscopic examination of the eyes may permit the diagnosis. In the absence of these corneal

changes, and of a family history of Wilson's disease, or in hepatitic present-
ations, laboratory methods are required for the diagnosis.

(2) Laboratory: A decreased plasma caeruloplasmin level (in the absence of
malnutrition, sprue or nephrosis) is virtually diagnostic. Urinary copper
excretion is usually increased. Liver biopsy shows cirrhosis, and measurement of the increased copper content of the biopsy specimen indicates the
diagnosis.

Therapy – Chelation of the excess copper with D-penicillamine or dimercaprol
(with subsequent renal excretion of the chelate) and the intake of a diet with
a low copper content, will halt the progress of the disorder[19]. Such treatment
may produce some improvement in the neurological manifestations.

4.6.2 Copper deficiency: Menkes' kinky (steely) hair disease (Trichopoliodystrophy)

This rare disorder of infancy, due to dietary copper malabsorption, alters hair
growth and disturbs brain development and function. The topic has been
reviewed by French[20] and Sass-Kortsak and Bearn[16].

Biochemical abnormality – Failure to absorb dietary copper in Menkes' disease
leads to copper deficiency with low serum levels of copper and caeruloplasmin.
In fact copper does absorb into intestinal mucosal cells in the disorder, but
is not transported through them to enter the circulation.
  A number of enzymes contain copper. These include:

(1) Lysyl oxidase, required for collagen and elastin maturation.

(2) Tyrosinase, required for the conversion of tyrosine to melanin.

(3) Cytochrome oxidase, the terminal enzyme of the mitochondrial electron
transport pathway (Section 2.4).

(4) Monoamine oxidase, involved in neurotransmitter degradation (Sections
5.3.1 and 5.4.1).

(5) Superoxide dismutase.

(6) Dopamine-β-hydroxylase, which catalyses the conversion of dopamine to
noradrenaline (Section 5.3.1).

(7) Tryptophan pyrrolase, involved in tryptophan metabolism to N-formyl
kynurenine (Section 2.5.1.3).

(8) Cysteamine, which catalyses the conversion of 6-mercaptoethylamine to
hypotaurine.

If the disorder of copper absorption is present from birth it might decrease
the activities of the above enzymes at a time of rapid tissue growth. Thus it

211

might alter brain development, reduce brain catecholamine levels[21] and interfere with the formation of elastic tissue in arterial walls.

Aetiology – Menkes' disease is inherited as an X-linked recessive trait.

Structural pathology – The internal elastic laminae of arteries are beaded and split in Menkes' disease[22]. Intracranial arteries are dilated and tortuous. Subdural haematomas or hygromas have occurred in some cases. The cerebrum is atrophic and shows foci of cortical neuronal loss with gliosis. Similar areas occur in the basal ganglia and cerebellar cortex and the dorsal spinocerebellar tracts degenerate[23]. There is widespread loss of brain myelin[24]. In the skin the appearance of pili torti is seen[25].

Clinical features – Affected (male) infants, often born prematurely, fail to make developmental progress or to grow normally. Hypothermia often occurs. The scalp hair is short, coarse, sparse and colourless. They are hypotonic and develop seizure disorders and usually die in the second or third year of life. Some have large bladder diverticulae.

Diagnosis
(1) Clinical: It is unlikely that Menkes' disease will be diagnosed clinically unless the characteristic hair appearances are noted and their significance is appreciated. A positive family history is helpful diagnostically.

(2) Laboratory: Low serum copper and copper oxidase levels (relative to normal values for age) help confirm the diagnosis, though such reduced levels are not pathognomic. Low levels also occur in Wilson's disease, though in that disease tissue copper levels are raised, whereas in Menkes' disease they are low.

Therapy – Parenteral copper administration appeared of benefit in a single case of Menkes' disease in whom the diagnosis was made early in the course of the disorder.

4.7 AMMONIUM

The subject of altered circulating NH_4^+ concentrations is discussed in Chapter 3.

4.8 URAEMIA

Uraemia resulting from renal insufficiency is a common disorder which may disturb function at several levels of the nervous system. The associated biochemical abnormalities are multiple, and the chemical pathogenesis of the neurological disturbance is not fully understood.

Biochemical abnormality – Renal insufficiency may produce a number of biochemical abnormalities, as listed below:

(1)   Altered water balance, commonly leading to dehydration (unless anuria develops).

(2)   Hyper- or hypo-natraemia, depending on the dietary Na^+ load.

(3)   Acidosis.

(4)   Hyperkalaemia.

(5)   Hypocalcaemia.

(6)   Hypomagnesaemia.

(7)   Sulphate and phosphate retention.

(8)   Raised blood urea, uric acid and creatinine levels.

(9)   Hypoproteinaemia.

In addition anaemia, hypertension and a bleeding tendency are commonly present. These latter factors, plus altered fluid and electrolyte concentrations in extracellular fluid, all may individually alter brain function. When present collectively, as in uraemia, the pattern of resultant brain dysfunction may be varied, and it may be difficult to attribute particular facets of the disturbance to individual chemical alterations.

Aetiology – Renal insufficiency may be due to:

(1)   Prerenal factors – circulatory insufficiency.

(2)   Renal factors – renal parenchymal disease.

(3)   Postrenal factors – urine outflow obstruction.

Structural pathology – The neuropathological changes described in uraemia have varied[26]. The brain may be grossly normal, or oedematous. Sometimes there has been neuronal loss from the cerebral and the cerebellar cortices. The white matter in some cases has shown focal demyelination. In addition, in some cases there have been hypertensive or haemorrhagic lesions. Peripheral nerves have shown demyelination, which may partly reverse after correction of the disorder[27].

Clinical features – The neurological disturbances associated with uraemia include mental changes, personality alterations, varying degrees of clouding of consciousness ranging from mild drowsiness to coma, epilepsy, and sometimes focal neurological signs and peripheral polyneuropathy.

In patients receiving dialysis therapy for uraemia a 'dialysis dementia' syndrome may develop, with slowly progressive dementia, alterations of mood and personality, speech disorders, seizures and coma, and diffuse paroxysmal e.e.g. changes[28]. The condition is further discussed by Glaser[29].

213

Diagnosis

(1) Clinical: There are no individual clinical features which are pathognomonic of uraemia. However, certain constellations of clinical symptoms can suggest the diagnosis, particularly if there are indications that a disorder is present which may impair renal function.

(2) Laboratory: Measurement of blood urea or serum creatinine levels confirms the diagnosis of uraemia. Measurement of levels of the individual substances whose metabolism is disturbed, e.g. H^+, Ca^{2+}, K^+, is helpful in assessing the extent of the disorder in individual patients.

Therapy – If possible, the cause of uraemia should be corrected. Disturbances of individual electrolytes may need to be remedied and any arterial hypertension that is present should be treated.

REFERENCES

1 Fishman, R. A. (1976). Neurological manifestations of hyponatremia. In Vinken, P. J. and Bruyn, G. W. (eds.) *Handbook of Clinical Neurology*. Vol. 28, pp. 495–505. (Amsterdam: North Holland)

2 Satran, R. and Griggs, R. C. (1979). Metabolic encephalopathy. In Tyler, H. R. and Dawson, D. M. (eds.) *Current Neurology*. Vol. 2, pp. 474–505. (Boston: Houghton Mifflin)

3 Swanson, P. D. (1976). Neurological manifestations of hypernatraemia. In Vinken, P. J. and Bruyn, G. W. (eds.) *Handbook of Clinical Neurology*. Vol. 28, pp. 443–461. (Amsterdam: North Holland)

4 Young, R. S. K. and Truax, B. T. (1979). Hypernatraemic haemorrhagic encephalopathy. *Ann. Neurol.*, **5**, 588–591

5 Sparacio, R. R., Anziska, B. and Schutta, H. S. (1976). Hypernatremia and chorea: a report of two cases. *Neurology*, **26**, 46–50

6 Reynolds, E. H. (1976). Neurological manifestations of potassium imbalance. In Vinken, P. J. and Bruyn, G. W. (eds.) *Handbook of Clinical Neurology*, Vol. 28, pp. 463–494. (Amsterdam: North Holland)

7 Engel, A. G. and Lambert, E. H. (1969). Calcium activation of electrically inexcitable muscle fibres in primary hypokalemic periodic paralysis. *Neurology*, **19**, 851–858

8 Pearson, C. M. (1976). The periodic paralyses. In Vinken, P. J. and Bruyn, G. W. (eds.) *Handbook of Clinical Neurology*. Vol. 28, pp. 581–601. (Amsterdam: North Holland)

9 Brooke, M. H. (1977). *A Clinician's View of Neuromuscular Diseases*. (Baltimore: Williams and Wilkins)

10 Lewis, E. D., Griggs, R. C. and Moxley, R. T. (1979). Regulation of plasma potassium in hyperkalemic periodic paralysis. *Neurology*, **29**, 1131–1137

11 Davis, F. A. and Schauf, U. (1976). Neurological manifestations of calcium imbalance. In Vinken, P. J. and Bruyn, G. W. (eds.) *Handbook of Clinical Neurology*. Vol. 28, pp. 527–542. (Amsterdam: North Holland)

12 Paymaster, N. J. (1976). Magnesium metabolism: a brief review. *Ann. R. Coll. Surg. Engl.*, **58**, 309–314

13 Durlach, J. (1976). Neurological manifestations of magnesium imbalance. In Vinken, P. J. and Bruyn, G. W. (eds.) *Handbook of Clinical Neurology*. Vol. 28, pp. 545–579. (Amsterdam: North Holland)

14 Lockman, L. A. (1976). Neurological aspects of acid–base metabolism. In Vinken, P. J. and Bruyn, G. W. (eds.) *Handbook of Clinical Neurology*. Vol. 28, pp. 507–525. (Amsterdam: North Holland)

214

15 Folbergrová, J., Norberg, K., Quistorff, B. and Siesjo, B. K. (1975). Carbodydrate and amino-acid metabolism in rat cerebral cortex in moderate and extreme hypercapnia. *J. Neurochem.*, **25**, 457–462

16 Sass-Kortsak, A. and Bearn, A. G. (1978). Hereditary disorders of copper metabolism (Wilson's disease (hepatolenticular degeneration) and Menkes' disease (kinky-hair or steely hair syndrome)). In Stanbury, J. B., Wyngaarten, J. B. and Fredrickson, D. S. (eds.) *The Metabolic Basis of Inherited Disease.* 4th Edn., pp. 1098–1126. (New York: McGraw-Hill)

17 Schulman, S. (1968). Wilson's disease. In Minckler, J. (ed.) *Pathology of the Nervous System.* Vol. 1, pp. 1139–1152. (New York: McGraw-Hill)

18 Smith, W. T. (1976). Intoxications, poisons and related metabolic disorders. In Blackwood, W. and Corsellis, J. A. N. (eds.) *Greenfield's Neuropathology.* 3rd Edn., pp. 148–193. (London: Arnold)

19 Barbeau, A. (1981). Treatment of Wilson's disease. In Barbeau, A. (ed.) *Disorders of Movement.* pp. 209–220. (Lancaster: MTP Press)

20 French, J. H. (1977). X-chromosome-linked copper malabsorption. In Vinken, P. J. and Bruyn, G. W. (eds.) *Handbook of Clinical Neurology.* Vol. 29, pp. 279–304. (Amsterdam: North Holland)

21 Morgan, R. F. and O'Dell, B. L. (1977). Effect of copper deficiency on the concentrations of catecholamines and related enzyme activities in the rat brain. *J. Neurochem.*, **28**, 207–213.

22 Grover, W. D., Johnson, W. C. and Henkin, R. I. (1979). Clinical and biochemical aspects of trichopoliodystrophy. *Ann. Neurol.*, **5**, 65–71

23 Iwata, M., Hirano, A. and French, J. H. (1979). Degeneration of the cerebellar system in X-chromosome-linked copper malabsorption. *Ann. Neurol.*, **5**, 542–549

24 Williams, R. S., Marshall, P. C. and Caviness, V. S. Jr. (1978). The cellular pathology of Menkes steely hair syndrome. *Neurology*, **28**, 575–583

25 Aguilar, M. J., Chadwick, D. L., Okuyama, K. and Kamoshita, S. (1966). Kinky hair disease. 1. Clinical and pathological features. *J. Neuropathol. Exp. Neurol.*, **25**, 507–522

26 Greenhouse, A. H. (1968). The neuropathology of renal disease. In Minckler, J. H. (ed.) *Pathology of the Nervous System.* Vol. 1, pp. 1029–1042. (New York: McGraw-Hill)

27 Bolton, C. F. (1976). Electrophysiologic changes in uremic neuropathy after successful renal transplantation. *Neurology*, **26**, 152–161

28 Noriega-Sanchez, A., Maldonado-Martinez, M. and Haiffe, R. M. (1978). Clinical and electroencephalographic changes in progressive uremic encephalopathy. *Neurology*, **28**, 667–669

29 Glaser, G. H. (1978). Progressive dialysis encephalopathy (dialysis dementia). In Matthews, W. B. and Glaser, G. H. (eds.) *Recent Advances in Clinical Neurology.* Vol. 2, pp. 14–18. (Edinburgh: Churchill Livingstone)

5
Disturbances of synaptic transmission

5.1 PRINCIPLES OF NEUROTRANSMITTER BIOLOGY

There appears to be more variety in the chemical substances involved in synaptic transmission than in those involved in any other single stage of neurophysiological activity. This fact is utilized in clinical neuropharmacology, where an attempt is made to alter specific aspects of neural function selectively by pharmacological manipulation of individual neurotransmitters. Unlike the disorders so far considered in this book, disorders of neurotransmitter function all appear to be consequences of other disease processes. The natures of the primary disease processes are usually poorly understood. In general, disorders of synaptic transmission are more frequent and more important clinically than most of the disorders of other neurochemical mechanisms so far dealt with in this book.

The established neurotransmitters are basic molecules (acetylcholine, the two catecholamines dopamine and noradrenaline, and the indolealkylamine 5-hydroxytryptamine, i.e. serotonin). Certain neutral amino acids (glycine, γ-aminobutyrate, taurine and β-alanine) are probable or possible inhibitory transmitters, while certain acidic amino acids (aspartate, cysteine, homocysteine and glutamate) are possible excitatory transmitters[1]. In addition, there is some evidence that the amine histamine may be a neurotransmitter, at least in the hypothalamus[2].

Certain molecules within the nervous system appear to act as modulators of neurotransmission rather than as neurotransmitters themselves (e.g. prostaglandins and certain oligopeptides, including endorphins, and purine nucleotides and nucleosides[3]). It should be pointed out that there seems to be no absolute distinction between neurotransmitters, neuromodulators and neurohormones. The one molecule (e.g. dopamine) may serve a neurotransmitter role at some sites in the nervous system, and a neurohormonal role at others[4]. Neurotransmitters, neuromodulators and neurohormones are all synthesized within neurons, concentrated in presynaptic nerve terminals, and released in response to nerve ending depolarization. Released neurotransmitters act directly but transiently to increase (inhibition) or decrease (excitation) the polarization of the subsynaptic membrane of neurons in synaptic contact with the neuron from which the transmitter is released. Neuromodulators, after release, alter local cell membrane conductance (and thus polarization) indirectly, and more gradually and more diffusely, by altering neurotransmitter-coupled subsynaptic mechanisms. Neurohormones act at a distance from their sites of release, after carriage through extracellular fluid.

The amine neurotransmitters are synthesized within neurons. The catecholamines are formed mainly within perikarya and are transported along axons to axon terminals where they accumulate in storage granules. Acetylcholine is formed mainly in the axon terminals where it and 5-hydroxytryptamine are also kept within storage granules. Within the body, the amine neurotransmitters are found mainly in the central and peripheral nervous systems, where they are concentrated in particular neuronal pathways. However, the catecholamines also occur in the adrenal medulla, and 5-hydroxytryptamine (serotonin) is found in platelets and the intestinal wall. In contrast, most of the amino-acid neurotransmitters are ubiquitous tissue molecules. However γ-aminobutyrate (GABA) and taurine have distributions largely restricted to the nervous system, while glutamate occurs in higher concentration in the nervous system than in other tissues. The sites of amino-acid neurotransmitter formation within neurons do not appear to be firmly established, and the amino-acid transmitter molecules do not seem to be contained in specific storage granules within axon terminals.

In response to action potentials which pass along axons and reach axon terminals, neurotransmitter molecules are released into the synaptic cleft adjacent to the axon terminal. Neurotransmitter release is Ca^{2+} dependent. The release seems to be determined by the influx of Ca^{2+} which occurs in parallel with Na^+ influx when the neuronal cell membrane becomes depolarized by an action potential. There are suggestions that the Ca^{2+} entry facilitates protein carboxymethylation which expedites neurotransmitter release from storage granules after they fuse with the cell membrane of the terminal axon[5].

The released neurotransmitter molecules may become attached to certain molecules on the subsynaptic dendritic or somal membrane of another neuron. A neurotransmitter may bind specifically to more than one chemical and functional type of receptor[6]. Neurotransmitter binding to receptors alters the polarization of the subsynaptic membrane around the receptors. If the change in polarization is in the direction of depolarization (i.e. a postsynaptic excitatory potential is produced) and if the potential achieves sufficient magnitude, a propagating action potential may be initiated. The chemical mechanisms whereby the binding of neurotransmitter to receptor are transduced into altered cell membrane polarization are not understood for some of the known and putative neurotransmitters. Most information is available for the catecholamine neurotransmitters. At some, though probably not all, sites of catecholaminic neurotransmission the subsynaptic receptor appears to be either the enzyme adenylate cyclase or the enzyme guanylate cyclase. Bonding of neurotransmitter to receptor activates the enzyme and sets in train the following series of events.

adenosine / guanosine } triphosphate —(ADENYLATE CYCLASE / GUANYLATE CYCLASE)→ cyclic { adenosine / guanosine } monophosphate

cAMP or cGMP → PROTEIN KINASE activation

protein + ATP —(PROTEIN KINASE)→ phosphorylated protein + AMP

Cell membrane protein phosphorylation alters the permeability of ion channels in the membrane. This membrane permeability change permits ion flux changes which alter local ion concentration gradients across the cell membrane. This changes the polarization of the membrane.

The action of each batch of released neurotransmitter molecules terminates as the neurotransmitter concentration falls in the synaptic cleft. (The binding of neurotransmitter to receptor appears to be reversible, so that bound transmitter dissociates from its receptors when the local transmitter concentration falls.) In the case of acetylcholine the local fall in synaptic neurotransmitter concentration appears due to enzymatic hydrolysis of the molecule. For all the other known neurotransmitters the fall in concentration appears due to active re-uptake of the neurotransmitter into the axon terminal from which it was released, even though chemical degradative mechanisms may also be present locally.

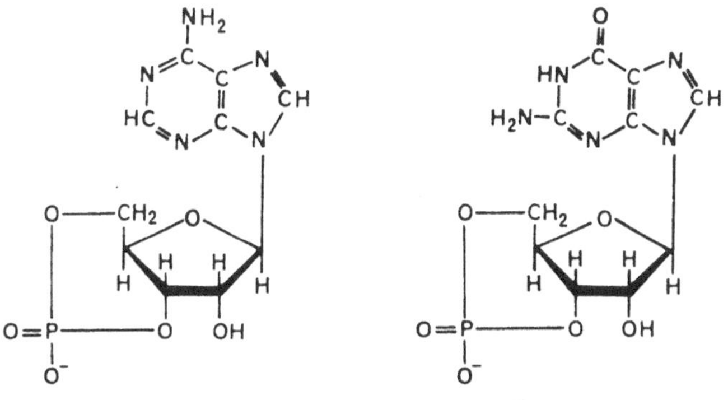

.cyclic adenosine monophosphate

.cyclic guanosine monophosphate

Certain mechanisms exist governing the amount of neurotransmitter release. Rising noradrenaline or dopamine concentrations in synapses may exert a feedback inhibition at an earlier stage of neurotransmitter synthesis. Released noradrenaline may activate presynaptic[7] as well as postsynaptic receptors. The presynaptic receptor activation may modulate noradrenaline release. Presynaptic receptors with similar autoregulatory functions exist at certain cholinergic synapses.

Neurotransmitter biology is well discussed in the monograph of Cooper, Bloom and Roth[8].

5.2 ACETYLCHOLINE

5.2.1 Acetylcholine biology

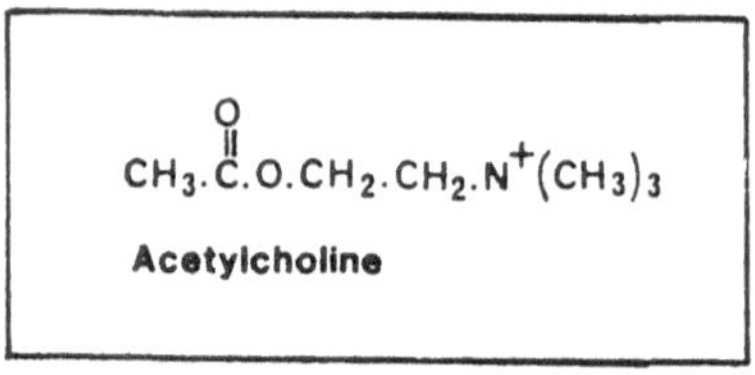

$$CH_3.\overset{O}{\overset{\|}{C}}.O.CH_2.CH_2.N^+(CH_3)_3$$

Acetylcholine

Synthesis – Acetylcholine is synthesized in neuronal cytoplasm, probably mainly in axon terminals, from choline and acetyl Coenzyme A.

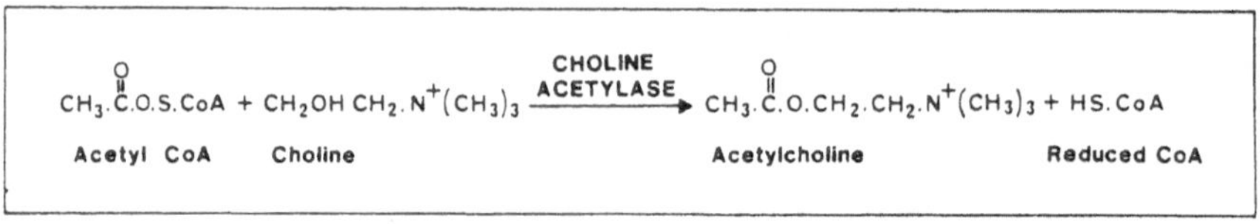

$$CH_3.\overset{O}{\overset{\|}{C}}.O.S.CoA + CH_2OH\,CH_2.N^+(CH_3)_3 \xrightarrow[\text{ACETYLASE}]{\text{CHOLINE}} CH_3.\overset{O}{\overset{\|}{C}}.O.CH_2.CH_2.N^+(CH_3)_3 + HS.CoA$$

Acetyl CoA   Choline   Acetylcholine   Reduced CoA

The enzyme which catalyses the reaction, choline acetylase (or choline acetyltransferase) is present in cytoplasm, mainly in nerve endings. Acetyl CoA is derived from pyruvate metabolism. The mechanism whereby the acetyl CoA moves from mitochondria to cytoplasm prior to acetylcholine synthesis is not established. Sterling and O'Neill[9] suggested that citrate may carry acetyl groups out of mitochondria into the cytosol. Choline molecules may be taken up from the blood, from the synaptic cleft (where they have been produced by acetylcholine hydrolysis) or they may be formed from the degradation of phosphatidyl choline. The brain cannot synthesize choline itself[9], but can take the preformed molecule up via an active transport system[10,11].

Storage – Acetylcholine is contained within storage granules inside axon terminals.

Release – Acetylcholine release in response to an action potential is Ca^{2+} dependent, as explained above.

Receptor action – Pharmacologically (and chemically), there are two well established classes of acetylcholine receptor. Muscarinic cholinergic receptors (blocked by atropine) occur in the central nervous system and, in the peripheral nervous system, at parasympathetic synapses. Nicotinic cholinergic receptors occur in sympathetic ganglia, and at myoneural junctions of skeletal muscle. Nicotinic cholinergic receptors are not blocked by atropine. A certain amount of information is available concerning the chemical nature of the nicotinic acetylcholine receptor of skeletal muscle[12]. The receptor appears to be an acidic glycoprotein which contains acetylcholine binding sites and also a cation-specific membrane channel which is 'opened' when acetylcholine molecules occupy the receptor binding sites. Subsequent Na^+ influx along the ion channel appears to initiate muscle contraction. Probably analogous molecular events occur at central nervous system acetylcholine receptors. Intimate details of the chemical changes involved in the binding of acetylcholine to its receptor are known[13].

Hydrolysis – Acetylcholine is hydrolysed by the relatively specific enzyme acetylcholinesterase, the distribution of which is largely restricted to neural tissue (where it occurs in the microsomes and mitochondria). Acetylcholine may also be hydrolysed by non-specific esterases (pseudocholinesterases) which occur in many tissues and in the plasma. Details of the chemical mechanisms involved in the enzymatic catalysis of the hydrolysis of acetylcholine are known[13]. The choline released by the hydrolysis in the synaptic cleft undergoes

$$CH_3.\overset{O}{\overset{\|}{C}}.O.CH_2.CH_2.\overset{+}{N}(CH_3)_3 + H_2O \xrightarrow{\text{ESTERASE}} CH_3COOH + N^+(CH_3)_3.CH_2.CH_2OH$$

acetylcholine   acetic acid   choline

subsequent Na^+-dependent active re-uptake into axon terminals. Here it is used for resynthesis of acetylcholine (Figure 5.1). The active transport system for choline is found only in cholinergic neurons[14].

5.2.2 Disorders related to altered acetylcholine function

As mentioned earlier (Section 2.2.5), biochemical abnormalities which cause a decreased metabolic flux through the anaerobic glycolytic pathway (e.g. hypoglycaemia, hypoxia), and deficient activity of the pyruvate dehydrogenase complex (as in thiamine deficiency and certain instances of Friedreich's ataxia and related spinocerebellar degenerations) may cause deficient acetyl CoA production. This deficiency leads to reduced acetylcholine synthesis. It is believed that central acetylcholine deficiency may be responsible for some of the more important manifestations of these disorders. It is also possible that acetylcholinergic transmission in the central nervous system may be altered in certain forms of epilepsy, and several anti-epileptic drugs are known to alter

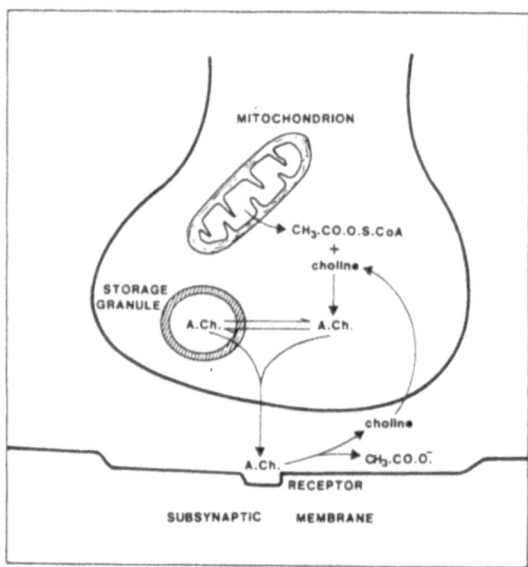

Figure 5.1   Diagram of events at a cholinergic axon terminal

brain acetylcholine concentrations[15]. However, there is no unequivocal evidence of a relation between epilepsy and altered acetylcholine function in the brain. The neurological disorder most unambiguously related to altered acetylcholinergic function in man is myasthenia gravis.

5.2.2.1 Myasthenia gravis

Myasthenia gravis is a not infrequent disorder (prevalence $1:20000$ of the population). It is potentially a very serious condition which is due to an inadequacy of cholinergic transmission at the neuromuscular junctions of skeletal muscle[16].

Biochemical abnormality – Although other theories have held sway in the past, it is now generally accepted that myasthenia gravis is an autoimmune disorder[17]. Antibodies form to nicotinic type acetylcholine receptors. These antibodies destroy some receptors and compete with acetylcholine to bind to the surviving receptors[12]. The net result of these processes is that fewer receptors are activated by the acetylcholine molecules released into the myoneural junctions after each nerve impulse. Accordingly, skeletal muscle contraction is weakened. There appears to be no alteration in the amount of acetylcholine released from axon terminals in response to each nerve impulse, and no significant increase in acetylcholine degradation. The essential problem in the disease is the competition between acetylcholine and antibody for a decreased population of acetylcholine receptors Circulating antibodies to nicotinic acetyl-choline receptors are present in considerably raised amounts in plasma.

Aetiology – The factor or factors which initiate the abnormal antibody pro-duction in myasthenia gravis are obscure. In myasthenia gravis other auto-immune disorders (e.g. polymyositis, thyroiditis, arthritis, diabetes mellitus, pernicious anaemia) occur more frequently than would be expected on a chance basis. This suggests the possibility that some myasthenics may be predisposed to form antibodies to their own tissues. The main site of abnormal antibody formation is probably the thymus.

Structural pathology – Some 10% of myasthenics have thymomas. The great majority of the remainder have prominent thymic medullary germinal centres, which probably represent sites of antibody formation. There may be small lymphocyte collections (lymphorrhages) between skeletal muscle fibres, and in advanced cases skeletal muscle atrophy with myopathic changes may occur. At electron microscopy the morphology of the subsynaptic membrane of skeletal muscle is altered, with a simplification of the architecture[18].

Clinical features – Myasthenia gravis causes the pathologically rapid develop-ment of weakness of skeletal muscle on repeated contraction, with rapid recovery after rest. The disorder may be localized (if so, often being restricted to the external ocular muscles) or it may be generalized. Life may be en-dangered if the bulbar or respiratory muscles are involved. Tendon reflexes tend to be preserved, and there is no muscle wasting till the disorder reaches a late stage. Temporary myasthenia may occur in the new born offspring of myasthenic mothers.

Diagnosis
(1) The abnormally rapid development of weakness and muscle fatigue on repeated or sustained contraction, with recovery on rest, is characteristic. Temporary correction of the weakness by the use of agents (edrophonium, neostigmine) which delay acetylcholine breakdown (by inhibiting acetyl-cholinesterase) helps confirm the diagnosis.

(2) Laboratory: The diagnosis of myasthenia may be suggested by finding an abnormal susceptibility of neuromuscular transmission to neuromuscular blocking agents, e.g. tubocurarine (though this procedure may be hazar-dous if facilities for assisted ventilation are not available). The diagnosis is confirmed by the bedside pharmacological tests mentioned above which have the effect of prolonging acetylcholine action. Electromyography shows a decremental response of muscle action potentials to repetitive nerve stimulation at 3–5 impulses per second. Acetylcholine receptor antibodies are present in plasma in some 90% of cases of myasthenia gravis.

Treatment – There are two main approaches to the treatment of myasthenia gravis. The two approaches may be employed simultaneously.

223

(1) The amount of anticholinesterase receptor antibody can be decreased by:
   (a) Removing its site of formation, namely the thymus,
   (b) Suppressing antibody formation with high dosage glucocorticoids or other immunosuppressant agents, or
   (c) Removing circulating antibody by plasmapheresis or thoracic duct drainage.

(2) The inactivation of released acetylcholine at axon terminals may be delayed by giving anticholinesterase agents, e.g. neostigmine, pyridostigmine.

The practical management of the myasthenic patient requires considerable therapeutic skill and some pharmacological insight. The above account does little more than outline the underlying principles of therapy.

5.2.2.2 Myasthenic syndrome (Eaton–Lambert syndrome)

This disorder, often but not necessarily associated with carcinoma elsewhere in the body, is less frequent than myasthenia gravis but resembles it clinically. In the myasthenic syndrome the deficiency of acetylcholine activity at skeletal muscle myoneural junctions appears to arise from deficient acetylcholine release from axon terminals, rather than from competition between an abnormal antibody and acetylcholine for postsynaptic receptors (as in myasthenia gravis).

Abnormal biochemistry – Unlike myasthenia gravis, there is no known disturbance in antibody production in the myasthenic syndrome[19]. Through mechanisms which are not yet understood, neoplasia elsewhere in the body causes the release of a decreased number of acetylcholine molecules from motor axon terminals in response to each action potential which reaches the axon terminal. In botulism, clostridium botulinum toxin appears to have a reasonably similar effect[20], probably by interfering with the provision of the Ca^{2+} essential for acetylcholine release[21]. Decreased acetylcholine concentrations in the synaptic clefts in response to each nerve impulse cause decreased voluntary muscle power, and abnormally rapid fatigue. However, power may increase after repeated axonal discharge, apparently because more acetylcholine is then released in response to each action potential. In the myasthenic syndrome it seems that there is no defect in acetylcholine synthesis, or in choline re-uptake. The defect lies in the release of preformed acetylcholine.

Aetiology – Although neoplasia somewhere in the body is usually present in the myasthenic syndrome, the molecular mechanisms through which the neoplasm causes the syndrome are obscure.

Structural pathology – It is difficult to obtain clear descriptions of the structural changes in muscle which occur in the myasthenic syndrome.

Clinical features – The disorder presents as a myasthenia-like condition, often in a person with known or occult malignancy. Power may increase to some extent when there are repeated attempts at voluntary muscle contraction. The response to anticholinesterases is less complete than in myasthenia gravis, and may be lost as time passes. The e.m.g. shows an 'incremental' response to nerve stimulation at 3–5 impulses per second.

Diagnosis
(1) Clinical: Although the circumstantial clinical evidence may be suggestive, diagnosis of the myasthenic syndrome depends on electromyography.

(2) Laboratory: The diagnostic e.m.g. finding (namely an incremental response) is mentioned above.

Treatment – Production of increased acetylcholine concentrations in the synaptic cleft by administration of anticholinesterases often yields incomplete restoration of power in the myasthenic syndrome. Further, the response to therapy may diminish with continued use of the drugs. Treatment with guanine, or germine diacetate, agents which increase the quantities of acetylcholine released from nerve terminals in response to each action potential, appears to yield better results. Such therapy is also said to be useful in botulism.

5.2.2.3 *Pseudocholinesterase deficiency*

A genetically-determined deficiency in pseudocholinesterase enzymes, which occurs in 1 in 2000 Europeans, can lead to prolonged paralysis following anaesthesia in which the neuromuscular blocking agent succinylcholine is given. Otherwise this enzyme deficiency causes no clinical problem.

Abnormal biochemistry – Succinylcholine is a depolarizing neuromuscular junction blocking agent which is used as a muscle relaxant in anaesthesia. Its duration of action is brief, due to its rapid hydrolysis by plasma and tissue pseudocholinesterase. When these enzymes are deficient, the action of the drug becomes greatly prolonged.

Aetiology – Pseudocholinesterase deficiency is inherited but the genetic factors concerned are complex, with at least two chromosomal loci being involved. Details of the genetics are provided by Harris[22]. Pseudocholinesterase may also be deficient in severe hepatic disease, where there presumably is failure of enzyme synthesis.

Structural pathology – The disorder does not cause morphological changes.

Clinical features – Pseudocholinesterase deficiency causes prolonged paralysis of skeletal muscles (including bulbar and respiratory muscles) after procedures in which succinylcholine has been used as a neuromuscular blocking agent.

Diagnosis

(1) Clinical: The story is characteristic. The possibility that myasthenia has been uncovered by the use of a blocking agent can be excluded by injecting an anticholinesterase agent, and ascertaining whether power improves (as it should in myasthenia).

(2) Laboratory: Plasma cholinesterase levels are reduced in patients with pseudocholinesterase deficiency.

Treatment – Avoidance of succinylcholine will prevent future anaesthetic difficulties. There is no contraindication to the use of competitive blockers of neuromuscular transmission as muscle relaxants.

5.3 CATECHOLAMINES

5.3.1 Catecholamine biology

Synthesis – The two catecholamine neurotransmitters relevant to clinical neurology are dopamine and noradrenaline. It is convenient to consider these substances together. Noradrenaline is an alkylamine side-chain oxidation product of dopamine. Both are derived from the amino acid phenylalanine,

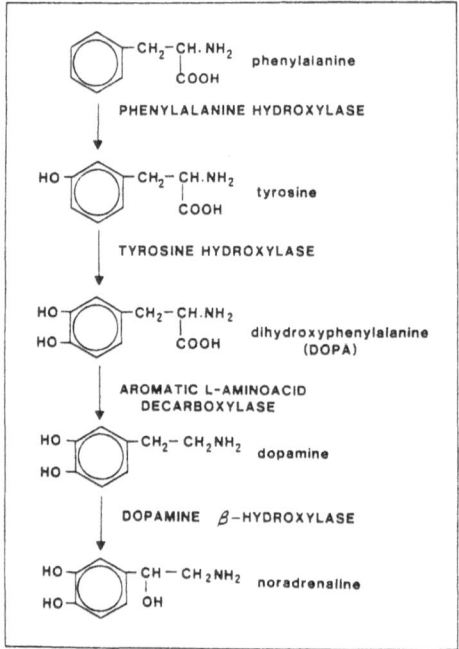

Figure 5.2 Catecholamine biosynthesis pathway

which is oxidized to tyrosine in the liver (Figure 5.2). Tyrosine enters the blood stream and is actively transported from blood into the brain across the blood–brain barrier. In neurons tyrosine is oxidized to dihydroxyphenylalanine in a reaction catalysed by the enzyme tyrosine hydroxylase. Dihydroxyphenyl-alanine (dopa) is subsequently decarboxylated to dopamine. This de-carboxylation is catalysed by aromatic L-amino-acid decarboxylase (col-loquially known as dopa decarboxylase), which has pyridoxal phosphate as a cofactor. Dopamine is then oxidized to noradrenaline, the reaction being catalysed by the enzyme dopamine-β-hydroxylase.

It should be pointed out that there is an analogous series of phenolic amines derived from tyrosine (Figure 5.3). Some of these compounds occur in the nervous system, and are candidates for a neurotransmitter role (e.g. octo-pamine) though this role is not yet established.

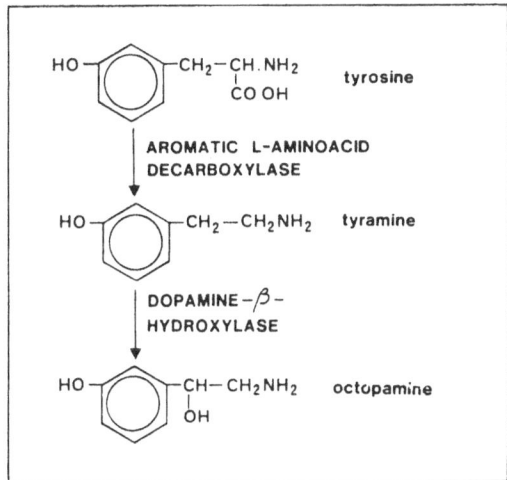

Figure 5.3   Biosynthesis of phenolic amines

Catecholamines are synthesized mainly in the cell bodies of neurons, in the cytosol. Catecholamine concentration exerts a feedback inhibitory control on tyrosine hydroxylase, the rate-limiting enzyme in catecholamine formation. After synthesis, catecholamines are transported along axons to axon terminals, where they are stored.

Storage – Dopamine and noradrenaline are stored in granules, (or vesicles) in axon terminals. Dopaminergic and noradrenergic neurons appear distinct and their transmitter storage and release mechanisms differ in points of detail[23]. Highest dopamine concentrations occur in the striatum, the septal area and the olfactory tubercule, whereas noradrenaline levels are highest in the peri-pheral sympathetic nerves, and in the hypothalamic region and locus coeruleus.

Release – Catecholamines undergo Ca^{2+} dependent release from nerve terminals in response to action potentials which reach the terminal axons.

Receptor actions – Recent work[24] has provided evidence of at least two populations of postsynaptic dopamine receptor in the brain, and there is now some evidence for at least a third type of dopamine receptor[25,26]. The so-called D_1 receptors are linked to adenyl cyclase, whereas the D_2 receptors are not. There are also presynaptic D_1 receptors in the substantia nigra, and presynaptic D_2 receptors in the corpus striatum. Certain drugs are reasonably specific agonists (apomorphine, bromocriptine) or antagonists (metoclopramide) at D_2 receptors, though as yet there appear to be few drugs which act specifically at D_1 receptors.

Two classes of noradrenergic receptors (α and β) are well established and there may be two subtypes of β-noradrenergic receptors within the nervous system[27]. Both types of noradrenergic receptors are linked to adenyl cyclase, and are differentiated with respect to their specificities towards the agonist and antagonist effects of certain drugs. Presynaptic noradrenergic receptors also exist[28]. Activation of these receptors can inhibit noradrenaline release.

The way in which noradrenergic receptor activation is transduced into

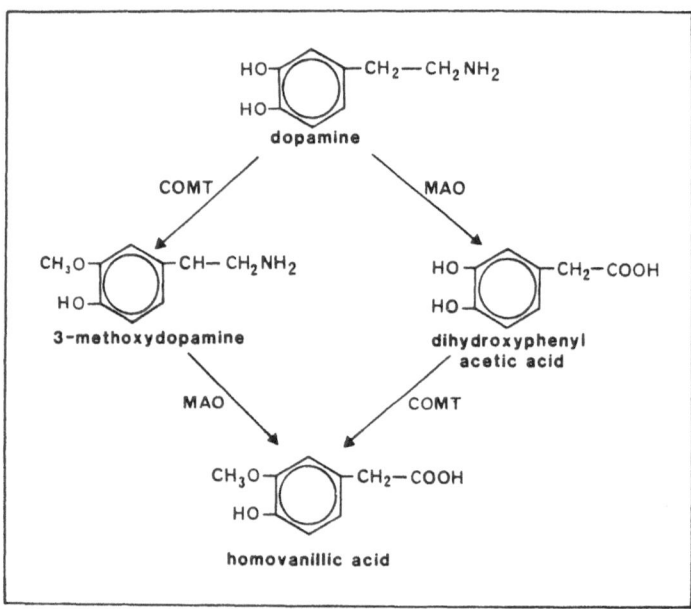

Figure 5.4 Dopamine degradation pathway. COMT = Catechol-*O*-methyl transferase; MAO = Monoamine oxidase. The structures of the intermediate aldehyde products of monoamine oxidase activity are not shown

altered ion channel permeability and membrane polarization was discussed earlier in this chapter (Section 5.1).

Re-uptake – After release into the synaptic cleft the actions of catecholamine neurotransmitters are terminated mainly by re-uptake of the intact amines rather than by their metabolic degradation. The re-uptake systems for both catecholamines are high affinity active transport ones which are stereochemically relatively specific. These mechanisms are dependent on extracellular Na^+ concentrations[14].

Degradation – Two enzymes, monoamine oxidase and catechol-*O*-methyl transferase, catalyse the metabolic degradations of dopamine and noradrenaline. The metabolic pathways are shown in Figures 5.4 and 5.5. Homovanillic acid and dihydroxyphenylacetic acid are the main metabolites of dopamine. In the

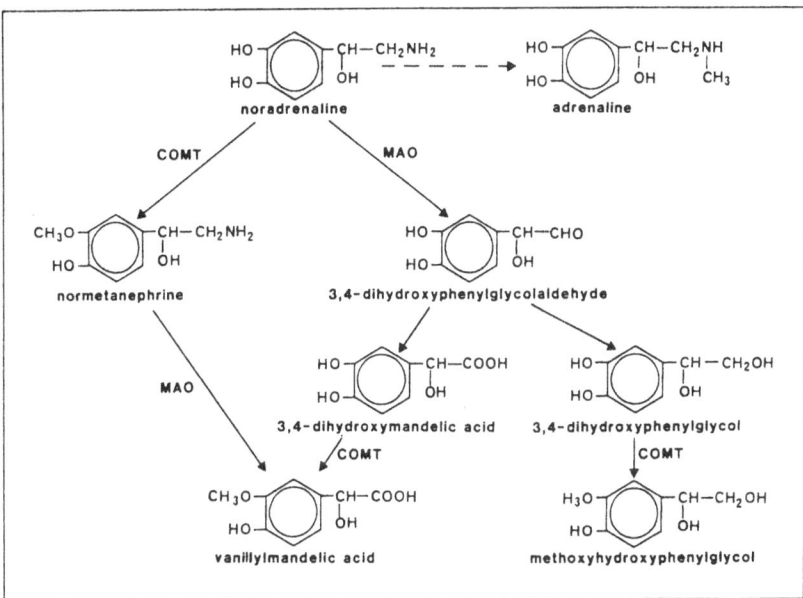

Figure 5.5 Noradrenaline degradation pathway. COMT = Catechol-*O*-methyl transferase; MAO = Monoamine oxidase. In the Figure, with the exception of 3,4-dihydroxyphenylglycol aldehyde, the various aldehyde oxidation products of reactions catalysed by monoamine oxidase are not shown, though the further reaction stages of these products are shown

peripheral nervous system vanillylmandelic acid is the major metabolite of noradrenaline, whereas in the central nervous system the dominant metabolite is methoxyhydroxyphenylglycol[29].

5.3.2 Disorders related to altered catecholamine function

There is reasonably good evidence that most of the essential features of Parkinsonism are related to striatal dopamine deficiency, and that striatal dopamine excess is largely responsible for producing chorea. There is also reasonable evidence that abnormal basal ganglia dopaminergic activity is related to the pathogenesis of tardive dyskinesia. A relationship between dystonia and between some forms of myoclonus and altered brain dopamine function has been suggested, but is less definite. Certain lines of evidence would incriminate excessive limbic system dopaminergic activity in the production of schizophrenic thought disorder.

There is some evidence that a relationship exists between diminished brain noradrenergic activity and endogenous depression, though in this regard serotonin depletion may also play a significant role. Mania, in many ways the phenomenological opposite to endogenous depression, may be related to central nervous system noradrenergic overactivity. Noradrenergic overactivity, central and also peripheral, appears involved in producing some of the somatic manifestations of anxiety. There is some reason to suspect that noradrenergic overactivity may also be related to the production of essential tremor. In addition, there is some rather tenuous evidence of a relationship between raised brain noradrenaline levels and protection against epilepsy, but this matter is so indefinite that it does not seem worth pursuing in the present context.

It should be recognized that the relationship between altered central catecholamine activity and several of the disorders mentioned above is provisional. Details of the relevant biochemical disturbances may not always be adequately worked out to permit a satisfactory understanding of the pathogenesis of the disorders. In some cases the strongest evidence for a relation between a disorder and altered catecholaminergic neurotransmission is derived from the response of the disorder to drugs which are believed to affect central catecholaminergic function in a specific way. The fallacies inherent in arguing from supposedly specific pharmacological responses to aetiology and pathogenesis are generally recognized.

5.3.2.1 Striatal dopamine deficiency: Parkinson's disease

Dopamine depletion in the corpus striatum appears to be responsible for the major disturbance of parkinsonism, namely consequences of rigidity and bradykinesis. Parkinsonism, a syndrome, is one of the more common neurological disorders.

Abnormal biochemistry – Dopamine concentrations in the human corpus striatum are decreased in parkinsonian brains[30]. This decrease appears to be a result of degeneration of a dopaminergic nigro-striatal pathway which normally inhibits certain neurons in the striatum (Figure 5.6). When this

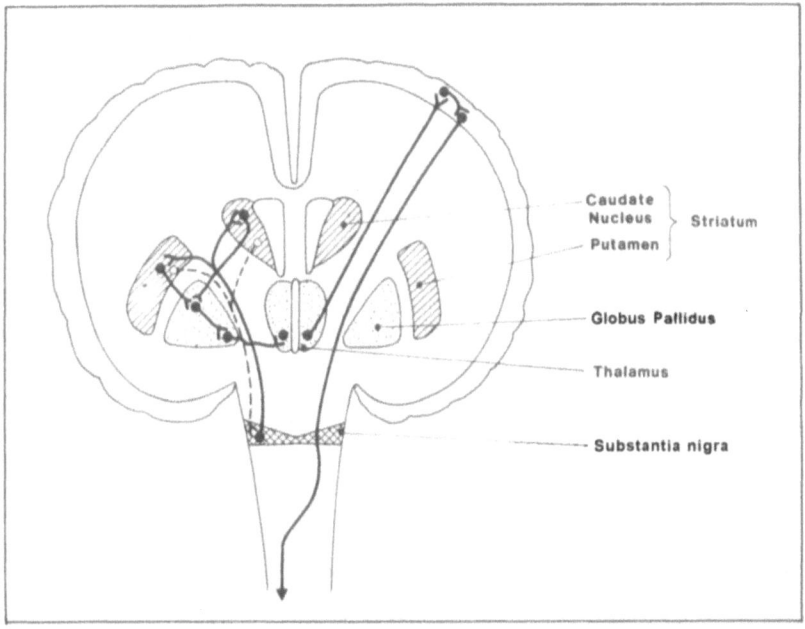

Figure 5.6 Neural pathways involved in parkinsonism. The striato-nigral pathway (GABAergic) is shown in dashed line, the nigro-striatal pathway (dopaminergic) in a continuous line. Connexions within the basal ganglia and thalamus are shown in the left side of the figure only, and thalamo-cortico-spinal connexions on the right side, purely for simplicity of display

inhibition is lessened, there is relative overactivity in pathways extending from the inner portion of the globus pallidus to the ventro-lateral nucleus of the thalamus, thence to the motor cortex, and finally, via the cortico-bulbar and cortico-spinal tracts, to bulbar and spinal motor neurons. Overactivity in this system produces rigidity and bradykinesis. The collective effects of these functional disturbances explain many of the features of parkinsonism, e.g. the flexed posture, the lack of involuntary associated movements, the drooling of saliva, the lack of facial expression. These functional deficits can be reversed, at least in part, by giving dopamine agonist drugs, or by giving the dopamine precursor L-dihydroxyphenylalanine (L-dopa). This amino acid, unlike dopamine, can cross the blood–brain barrier (by virtue of an active transport process). Presumably some of the L-dopa load is then decarboxylated to dopamine in surviving nigral neurons in which the underlying disease process may have already rendered dopamine formation deficient. There is also a suggestion that the decarboxylation may occur in non-aminergic striatal interneurons or efferent neurons, which are known to contain dopa-decarboxylase[31]. It had earlier been suggested that the decarboxylation of L-dopa may also occur in the serotoninergic neurons of the striatum (serotonin and dopamine

231

are both formed in decarboxylation reactions catalysed by the same enzyme, aromatic-L-amino-acid decarboxylase). If this were so, the dopamine so formed would act as a false neurotransmitter in relation to the serotoninergic neurons in which it was synthetized. However, it would be released in the vicinity of striatal neurons with postsynaptic dopamine receptors, and it would presumably act at these receptors. This suggested site of dopamine formation in serotoninergic neurons now seems unlikely[31].

It should be noted that the resting tremor of parkinsonism does not appear to be mediated simply through altered dopamine activity in the nigro-striatal pathway. The pathway mediating tremor probably runs from the red nucleus to the olive, then to the cerebellar cortex, to the dentate nucleus of the cerebellum, then back to the red nucleus and field of Forel, and thence influences the ventro-lateral nucleus of the thalamus[32]. The neurotransmitter (or neurotransmitters) involved in this pathway are not yet identified. There is no known pharmacological agent which will reliably correct true parkinsonian tremor. It should also be noted that the biochemical basis is not yet established for some of the minor autonomic manifestations of parkinsonism, e.g. the abnormalities of alimentary tract motility, the altered sweating and sebum secretion. The clinical expressions of striatal dopamine depletion are somewhat less than the full gamut of clinical manifestations of idiopathic parkinsonism.

Aetiology – Parkinsonism is a syndrome. The following are its major aetiologies:

(1) Idiopathic (paralysis agitans),
(2) Drug (dopamine antagonist) induced,
(3) Postencephalitis lethargica (now becoming very rare),
(4) Occurring as a facet of certain syndromes, e.g. olivo-ponto-cerebellar atrophy, Shy–Drager syndrome,
(5) Poisoning with carbon monoxide or manganese,
(6) Miscellaneous, e.g. following trauma, cerebral tumour, neurosyphilis.

It seems doubtful whether 'arteriosclerotic' parkinsonism exists as a genuine pathological entity[33]. The great majority of contemporary cases of parkinsonism are idiopathic, or iatrogenic.

Structural pathology – Macroscopically, the parkinsonian brain shows diffuse cerebral cortical atrophy, with notable depigmentation of the substantia nigra[34,35]. Microscopically, the most consistent change is neuronal loss in the substantia nigra, with replacement gliosis. There may also be neuronal loss in various parts of the basal ganglia, the thalamus, hypothalamus and neocortex, and with selective neuronal fallout in certain brain stem nuclei, e.g. the locus coeruleus and the dorsal vagal nucleus. In idiopathic cases Lewy bodies occur in affected areas of the brain and elsewhere in the nervous system.

Clinical features – The clinical features of parkinsonism, e.g. the slowness of movement, flexed posture, expressionless facies, characteristic resting tremor, plastic rigidity, loss of arm swinging when walking, are too well known to warrant detailed description here.

Diagnosis
(1) Clinical: The diagnosis of Parkinson's disease rests solely on recognition of the clinical features, and then searching for the aetiology.

(2) Laboratory: Laboratory investigation is of use in determining the aetiology of parkinsonism, but not in diagnosing the disorder.

Treatment – There are three main approaches to the treatment of Parkinson's syndrome.

(1) Dopaminergic activity in the striatum may be increased by giving dopamine agonists, e.g. amantadine, bromocriptine, or the dopamine precursor levodopa. Levodopa (L-dopa) may be given alone, or with bense-razide or carbidopa, both inhibitors of the enzyme aromatic L-amino-acid decarboxylase which do not themselves cross the blood–brain barrier. Simultaneous use of the inhibitors reduces 'wasteful' extraneural decar-boxylation of L-dopa, and consequently reduces adverse effects from excess circulating dopamine (mainly nausea and vomiting).

(2) Centrally acting anticholinergic drugs may be used. There is some evidence that a relative excess of striatal acetylcholinergic activity occurs in parkin-sonism, though the disorder is principally due to striatal dopamine depletion. Anticholinergic agents are much less effective than dopamine agonists or L-dopa itself. Further, some anticholinergics, e.g. benztropine, also appear to have dopamine agonist effect.

(3) Surgical lesions may be made in the ventro-lateral nucleus of the thalamus on both sides. These lesions decrease the excessive inflow of pallido-thalamo-cortical impulses which is responsible for parkinsonian rigidity. The lesions also relieve parkinsonian tremor, but not bradykinesis.

5.3.2.2 Striatal dopamine excess: chorea

Chorea is in many ways the biochemical and functional converse of parkinson-ism. Chorea is a less common neurological disorder than parkinsonism, and is related to a relative excess of dopaminic activity in the corpus striatum.

Biochemical abnormality – Biochemical data are available mainly for Hunting-ton's chorea, but it seems likely that similar molecular mechanisms would apply in rheumatic and the other rare types of chorea. In brains from cases of Huntington's chorea, dopamine concentrations are increased in the striatum and nucleus accumbens and γ-aminobutyrate (GABA) concentrations are reduced[36], as is activity of glutamate decarboxylase, the enzyme catalysing

GABA synthesis. To some extent the increased dopamine relative concentrations may result from striatal neuron loss with dopaminergic nigro-striatal axon terminals in the striatum remaining intact. The loss of GABA-ergic neurons of the striato-nigral pathway (with their cell bodies in the striatum) may cause diminished GABA inhibition of nigro-striatal dopaminergic neurons (originating in the substantia nigra), thus leading to increased dopamine activity in the striatum (Figure 5.6). Consequently there is excess inhibition of the same neural circuits which are disinhibited in parkinsonism. As a result of the underinhibition there is the hypotonia and the excessive involuntary associated movements of chorea. The fact that pharmacological measures[37] aimed at increasing brain GABA content fail to alter chorea argues against the loss of the GABA-ergic striato-nigral pathway playing any major role in producing excessive striatal dopamine activity[38,39]. In contrast, blockading striatal dopamine receptors reverses chorea effectively.

Other biochemical abnormalities are found in Huntington's disease though their pathogenesis and causative roles in the disorder are obscure. Red blood cell membrane Na^+, K^+-adenosine triphosphatase activity is increased[40], platelet monoamine oxidase activity may be decreased[41], γ-hydroxybutyrate levels are raised in the caudate nucleus and substantia nigra[42], choline acetyltransferase activity is reduced in the basal ganglia[43], and there are decreased homocarnosine levels in the lenticular nuclei and cerebellar cortex[44].

Aetiology – The main varieties of chorea are:

(1) Rheumatic (Sydenham's),
(2) Huntington's (autosomal dominant inheritance).

In addition chorea is occasionally associated with systemic lupus erythematosus, oral contraceptive intake[45] and levodopa overdosage, and a senile variety of chorea is often recognized. Hemiballismus, a violent hemichorea, is nearly always due to cerebrovascular disease.

Structural pathology – The structural pathology of rheumatic chorea is not well documented, as death during the active stage of the disorder must be rare. In Huntington's disease[46,47] the caudate nucleus and putamen are greatly shrunken due to neuronal fall-out, and there is replacement gliosis. The cerebral cortex is diffusely atrophic (probably correlating with the dementia of the disorder). The substantia nigra is deeply pigmented, but atrophic.

In hemiballism, the subthalamic nucleus on the side opposite to the dyskinesis often contains a haemorrhagic lesion or an infarct[48].

Clinical features – Chorea is manifested as hypotonia with variable, irregular, brief involuntary movements of the face, tongue, limbs and trunk. There are excessive arm and trunk movements when walking, with purposeless, restless movements of the face, head, neck and limbs when the patient is sitting or lying. In Huntington's disease there is also a progressive dementia.

234

Diagnosis

(1) Clinical: Chorea is diagnosed on the basis of the patient's appearance.

(2) Laboratory: No laboratory investigation provides diagnostic information.

Treatment – Chorea results from relative dopamine overaction in the striatum. The involuntary movements can be decreased by depleting nigro-striatal dopaminic neurons of their neurotransmitter (by giving the drugs tetrabenazine or reserpine, which prevent neuronal dopamine storage) or by blockading striatal postsynaptic dopamine receptors with phenothiazine or butyrophenone drugs. No known treatment helps the intellectual failure of Huntington's disease. A more detailed account of therapy is given by Goetz *et al.*[49]

5.3.2.3 Tardive dyskinesia

Tardive dyskinesia is almost entirely an iatrogenic disorder. It occurs rather frequently in contemporary practice. The dyskinesis comprises choreiform involuntary movements, usually quite localized in their distribution and often induced by a reduction in dose of dopamine receptor blocking agents. Recent reviews of the topic include those of Baldessarini and Tarsy[50] and Gerlach[51].

Abnormal biochemistry – Tardive dyskinesia nearly always develops in patients, particularly older patients, who have been taking dopamine receptor blocking agents (phenothiazine or butyrophenone derivatives) in high dosage for long periods. Often drug dosage has been reduced shortly before the movements appear. The involuntary movements are relieved by increasing the degree of dopamine receptor blockade, or by producing central neuronal dopamine depletion. Tardive dyskinesia represents the paradoxical situation of a disorder being treated by prescribing the agent which is its cause. The pathogenesis of the disorder may be explained by prolonged severe dopamine receptor blockade causing the development of increased numbers of dopamine receptors on striatal neurons[52]. In experimental animals such a change is known to occur in β-adrenergic receptors after receptor blockade[53]. When the degree of dopamine receptor blockade is lessened, e.g. by dosage reduction, there is an abnormally large number of receptors available to bind the dopamine released in the basal ganglia in response to neuronal activity. Consequently, an abnormally great striatal dopamine response occurs, analogous to the phenomenon of denervation hypersensitivity[54]. Increased striatal dopamine activity is accepted as the cause of involuntary movements in chorea (Section 5.3.2.2).

The above explanation is based on a mixture of experimental evidence and argument from analogy. With the recognition of multiple classes of central pre- and postsynaptic dopamine receptors[24,26], attempts are now being made to explain tardive dyskinesia in terms of differential disturbance of different dopamine receptors[55]. However, the experimental proof of such hypotheses is incomplete.

Aetiology – Most cases of tardive dyskinesia occur in older patients who have received prolonged high dosage phenothiazine or butyrophenone therapy. Rarely cases have followed intake of other drugs (e.g. centrally acting anticholinergics, phenytoin) or have occurred in aged persons with no known exposure to dopamine antagonists or dopamine depleting agents.

Structural pathology – No impressive or consistent structural neuropathological changes have been found in the brains of persons with tardive dyskinesia[50].

Clinical features – Although the involuntary movements of tardive dyskinesia resemble those of chorea more than any other condition, the movements tend to be more stereotyped than in chorea and they are usually very much more localized, commonly being restricted to the face, mouth and tongue (buccolingual or oro-bucco-lingual dyskinesia).

Diagnosis – The diagnosis of tardive dyskinesia is purely clinical. No laboratory investigation is of value.

Treatment – As explained above, the only treatment which can consistently control tardive dyskinesia is to increase the degree of central dopamine antagonism by presciding dopamine receptor blocking agents, or dopamine store depleting agents. Unfortunately, such therapy perpetuates the situation which caused the disorder. The therapy of the disorder is discussed more extensively by Paulson[56]. There is a recent report that the acetylcholine precursors choline and lecithin are of benefit[57].

5.3.2.4 *Other dyskinesias related to dopamine dysfunction*

Overdosage with L-dopa not infrequently causes localized forms of intermittent torsion dystonia, commonly occurring in one or both lower limbs. These involuntary movements are sometimes relieved by L-dopa dosage reduction. Generalized torsion dystonia (dystonia musculorum deformans), an inherited progressive disorder with no known structural pathological basis, is reported as sometimes responding, temporarily, to L-dopa therapy[58]. Apparently occurring as an idiosyncratic reaction, certain patients develop temporary localized dystonic reactions, often oculogyria, in response to single doses of certain phenothiazine dopamine receptor-blocking agents, e.g. prochlorperazine. These clinical observations suggest a relation between altered central dopamine activity and certain forms of dystonia. However, it is difficult to yet develop any coherent hypothesis from the data.

Myoclonus occasionally follows L-dopa overdosage, but different forms of myoclonus probably have different aetiologies. Myoclonus is discussed further in relation to serotoninergic dysfunction (Section 5.4.2.3).

5.3.2.5 Meige's disease

This disorder comprises symmetrical dystonic contractions of the face, particularly the periocular region, with spasms of mouth retraction and jaw opening. The tongue, palate, pharynx and neck muscles may be involved in the spasmodic dystonia[59]. The occurrence of the condition is largely confined to the elderly. The condition is considered related to a preponderance of striatal dopamine activity[60]. Dopamine antagonists confer benefit on sufferers, and L-dopa and anticholinergic agents make the dyskinesia worse[61].

5.3.2.6 Dopamine and schizophrenia

There are certain items of evidence consistent with the suggestion that schizophrenic thought and behaviour disorder are mediated through increased dopamine activity in the limbic system. The case incriminating dopamine is not so strong that it seems justified to present the subject of schizophrenia in the systematic fashion in which nearly all disease entities have been dealt with in this book. However, schizophrenia is such an important disorder that some mention of its possible neurochemical basis appears warranted. The dopamine hypothesis of schizophrenia is heavily dependent on the fact that dopamine antagonists, if given in sufficient dosage, relieve the thought disturbance and behaviour alterations of the disorder. The effect seems to be mediated through D_2 receptors[62]. Although it has long been suspected on clinical grounds that there is a relation between temporal lobe dysfunction and schizophrenia, no consistent structural or biochemical abnormalities had been described in the relevant brain regions in schizophrenics until quite recently. Now, in schizophrenic brains, increased dopamine levels have been demonstrated in specific areas of the limbic system, in the nucleus accumbens and in the anterior perforated substance, but not in the basal ganglia[63]. This differential distribution of altered dopamine contents appeared unlikely to be a consequence of prior neuroleptic therapy. Serum dopamine-β-hydroxylase levels are known to be reduced in the disorder[64].

5.3.2.7 Noradrenaline and affective illness

Bipolar depression (endogenous depression: manic-depressive psychosis) is a common and serious psychiatric disorder. Mania is a less common but equally serious one. In both depression and mania there is some evidence that the biochemical pathogenesis involves altered central nervous noradrenergic function, though the role of noradrenaline in the disorder is certainly not yet established beyond reasonable doubt.

Biochemical abnormalities – There is rather better evidence that altered central noradrenergic activity is related to the pathogenesis of endogenous depression and its behavioural converse, mania, than there is for a relation between

237

dopamine and schizophrenia. Much of the evidence in relation to depression is pharmacological[65], and a certain amount of it could apply to serotonin as much as to noradrenaline.

Monoamine oxidase inhibitor intake raises brain contents of various amines, including noradrenaline and serotonin, by interfering with the degradations of these substances. Monoamine oxidase inhibitors correct the mood disorder of endogenous depression. Amphetamines, which through several different mechanisms of action have the final common effect of making more noradrenaline available at postsynaptic central nervous system adrenoreceptors, also have some mood elevating effects, though they are not efficient antidepressant agents. The main antidepressant agents in current use, the tricyclic derivatives, are capable of fully reversing the mood disorder of depression. Some of the tricyclic drugs are tertiary amine derivatives, and appear to act mainly by preventing re-uptake of serotonin released at nerve terminals, thus increasing serotonin concentrations at postsynaptic receptors. However, these tertiary amine agents are N-dealkylated in the human liver. Their secondary amine metabolites, which are effective antidepressants when given directly to man, act by decreasing presynaptic noradrenaline re-uptake into axon terminals. This diminished re-uptake makes available increased amounts of noradrenaline for attachment to postsynaptic receptors in the central nervous system. Thus three different types of drugs with antidepressant actions have in common the property of increasing noradrenaline availability at postsynaptic receptors. Further, reserpine, an amine depleting agent in central and peripheral neurons (and in platelets), was known as a cause of depression when it was used to treat hypertension. Lithium, used to prevent mania in manic-depressive psychosis, appears to decrease noradrenaline availability at synapses. Lithium seems to diminish the release of noradrenaline from axon terminals in response to action potentials. It may also enhance the re-uptake of released noradrenaline. However, in chronic use lithium also interferes with the receptor binding of serotonin and γ-aminobutyrate[66].

It is often unwise to argue too strongly from pharmacological data to biochemical mechanisms. Drugs may have more than one known chemical action, and may also have as yet unrecognized ones. The tricyclic antidepressants are also central acetylcholine receptor blocking agents. In man there is an unexplained delay of some days between the time of peak plateau tricyclic antidepressant drug concentrations and onset of antidepressant effect. In animals it is known that noradrenaline receptor-blocking drugs, and altered noradrenaline concentrations, can change brain noradrenaline receptor numbers. Therefore the tricyclic antidepressants may not relieve depression by a direct effect on noradrenaline, but by an indirect effect on noradrenaline receptors[55]. Certain clinically effective antidepressant drugs, e.g. iprandole, do not appear to affect central catecholamine (or serotonin) function[67,68]. Thus the pharmacological evidence for the noradrenergic hypothesis of affective disorders is not unambiguous, though on the whole it does point in the direction of a role for noradrenaline deficiency. The possibility of a defective response at postsynaptic noradrenaline receptors in depression has been suggested[69].

Evidence is also beginning to appear suggesting a possible role for phenolic amines in depression[68]. Some non-pharmacological evidence is also consistent with the noradrenergic hypothesis of depression. Urinary levels of methoxy-hydroxyphenylglycol, the main product of central nervous system noradrenaline breakdown (Section 5.3.1), tend to be reduced in patients with untreated endogenous depression, and to rise if patients with bipolar depression move into a hypomanic phase. However, it should be mentioned that some workers deny the reported urinary methoxyhydroxyphenylglycol changes in depression[70].

Aetiology – The aetiology of bipolar depression is inadequately understood, but may be a primary disturbance in brain chemistry.

Structural pathology – There are no structural pathological changes consistently associated with bipolar depressive illness in man.

Clinical features – The symptoms of endogenous depression are protean, though the more common ones include flatness of affect, gloom, ideas of worthlessness, loss of appetite, weight loss, insomnia and thoughts of suicide. Secondary anxiety and sometimes delusional reactions may occur. The disorder arises without adequate provocation. Even if untreated, after a period from weeks to years, depression tends to resolve spontaneously. Such unipolar depression may subsequently relapse after a time. In the less common bipolar depression, at times the sufferer may move into a phase of heightened mood with a surfeit of energy. On occasions such hypomania may expand into frank mania with its disordered thinking, flight of ideas and extremely hyperactive behaviour.

Diagnosis
(1) Clinical: The diagnosis of affective illness is largely a matter of clinical recognition of the features of the disorder.

(2) Laboratory: No laboratory investigation is diagnostic, though there is said to be a correlation between reduced urine methoxyhydroxyphenylglycol excretion and response to tricyclic antidepressants.

Treatment – The main treatments currently used for endogenous depression are:

(1) Tricyclic antidepressants, which are generally thought to act by increasing neurotransmitter amine concentrations in synaptic clefts, (since these drugs block amine re-uptake into axon terminals).

(2) Monoamine oxidase inhibitors, which increase neurotransmitter amine concentrations by interfering with the metabolic degradation of these substances.

(3) Electroconvulsive therapy.

Development of hypomanic or manic phases may be prevented by continuous lithium therapy. In such therapy plasma lithium levels should be kept below 1.4mEq/l to avoid toxic manifestations. Major tranquillizers may be needed to control the manifestations of frank mania.

5.3.2.8 Essential tremor

The hypothesis that essential tremor is mediated by abnormal noradrenergic function depends almost entirely on pharmacological evidence.

Biochemical abnormality – The facts that propranolol, a centrally-active β-adrenergic blocking agent, relieves essential tremor, while intravenously infused noradrenaline tends to augment the tremor, suggest a relationship between essential tremor and noradrenergic over-activity. However, it may be premature to argue too closely from a single line of pharmacological evidence to a biochemical mechanism.

Aetiology – Essential tremor often appears inherited as an autosomal dominant trait, though sporadic cases occur.

Structural pathology – Essential tremor has no established structural pathological basis[71].

Clinical features – Essential tremor may develop at any age from infancy to senescence. There is an ataxic (action, intention) tremor of the upper limbs, without disturbance of tone or of rapid alternating movements. In more severe cases there is also a vertical or horizontal tremor of the head and neck. The legs are rarely involved. The disorder is often only very slowly progressive. Ethyl alcohol intake characteristically produces temporary relief of the tremor.

Diagnosis
(1) Clinical: The diagnosis of essential tremor is clinical.

(2) Laboratory: No laboratory investigation is of value.

Treatment – The β-adrenergic receptor blocking agent propranolol is the most effective available drug therapy[72]. Stereotaxic surgical destruction of the ventro-lateral nucleus of the thalamus on both sides is said to be effective, but it must rarely be carried out for this indication.

5.4 SEROTONIN

5.4.1 Serotonin biology

Synthesis – Serotonin (5-hydroxytryptamine) is derived from the amino acid tryptophan, which is first oxidized to 5-hydroxytryptophan. This, in turn, is decarboxylated to the amine in a reaction catalysed by the enzyme aromatic L-amino-acid decarboxylase (Figure 5.7). These biosynthetic reactions occur within neural tissue (and elsewhere in the body, e.g. in alimentary epithelium). Within neurons the synthesis takes place in the cell bodies. Serotoninergic neurons have a restricted distribution within the nervous system. Their cell bodies lie mainly in the midline of the pons and mid brain; they project to the medulla, spinal cord, diencephalon, cerebellum and forebrain, including the neocortex and striatum. Serotonin is stored in vesicles in the axon terminals of these neurons.

Release – Serotonin in axon terminals undergoes Ca^{2+} dependent release following depolarization of the nerve cell membrane by an action potential.

Receptor action – Serotonin released into synaptic clefts may bind to post-synaptic receptors. The biochemical mechanisms whereby activation of serotoninergic receptors is transduced into altered ion permeability of the neuronal subsynaptic membrane are incompletely understood.

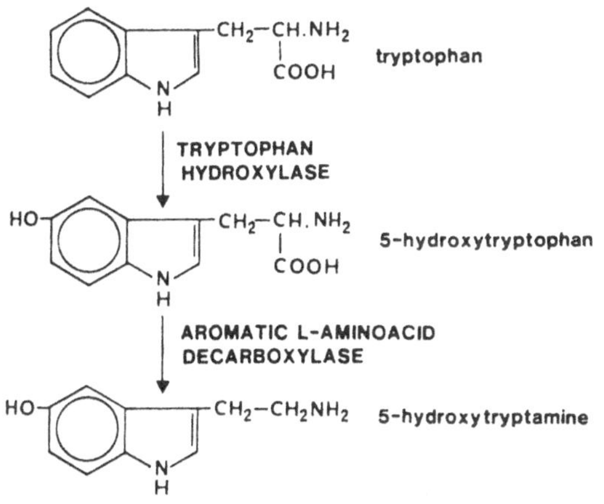

Figure 5.7   Serotonin (5-hydroxytryptamine) biosynthesis

Re-uptake – Serotonin from the synaptic cleft is taken up into the axon terminals of serotoninergic neurons by a Na^+ dependent, high-affinity process. There is also some tendency for serotonin to be taken up into the terminals of neighbouring dopaminic and noradrenergic neurons.

Degradation – Serotonin is metabolically degraded by an oxidation catalysed by the enzyme monoamine oxidase, forming an aldehyde derivative. This substance is converted into a carboxylic acid derivative, 5-hydroxyindoleacetic acid, in a further reaction catalysed by the enzyme aldehyde dehydrogenase,

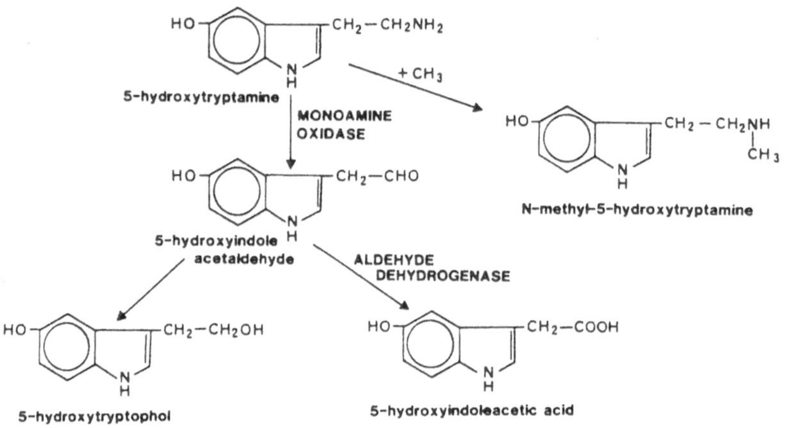

Figure 5.8 Pathways of serotonin metabolism

or else is reduced to 5-hydroxytryptophol. In the brain, serotonin can also be *N*-methylated, *S*-adenosylmethionine being the methyl donor (Figure 5.8).

5.4.2   Diseases related to altered serotonin function

5.4.2.1 Depression

When the noradrenergic hypothesis of depression was discussed in the previous section (5.3.2.7), it was mentioned that affective illness might also be related to disordered serotoninergic function. A good deal of the pharmacological evidence adduced in favour of the noradrenergic hypothesis could apply almost as well to serotonin. The tricyclic antidepressants with tertiary amine side chains, and the monoamine oxidase inhibitors, will both raise synaptic cleft serotonin concentrations in the same way in which they raise local noradrenaline concentrations. Platelet serotonin uptake is reduced in untreated depressed patients[70]. Lithium causes increased neuronal tryptophan uptake and a temporary increase in serotonin synthesis, with a residual decreased sensitivity of serotonin mechanisms to agents such as amphetamine. Some cases of depression respond to therapy with the serotonin precursor 5-hydroxytryptophan[73]. It may be that depressive illness involves deficiency of both noradrenaline and serotonin in the relevant synaptic clefts.

5.4.2.2 Sleep disorders

There seems to be a relation between sleep disturbance and brain serotonin and noradrenaline mechanisms[74]. Increased serotoninergic activity appears to promote sleep and increased noradrenaline activity favours wakefulness. Narcolepsy, a disorder functionally equivalent to excessive sleepiness, is benefited by amphetamine, which causes increased noradrenaline release from axon terminals. Narcolepsy is also helped by the serotonin receptor blocking agent methysergide[75]. Cataplexy, which seems to occur only in association with narcolepsy, is best treated with chlorimipramine, an agent which increases synaptic cleft serotonin and noradrenaline concentrations by blocking amine re-uptake into nerve terminals. Knowledge of the mechanisms involved in these sleep disorders is not sufficiently advanced to permit more detailed presentation of the material.

5.4.2.3 Myoclonus

Myoclonus is a relatively uncommon involuntary movement disorder which has many causes. It seems to arise from lesions at any level along the motor pathway from the neocortex to peripheral nerves. Some varieties arising in the cortex or thalamus appear to be epileptic in nature. In certain other varieties it now seems likely that deficient brain serotonin activity is involved. An induced increase in brain serotonin concentrations can correct the myoclonus which occurs in postanoxic action myoclonus (the Lance–Adams syndrome), a rare disorder which follows hypoxia-induced basal ganglia damage[76]. Raising brain serotonin concentration is also reported to correct palatal myoclonus[77]. This disorder is associated with changes in the inferior olives and may be secondary to vascular lesions of the central tegmental tract in the midbrain or pons. In both these disorders brain serotonin levels can be raised by giving the patient tryptophan or 5-hydroxytryptophan orally in high dosage, together with an inhibitor of aromatic L-amino-acid decarboxylase which does not cross the blood–brain barrier (i.e. carbidopa or benserazide). Analogously to the use of a decarboxylase inhibitor with L-dopa in parkinsonism, such therapy prevents the unwanted extraneural decarboxylation of 5-hydroxytryptophan and yet allows 5-hydroxytryptamine to form in the brain. In the case of such therapy the pharmacological evidence is less ambiguous than if the attempt had been made to manipulate central serotonin levels by inhibiting monoamine oxidase or by using amine re-uptake-blocking agents.

It is not yet clear whether brain serotonin depletion is involved in the biochemical pathogenesis of other varieties of myoclonus. The fact that clonazepam, a benzodiazepine anticonvulsant useful in several forms of myoclonic epilepsy, raises brain serotonin levels[78,79] encourages such speculation, particularly since the drug is also reported of use in postanoxic intention myoclonus[80].

5.4.2.4 Epilepsy

There is evidence in both experimental animals and man that successful anti-convulsant therapy with several different agents is associated with the production of raised brain serotonin levels[15]. The association, however, may be coincidental rather than causal.

5.4.2.5 Migraine

There is reasonably strong evidence connecting platelet serotonin with the pathogenesis of migraine. Speculation also exists that brain stem serotonin mechanisms may be involved in the pathogenesis of the disorder[81]. Migraine is so important a condition to the practising neurologist that the matter needs to be considered at some length.

Biochemical abnormality – Accounts of the biochemical pathogenesis of migraine are available, e.g. Lance[82], Bruyn[83]. The fully developed classical migraine attack comprises three phases, namely:

(1) The aura, due to intracranial vasoconstriction (there being associated scalp arterial constriction which is without clinical effect).

(2) The vascular headache phase, due to scalp arterial dilatation (there being more or less synchronous brain arterial dilatation).

(3) A subsequent phase of headache that is probably neurally mediated in its pathogenesis. This final phase appears to arise from excessive tightness of the epicranial aponeurosis resulting from reflex overcontraction of the posterior cervical muscles.

Only the first and second phases of migraine involve altered serotonin function.

Serotonin acts as a vasoconstrictor, particularly on intracranial arteries[84]. It is believed that, in the initial phase of a migraine attack, blood platelets release their stored serotonin. The released serotonin in plasma water causes vasoconstriction and this produces the aura phase of migraine. However, the released serotonin is now susceptible to degradation by the enzyme monoamine oxidase which is present in blood vessel walls. (There are two accepted subtypes of monoamine oxidase[85]; in platelets, the monoamine oxidase is of type B, which does not metabolize serotonin or noradrenaline[86].) The monoamine oxidase-catalysed serotonin degradation causes increased formation and increased urinary excretion of 5-hydroxyindoleacetic acid, and finally depletion of circulating serotonin in both whole plasma and plasma water[87]. Vaso-constriction can then no longer be maintained by circulating serotonin and vasodilation supervenes, causing the vascular headache element. There have been suggestions[88] that brain stem ischaemia during the vasoconstrictive phase may cause brain stem serotonin depletion. In animals, such depletion leads to increased awareness of pain because of diminished serotoninergic modulation of nocioceptive neurons in the substantia gelatinosa. Such an event in humans might contribute to the pain experienced during migraine attacks. One might also speculate that the factors responsible for platelet serotonin release could also activate neuronal serotonin release, which would be followed by nervous system serotonin depletion after local serotonin degradation took place.

There are several lines of evidence supporting the essential features of the serotonin hypothesis of migraine pathogenesis. Three of the more compelling lines are:

(1) The measured fall in total plasma serotonin levels during migraine attacks.

(2) The measured increase in urinary 5-hydroxyindoleacetic acid excretion during attacks.

(3) The reported relief of the vascular headache phase by the intravenous infusion of serotonin[89].

Secondary biochemical disturbances appear likely in migraine. Thus scalp ischaemia during the vasoconstrictive phase may lead to the formation of the nonapeptide bradykinin in and around vessels[90], and to the local formation of prostaglandins. In the presence of bradykinin, certain prostaglandins will

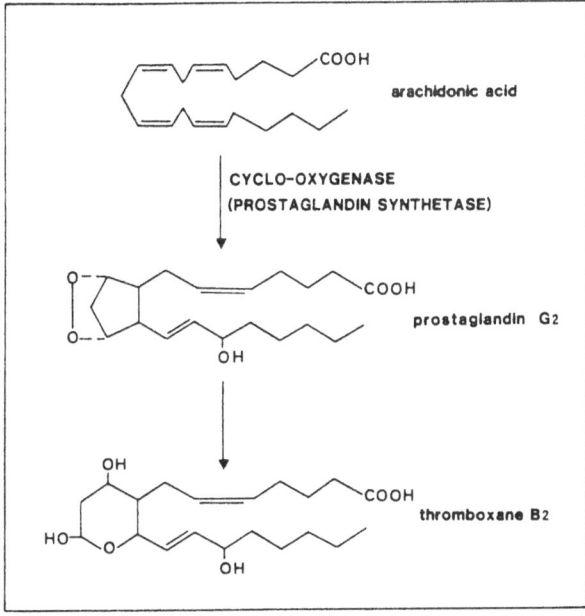

Figure 5.9 Thromboxane synthesis in platelets (in outline)

245

cause an increased input of impulses from peripheral pain fibres, and thus produce an experience of local pain.

The cause of the initial serotonin release from platelets in migraine, and the molecular mechanisms whereby certain chemical factors appear capable of provoking migraine attacks, should also be considered.

The tendency to migraine appears inherited and there is evidence that migraine sufferers have abnormalities of certain platelet functions[91]. Platelet monoamine oxidase activity (type B) is reduced in migraine sufferers, and falls further during migraine attacks[92-94]. Platelet aggregation increases during migraine attacks[95]. There is also evidence of an abnormal circulating factor in plasma during migraine attacks. The suggestion has arisen that the prostaglandin derivatives thromboxanes (Figure 5.9), which promote platelet aggregation[96] and subsequent serotonin release, may be the circulating factor which alters the behaviour of migraineurs' platelets. The platelet monoamine oxidase changes may possibly be merely coincidental (since platelet monoamine oxidase is not involved in serotonin metabolism). A number of migraine-provoking factors, e.g. emotional stress, falling plasma oestrogen levels premenstrually, high dietary fat intake, may activate prostaglandin and thromboxane formation (Figure 5.9). It has been suggested that much of the prostaglandin synthesis and release in migraine may occur in the lungs[97]. Associated thromboxane formation may then cause platelet aggregation and serotonin release, thus initiating the essential sequence of chemical events of migraine. Other provoking factors, e.g. intake of the drug reserpine and high dietary intake of certain amines (e.g. phenylethylamine, in chocolate) may also activate the essential biochemical sequence of migraine and thus induce attacks. Reserpine may have its effect by causing platelet storage granules to release serotonin. Phenylethylamine, metabolic breakdown of which will be decreased in migraineurs, whose platelet monoamine oxidase (type B) is defective, may competitively displace platelet serotonin from storage granules.

The biochemical pathogenesis of migraine may be schematized as in Figure 5.10.

Aetiology – Although the tendency to migraine appears inherited, the genetics are not well worked out, partly because of lack of a generally agreed definition of migraine.

Structural pathology – There is no structural pathology associated with migraine. Exceptionally, cerebral ischaemia from migraine can be severe enough to produce brain infarction[98].

Clinical features – The typical picture of migraine is well known though the manifestations of the disorder are protean. The essential feature is recurrent episodic headache, often unilateral, commonly throbbing in character and associated with photophobia, nausea and vomiting. An aura, usually a visual disturbance or mood alteration, occurs in a substantial minority of cases ('classical' migraine). Factors provoking attacks have been mentioned above.

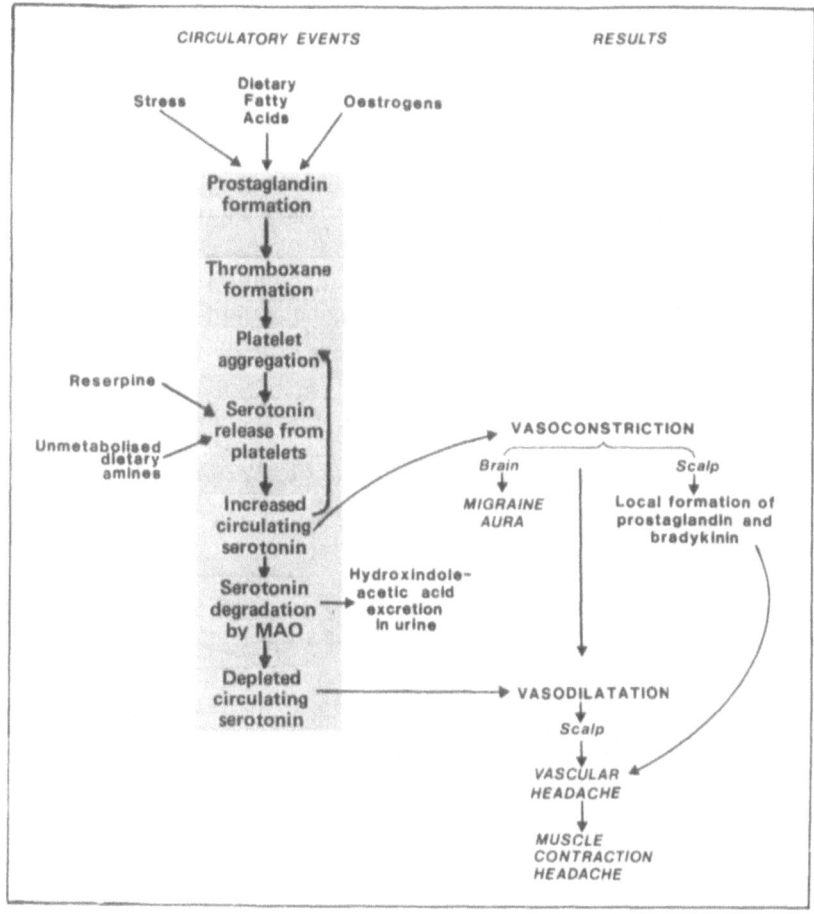

Figure 5.10 Possible biochemical mechanisms of migraine. The essential chemical events are shaded over

Diagnosis
(1) Clinical: The diagnosis of migraine depends on recognition of the history. There are no diagnostic physical findings.

(2) Laboratory: No laboratory investigation provides diagnostic information.

Treatment
(1) Acute attacks: At the onset of acute attacks of migraine, preferably in the aura phase, oral aspirin therapy may terminate the attack by interfering with further thromboxane synthesis, and thus preventing additional serotonin release. Aspirin is a very effective inhibitor of platelet cyclo-

247

oxygenase, and individual molecules of thromboxane have only a very transient existence so that thromboxane synthesis needs to be continuous to maintain increased platelet aggregability and serotonin release.

At a slightly later stage in the attack, the vasoconstrictor ergotamine may benefit the developing vascular headache element. Various tissue cyclo-oxygenase inhibitors (e.g. paracetamol, indomethacin) may be used to prevent secondary local prostaglandin synthesis in tissues. Local heat and massage may relieve the late stage headache due to excessive posterior neck muscle contraction.

(2) Prevention: Drugs which alter serotonin metabolism provide some protection against migraine attacks. Such drugs include antiserotonin agents (e.g. pizotifen, methysergide) and also amitriptyline and propranolol, which block serotonin re-uptake into platelets[99].

5.4.2.6 Migrainous neuralgia: cluster headache

This disorder is closely allied to migraine, but is much less common and probably has a different genetic background and sex dominance. It has a most characteristic time pattern and constellation of clinical features[100]. The disorder has been reported as occurring in relation to increased circulating histamine levels[101], though this work needs further confirmation.

5.5 γ-AMINOBUTYRIC ACID

5.5.1 γ-Aminobutyrate biology

Synthesis – γ-Aminobutyrate (GABA) is an amino acid which is both a metabolic intermediate and a neurotransmitter. Its distribution in the body

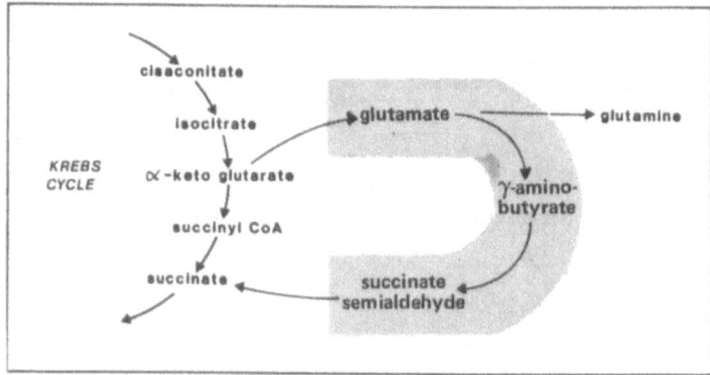

Figure 5.11 The GABA shunt (shaded over) seen in relation to relevant portions of the Krebs cycle

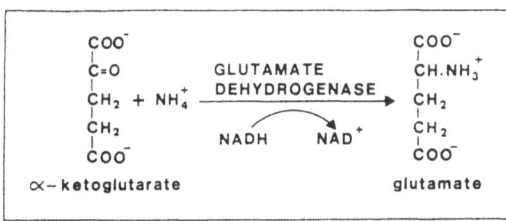

is almost entirely confined to the central nervous system and its appendage, the retina. GABA is a component of the GABA shunt (Figure 5.11), a metabolic bypass of the Krebs cycle between α-ketoglutarate and succinate. The extent of metabolic diversion from the Krebs cycle through the GABA shunt remains uncertain. In the first stage of the GABA shunt α-ketoglutarate is reduced to glutamate, the reaction being catalysed by the enzyme glutamate dehydrogenase. Glutamate is then decarboxylated to γ-aminobutyrate in a reaction catalysed by glutamate decarboxylase, with pyridoxal phosphate as cofactor.

γ-Aminobutyrate next undergoes a transamination reaction with α-ketoglutarate, succinic semialdehyde being formed. This substance is then oxidized to succinate, in a reaction catalysed by succinate semialdehyde dehydrogenase, and the succinate enters the Krebs cycle.

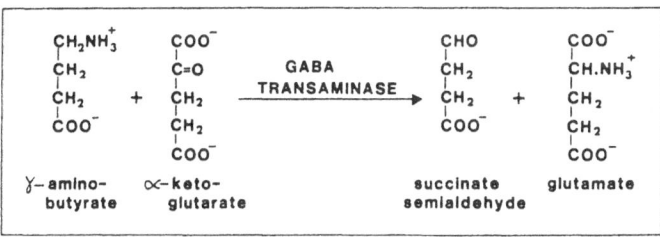

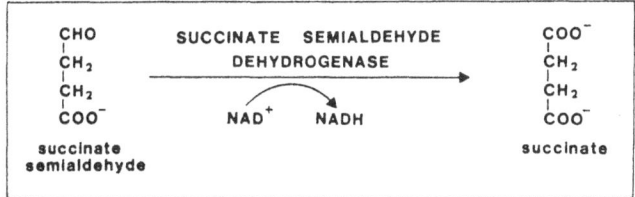

Storage – There appear to be no specific storage vesicles for GABA in axon terminals.

Release – GABA is released from axon terminals into synaptic clefts in response to presynaptic membrane depolarization.

Receptor action – GABA activates postsynaptic receptors to produce hyper-polarization of postsynaptic membranes. Thus it is an inhibitory neuro-transmitter. There appear to be multiple types of central nervous system GABA receptors[102].

Re-uptake – Synaptic GABA appears to be actively taken up into presynaptic terminals by a Na^+ dependent active transport process. The amino acid is also taken up into glia[103].

Degradation – The pathway for the metabolic conversion of GABA to succinate has already been described. Detailed ultracentrifugation and biochemical studies have produced data consistent with the possibility that GABA is synthesized presynaptically, and degraded postsynaptically. A number of alternative metabolic routes for GABA are known, including the formation[8] of carnitine $[(CH_3)_3.N^+.CH_2.CH(OH).CH_2.COO^-]$. The relevance of these latter pathways for human disease is uncertain.

5.5.2 Diseases related to altered γ-aminobutyrate function

GABA is well established as an inhibitory neurotransmitter. Inevitably the question of a relation between GABA depletion and epilepsy has arisen. Certainly epileptic phenomena can be produced in experimental animals by pharmacological measures designed to reduce brain GABA concentrations. Moreover, several anticonvulsant drugs effective in human epilepsy cause raised brain GABA concentrations when given to experimental animals, e.g. valproate. However, the only known instance of human epilepsy probably being related to brain GABA depletion is that which occurred when infants were fed pyridoxine-deficient powdered milk, and developed infantile spasms (hypsarrhythmia) which was reversed by supplying pyridoxine (Section 2.6.3.2). Pyridoxal phosphate is the coenzyme for other decarboxylations apart from the conversion of glutamate to GABA, so that the therapeutic response to pyridoxine is not itself proof that the outbreak of hypsarrhythmia was causally related to brain GABA depletion. The role, if any, of GABA in the great majority of cases of human epilepsy remains unclear.

The proven deficiency of striatal glutamate decarboxylase activity in the brains of persons with Huntington's disease has been discussed earlier (Section 5.3.2.2). As already pointed out, the involuntary movement disorder in this condition is not corrected by therapy designed to restore brain GABA levels.

Thus GABA so far appears not to have any well established major role in the biochemical pathogenesis of human disease.

5.6 DISORDERS INVOLVING GLYCINE FUNCTION

Neurological manifestations associated with non-ketotic hyperglycinaemia are mentioned in Section 2.5.2.2.1.

REFERENCES

1 Watkins, J. C. and Evans, R. H. (1981). Excitatory amino acid transmitters. *Annu. Rev. Pharmacol. Toxicol.*, **21**, 165–204
2 Schwartz, J-C., Pollard, H. and Quach, T. T. (1980). Histamine as a neurotransmitter in mammalian brain: neurochemical evidence. *J. Neurochem.*, **35**, 26–35
3 Fredholm, B. B. and Hedquist, P. (1980). Modulation of neurotransmission by purine nucleotides and nucleosides. *Biochem. Pharmacol.*, **29**, 1635–1643
4 Calne, D. B. (1979). Neurotransmitters, neuromodulators, and neurohormones. *Neurology*, **29**, 1517–1521
5 O'Dea, R. F., Viveros, O. H. and Diliberto, E. J. Jr. (1981). Protein carboxymethylation: role in the regulation of cell functions. *Biochem. Pharmacol.*, **30**, 1163–1168
6 Snyder, S. H. and Goodman, R. R. (1980). Multiple neurotransmitter receptors. *J. Neurochem.*, **35**, 5–15
7 Starke, K. (1981). Presynaptic receptors. *Annu. Rev. Pharmacol. Toxicol.*, **21**, 7–30
8 Cooper, J. R., Bloom, F. E. and Roth, R. H. (1978). *The Biochemical Basis of Neuropharmacology*. 3rd Edn. (New York: Oxford University Press)
9 Sterling, G. H. and O'Neill, J. J. (1978). Citrate as the precursor of the acetyl moiety of acetylcholine. *J. Neurochem.*, **31**, 525–530
10 Cornford, E. M., Braun, L. D. and Oldendorf, W. H. (1978). Carrier mediated blood–brain barrier transport of choline and certain choline analogs. *J. Neurochem.*, **30**, 299–308
11 Kuhar, M. J. and Murrin, L. C. (1978). Sodium-dependent, high affinity choline uptake. *J. Neurochem.*, **30**, 15–21
12 Lindstrom, J. and Dau, P. (1980). Biology of myasthenia gravis. *Annu. Rev. Pharmacol. Toxicol.*, **20**, 337–362
13 Michelson, M. J. and Danilov, A. F. (1971). Synaptic vesicles, specific granules, autopharmacology. 5.1. Cholinergic transmissions. In Bacq, Z. M. (ed.) *Fundamentals of Biochemical Pharmacology*. pp. 221–253. (Oxford: Pergamon)
13 Iversen, L. L. (1974). Uptake mechanisms for neurotransmitter amines. *Biochem. Pharmacol.*, **23**, 1927–1935
15 Eadie, M. J. and Tyrer, J. H. (1980). *Anticonvulsant Therapy. Pharmacological Basis and Practice*. 2nd Edn. (Edinburgh: Churchill-Livingstone)
16 Rowland, L. P. (1978). Myasthenia gravis. In Matthews, W. B. and Glaser, G. H. (eds.) *Recent Advances in Clinical Neurology*. Vol. 2, pp. 25–46. (Edinburgh: Churchill-Livingstone)
17 Simpson, J. A. (1977). Myasthenia gravis – validation of a hypothesis. *Scott. Med. J.*, **22**, 201–210
18 Engel, A. G. and Santa, I. (1971). Histometric analysis of the ultrastructure of the neuromuscular junction in myasthenia gravis and in the myasthenic syndrome. *Ann. N.Y. Acad. Sci.*, **183**, 46–63
19 Lindstrom, J. M. and Lambert, E. H. (1978). Content of acetylcholine receptor and antibodies bound to receptor in myasthenia gravis, experimental autoimmune myasthenia gravis and Eaton–Lambert syndrome. *Neurology*, **28**, 130–138
20 Gutmann, L. and Pratt, L. (1976). Pathophysiologic aspects of human botulism. *Arch. Neurol.*, **33**, 175–179

21 Wonnacott, S., Marchbanks, R. M. and Fiol, C. (1978). Ca^{2+} uptake by synaptosomes and its effect on the inhibition of acetylcholine release by botulinium toxin. *J. Neurochem.*, **30**, 1127–1134

22 Harris, H. (1970). *The Principles of Human Biochemical Genetics*. (Amsterdam: North Holland)

23 McMillen, B. A., German, D. C. and Shore, P. A. (1980). Functional and pharmacological significance of brain dopamine and norepinephrine storage pools. *Biochem. Pharmacol.*, **29**, 3045–3050

24 Kebabian, J. W. and Calne, D. B. (1979). Multiple receptors for dopamine. *Nature (London)*, **277**, 93–96

25 Seeman, P. (1980). Brain dopamine receptors. *Pharmacol. Rev.*, **32**, 230–313

26 Creese, I., Sibley, D. R., Leff, S. and Hamblin, M. (1981). Dopamine receptors: subtypes, localization and regulation. *Fed. Proc.*, **40**, 147–152

27 Minneman, K. P. and Molinoff, P. B. (1980). Classification and quantitation of β-adrenergic receptor subtypes. *Biochem. Pharmacol.*, **29**, 1317–1323

28 Starke, K., Taube, H. D. and Borowski, E. (1977). Presynaptic receptor systems in catecholaminergic transmission. *Biochem. Pharmacol.*, **26**, 259–268

29 Demet, E. M. and Halaris, A. E. (1979). Origin and distribution of 3-methoxy-4-hydroxyphenylglycol in body fluids. *Biochem. Pharmacol.*, **28**, 3043–3049

30 Ehringer, H. and Hornykewicz, O. (1960). Verteilung von Noradrenalin and Dopamin (3-Hydroxytryptamin) im Gehirn des Menschen und ihr Verhalten bei Erkrankungen des extrapyramidalen Systems. *Klin. Wochenschr.*, **38**, 1236–1239

31 Melamed, E., Hefti, F. and Wurtman, R. J. (1980). Nonaminergic striatal neurons convert exogenous L-dopa to dopamine in Parkinsonism. *Ann. Neurol.*, **8**, 558–563

32 Poirier, L. J., Filion, M., Larochelle, L. and Pecharde, J-C. (1976). Physiopathology of tremor and rigidity. In Birkmayer, W. and Hornykiewicz, O. (eds.) *Advances in Parkinsonism*. pp. 217–235. (Basle: Roche)

33 Eadie, M. J. and Sutherland, J. M. (1964). Arteriosclerosis in Parkinsonism. *J. Neurol. Neurosurg. Psychiatry*, **27**, 237–240

34 Alvord, E. C. Jr. (1968). The pathology of Parkinsonism. In Minckler, J. (ed.) *Pathology of the Nervous System*. Vol. 1, pp. 1152–1161. (New York: McGraw-Hill)

35 Oppenheimer, D. R. (1976). Diseases of the basal ganglia, cerebellum and motor neurons. In Blackwood, W. and Corsellis, J. A. N. (eds.) *Greenfield's Neuropathology*. 3rd Edn. pp. 608–651. (London: Arnold)

36 Bird, E. D. (1980). Chemical pathology of Huntington's disease. *Annu. Rev. Pharmacol. Toxicol.*, **20**, 533–551

37 Shoulson, I., Kartzinel, R. and Chase, T. N. (1976). Huntington's disease: treatment with dipropylacetic acid and gamma-aminobutyric acid. *Neurology*, **26**, 61–63

38 Perry, T. L., Wright, J. M., Hansen, S., Allan, B. M., Baird, P. A. and MacLeod, P. M. (1980). Failure of aminooxyacetic acid therapy in Huntington's disease. *Neurology*, **30**, 772–775

39 Manyam, B. V., Katz, L., Hare, T. A., Kaniefski, K. and Tremblay, R. D. (1981). Isoniazid-induced elevation of CSF GABA levels and effects on chorea in Huntington's disease. *Ann. Neurol.*, **10**, 35–37

40 Butterfield, D. A., Oeswein, J. Q., Prunty, M. E., Hisle, K. C. and Markesbery, W. R. (1978). Increased sodium plus potassium adenosine triphosphatase activity in erythrocyte membranes in Huntington's disease. *Ann. Neurol.*, **4**, 60–62

41 Mann, J. and Chiu, E. (1978). Platelet monoamine oxidase activity in Huntington's chorea. *J. Neurol, Neurosurg. Psychiatry*, **41**, 809–812

42 Ando, N., Gold, B. I., Bird, E. D. and Roth, R. H. (1979). Regional brain levels of γ-hydroxybutyrate in Huntington's disease. *J. Neurochem.*, **32**, 617–622

43 Bird, E. D., Gale, J. S. and Spokes, E. G. (1977). Huntington's chorea: postmortem activity of enzymes involved in cerebral glucose metabolism. *J. Neurochem.*, **29**, 539–545

44 Urquhart, N., Perry, T. L., Hansen, S. and Kennedy, J. (1975). GABA content and glutamic acid decarboxylase activity in brain of Huntington's chorea patients and control subjects. *J. Neurochem.*, **24**, 1071–1075

45 Nausieda, P. A., Koller, W. C., Weiner, W. J. and Klawans, H. L. (1979). Chorea induced by oral contraceptives. *Neurology*, **29**, 1605–1609

252

46 Dreese, M. J. and Netsky, M. G. (1968). Degenerative disorders of the basal ganglia. In Minckler, J. (ed.) *Pathology of the Nervous System*. Vol. 1, pp. 1185–1204. (New York: McGraw-Hill)

47 Corsellis, J. A. N. (1976). Ageing and the dementias. In Blackwood, W. and Corsellis, J. A. N. (eds.) *Greenfield's Neuropathology*. 3rd Edn., pp. 796–848. (London: Arnold)

48 Martin, J. P. (1959). Remarks on the functions of the basal ganglia. *Lancet*, **1**, 999–1005

49 Goetz, C. G., Weiner, W. J. and Klawans, H. L. (1981). Treatment of the choreas. In Barbeau, A. (ed.) *Disorders of Movement*. pp. 29–41. (Lancaster: MTP Press)

50 Baldessarini, R. J. and Tarsy, D. (1979). Relationship of the actions of neuroleptic drugs to the pathophysiology of tardive dyskinesia. *Int. Rev. Neurobiol.*, **21**, 1–45

51 Gerlach, J. (1979). Tardive dyskinesia. *Danish Med. J.*, **46**, 209–245

52 Snyder, S. H. (1979). Receptors, neurotransmitters and drug responses. *N. Engl. J. Med.*, **300**, 465–472

53 Weiss, B., Greenberg, L. and Cantor, E. (1979). Age-related alterations in the development of adrenergic denervation supersensitivity. *Fed. Proc.*, **38**, 1915–1921

54 Klawans, H. L. (1973). *The Pharmacology of Extrapyramidal Movement Disorders*. (Basel: Karger)

55 Creese, I. and Sibley, D. R. (1981). Receptor adaptations to centrally acting drugs. *Annu. Rev. Pharmacol. Toxicol.*, **21**, 357–391

56 Paulson, G. W. (1981). Treatment of tardive dyskinesia. In Barbeau, A. (ed.) *Disorders of Movement*. pp. 133–150. (Lancaster: MTP Press)

57 Wurtman, R. J., Hefti, F. and Melamed, E. (1981). Precursor control of neurotransmitter synthesis. *Pharmacol. Rev.*, **32**, 315–335

58 Eldridge, R., Kanter, W. and Koerber, T. (1973). Levodopa in dystonia. *Lancet*, **2**, 1027–1028

59 Tolosa, E. S. (1981). Clinical features of Meige's disease (idiopathic orofacial dystonia). A report of 17 cases. *Arch. Neurol.*, **38**, 147–151

60 Tolosa, E. S. and Lai, C. (1979). Meige disease: striatal dopaminergic preponderance. *Neurology*, **29**, 1126–1130

61 Casey, D. E. (1980). Pharmacology of blepharospasm-oromandibular dystonia syndrome. *Neurology*, **30**, 690–695

62 Snyder, S. H. (1981). Dopamine receptors, neuroleptics and schizophrenia. *Am. J. Psychiatry*, **138**, 460–464

63 Bird, E. D., Spokes, E. G. S. and Iversen, L. I. (1980). Dopamine and noradrenaline in post-mortem brain in Huntington's disease and schizophrenic illness. *Acta Psychiatr. Scand.*, **61** (Suppl. 280), 63–72

64 Fujita, K., Ito, T., Maruta, K., Teradaira, R., Beppu, H., Nakagami, Y., Kato, Y., Nagatsu, T. and Kato, T. (1978). Serum dopamine-β-hydroxylase in schizophrenic patients. *J. Neurochem.*, **30**, 1569–1572

65 Leonard, B. E. (1975). Neurochemical and neuropharmacological aspects of depression. *Int. Rev. Neurobiol.*, **18**, 357–387

66 Maggi, A. and Enna, S. J. (1980). Regional alteration in rat brain neurotransmitter systems following chronic lithium treatment. *J. Neurochem.*, **34**, 888–892

67 Zis, A. P. and Goodwin, F. K. (1979). Novel antidepressants and the biogenic amine hypothesis of depression. The case for iprindole and mianserin. *Arch. Gen. Psychiatry*, **36**, 1097–1107

68 Sandler, M., Ruthven, C. R. J., Goodwin, B. L., Reynolds, G. P., Rao, V. A. R. and Cohen, A. (1980). Trace amine deficit in depressive illness: the phenylalanine connexion. *Acta Psychiatr. Scand.*, **61** (Suppl. 280), 29–38

69 Sulser, F., Vetulani, J. and Mobley, P. L. (1977). Mode of action of antidepressant drugs. *Biochem. Pharmacol.*, **27**, 257–261

70 Coppen, A. and Wood, K. M. (1980). Peripheral serotonergic and adrenergic responses in depression. *Acta Psychiatr. Scand.*, **61** (Suppl. 280), 21–27

71 Herskovitz, E. and Blackwood, W. (1969). Essential (familial) tremor – a case report. *J. Neurol., Neurosurg. Psychiatry*, **32**, 509–511

72 Murray, T. J. (1981). Essential tremor. In Barbeau, A. (ed.) *Disorders of Movement*. pp. 151–170. (Lancaster: MTP Press)

73 Van Praag, H. M. and De Haan, S. (1980). Central serotonin deficiency – a factor which

increases depression vulnerability. *Acta Psychiatr. Scand.*, **61** (Suppl. 280), 89–95

74 Gillin, J. C., Mendelson, W. B., Sitaram, N. and Wyatt, R. J. (1978). The neuropharmacology of sleep and wakefulness. *Annu. Rev. Pharmacol. Toxicol.*, **18**, 563–579

75 Wyler, A. R., Wilkus, R. J. and Troupin, A. S. (1975). Methysergide in the treatment of narcolepsy. *Arch. Neurol.*, **32**, 265–268

76 Growdon, J. H., Shahani, B. T. and Young, R. R. (1976). L-5-Hydroxytryptophan in treatment of several different syndromes in which myoclonus is prominent. *Neurology*, **26**, 1135–1140

77 Magnussen, I., Dupont, E., Prange-Hansen, A. and De Fine Olivarius, B. (1977). Palatal myoclonus treated with 5-hydroxytryptophan and a decarboxylase inhibitor. *Acta Neurol. Scand.*, **55**, 251–253

78 Fennessy, M. R. and Lee, J. R. (1972). The effects of benzodiazepines on brain amines in the mouse. *Arch. Intern. Pharmacodyn.*, **197**, 37–44

79 Jenner, P., Chadwick, D., Reynolds, E. H. and Marsden, C. D. (1975). Altered 5HT metabolism with clonazepam, diazepam and diphenylhydantoin. *J. Pharm. Pharmacol.*, **27**, 707–710

80 Goldberg, M. A. and Dorman, J. D. (1976). Intention myoclonus successfully treated with clonazepam. *Neurology*, **26**, 24–26

81 Lance, J. W. (1981). Headache. *Ann. Neurol.*, **10**, 1–10

82 Lance, J. W. (1978). Migraine. In Matthews, W. B. and Glaser, G. H. (eds.) *Recent Advances in Clinical Neurology*. Vol. 2, pp. 145–161. (Edinburgh: Churchill-Livingstone)

83 Bruyn, G. W. (1976). The biochemical basis of migraine: a critique. In Klawans, H. (ed.) *Clinical Neuropharmacology*. Vol. 1, pp. 185–213. (New York: Raven Press)

84 Hardebo, J. E., Edvinsson, L., Owman, C. H. and Svendgaard, N-Aa. (1978). Potentiation and antagonism of serotonin effects on intracranial and extracranial vessels. *Neurology*, **28**, 64–70

85 Murphy, D. L. (1978). Substrate-selective monoamine oxidases-inhibitor, tissue, species and functional differences. *Biochem. Pharmacol.*, **27**, 1889–1893

86 Sandler, M., Reveley, M. A. and Glover, V. (1981). Human platelet monoamine oxidase activity in health and disease: a review. *J. Clin. Pathol.*, **34**, 292–302

87 Somerville, B. W. (1976). Platelet-bound and free serotonin levels in jugular and forearm venous blood during migraine. *Neurology*, **26**, 41–45

88 Sicuteri, F. (1976). Migraine, a central biochemical dysnociception. *Headache*, **16**, 145–159

89 Anthony, M., Hinterberger, H. and Lance, J. W. (1967). Plasma serotonin in migraine and stress. *Arch. Neurol.*, **16**, 544–552

90 Regoli, D. and Barabe, J. (1980). Pharmacology of bradykinin and related kinins. *Pharmacol. Rev.*, **32**, 1–46

91 Hanington, E. (1978). Migraine: a blood disorder. *Lancet*, **2**, 501–502

92 Sandler, M., Youdim, M. B. H. and Hanington, E. (1974). A phenylethylamine oxidising defect in migraine. *Nature* (London), **250**, 335–337

93 Glover, V., Sandler, M., Grant, E., Rose, F. C., Orton, D., Wilkinson, M. and Stevens, D. (1977). Transitory decrease in platelet monoamine-oxidase activity during migraine attacks. *Lancet*, **1**, 391–393

94 Sandler, M. (1978). Implications of the platelet monoamine oxidase deficit during migraine attacks. *Res. Clin. Stud. Headache*, **6**, 65–72

95 Masel, B. E., Chesson, A. L., Peters, B. H., Levin, H. S. and Alperin, J. B. (1980). Platelet antagonists in migraine prophylaxis. A clinical trial using aspirin and dipyridamole. *Headache*, **20**, 13–18

96 Moncada, S. and Vane, J. R. (1979). Pharmacology and endogenous roles of prostaglandin endoperoxides, thromboxane A_2, and prostacyclin. *Pharmacol. Rev.*, **30**, 293–331

97 Sandler, M. (1972). Migraine: a pulmonary disease? *Lancet*, **1**, 618–619

98 Neligan, P., Harriman, D. G. F. and Pearce, J. (1977). Respiratory arrest in familial hemiplegic migraine: a clinical and neuropathological study. *Br. Med. J.*, **2**, 732–734

99 Raskin, N. H. (1981). Pharmacology of migraine. *Annu. Rev. Pharmacol. Toxicol.*, **21**, 463–478

100 Symonds, C. P. (1956). A particular variety of headache. *Brain*, **79**, 217–232

101 Anthony, M. and Lance, J. W. (1971). Whole blood histamine and plasma serotonin in cluster headache. *Arch. Neurol.*, **25**, 225–231

102 Andrews, P. R. and Johnston, G. A. R. (1979). GABA agonists and antagonists. *Biochem. Pharmacol.*, **28**, 2697–2702

103 Iversen, L. L. and Kelly, J. S. (1975). Uptake and metabolism of γ-aminobutyric acid by neurones and glial cells. *Biochem. Pharmacol.*, **24**, 933–938

Index